MODELLING INTELLIGENT MULTI-MODAL TRANSIT SYSTEMS

MODELLING INTELLIGENT MULTI-MODAL TRANSIT SYSTEMS

Editors

Agostino Nuzzolo
Department of Enterprise Engineering
Tor Vergata University of Rome
Rome, Italy

William H. K. Lam
Department of Civil and Environmental Engineering
The Hong Kong Polytechnic University
Hong Kong

CRC Press is an imprint of the
Taylor & Francis Group, an **informa** business

A SCIENCE PUBLISHERS BOOK

CRC Press
Taylor & Francis Group
6000 Broken Sound Parkway NW, Suite 300
Boca Raton, FL 33487-2742

First issued in paperback 2020

© 2017 by Taylor & Francis Group, LLC
CRC Press is an imprint of Taylor & Francis Group, an Informa business

No claim to original U.S. Government works

ISBN-13: 978-1-4987-4353-2 (hbk)
ISBN-13: 978-0-367-78263-4 (pbk)

Library of Congress Cataloging-in-Publication Data

Names: Nuzzolo, Agostino, editor. | Lam, William H. K., editor.
Title: Modelling intelligent multi-modal transit systems / editors, Agostino
Nuzzolo Department of Enterprise Engineering, Tor Vergata University of
Rome, Rome, Italy, William H. K. Lam, Department of Civil and
Environmental Engineering, The Hong Kong Polytechnic University, Hong Kong.
Description: First edition. | Boca Raton, FL : CRC Press, [2016] | "A Science
Publishers book." | Includes bibliographical references and index.
Identifiers: LCCN 2016032377| ISBN 9781498743532 (hardback : alk. paper) |
ISBN 9781498743549 (e-book)
Subjects: LCSH: Local transit--Technological innovations. | Local
transit--Mathematical models. | Transportation--Technological innovations.
| Transportation--Mathematical models.
Classification: LCC HE147.7 .M64 2015 | DDC 388.401--dc23
LC record available at https://lccn.loc.gov/2016032377

Visit the Taylor & Francis Web site at
http://www.taylorandfrancis.com

and the CRC Press Web site at
http://www.crcpress.com

Preface

The key objectives of this book are to improve understanding of the role played by recent developments in Intelligent Transportation Systems (ITS) and Information Communication Technology (ICT) in addressing multi-modal transit problems; to outline the role of new ITS/ICT tools in enhancing the performance of multi-modal transit operations management and travel advice; to disseminate recent methods of multi-modal transit modelling, taking into account the new functions supplied by advanced ITS/ICT, to be applied for transit operations management and control; to present state-of-the-art approaches in transit modelling for transit design and planning, especially the activity-based approach and reliability-based approach; to analyse several methodological research issues and challenges connected to these new modelling approaches.

The contents of this book contents can be split into three, strictly related, main parts. The first part presents analyses of several methodological issues connected to the development of new tools supporting short-term forecasting (Chapter 1—*Introduction to Modelling Multi-modal Transit Systems in an ITS Context*) for transit operations control (Chapter 2—*New Applications of ITS to Real-time Transit Operations*) and for traveller information provision (Chapter 3—*A New Generation of Individual Real-time Transit Information Systems*).

The second part of the book explores some aspects of real-time multi-model transit modelling.

It starts from the general simulation framework (Chapter 4—*Real-time Operations Management Decision Support Systems: A Conceptual Framework*) and then investigates path choice modelling (Chapter 5—*Real-time Modelling of Normative Travel Strategies on Unreliable Dynamic Transit Networks: A Framework Analysis* and Chapter 6—*A Dynamic Strategy-based Path Choice Modelling in Real-time Transit Simulation*), dynamic routing (Chapter 7—*Time-dependent Shortest Hyperpaths for Dynamic Routing on Transit Networks*) and reverse assignment methods (Chapter 8—*Real-time Reverse Dynamic Assignment for Multiservice Transit Systems*).

The third part of this book finally reports some recent developments in transit modelling for planning (Chapter 9—*Optimal Schedules for Multi-modal Transit Services: An Activity-based Approach*) and design of multi-modal transit systems (Chapter 10—*Transit Network Design with Stochastic Demand*).

In summary, this book consists of 10 original chapters solicited to represent the broad base of contemporary themes in modelling multi-modal transit systems in the context of ITS and ICT. Scholars from Europe and Asia have contributed their knowledge to produce a unique compilation of recent developments in the field. Topics both in theory and innovative applications to multi-modal transit network design problems presented in this book are by no means exhaustive. However, they do provide general coverage of various important areas of Research and Development (R&D) on the theme of this book. The editors wish that this book will bring the up-to-date state-of-the-art methodologies of network modelling for intelligent multi-modal transit systems to the attention of researchers and practicing engineers, and that it will inspire and stimulate new R&D opportunities and efforts in the field particularly in view of the recent advancement in ITS and ICT. After all, it is hoped that this would make better the planning, design and operation of multi-modal transit systems and help promote their use to improve the effectiveness and efficiency of the multi-modal transit services in our cities.

The target audience of this book comprises academics and PhD students; researchers; students; transport and transit agency technicians, in relation to new opportunities of advanced ITS; decision support system (DSS) tool developers; transport professionals and other people who are interested in studying and implementing ITS in mass transit systems, especially in the ICT context, to support transit operations management and control and travel advice.

A. Nuzzolo
W. H. K. Lam

Contents

List of Figures

Introduction to Modelling Multimodal Transit Systems in an ITS Context

Agostino Nuzzolo

ABSTRACT

Transit network 'big data' collecting and processing and bi-directional communication between transit travellers and info centres are emerging as two factors which could improve both transit network planning and design and short-term forecasting of network states for transit operations control and traveller information provision. Several methodological issues connected to the development of models for real-time transit network simulation and forecasting of vehicle occupancy and crowding degree, for supporting such activities, are analyzed in this chapter. The main issues concern real-time transit system modelling, dynamic real-time strategy-based path choice and assignment models, individual path choice modelling and, finally, estimation of OD flows and aggregate calibration of model parameters through collected data.

Department of Enterprise Engineering, Tor Vergata University of Rome, via del Politecnico 1, 00133, Rome, Italy.
E-mail: nuzzolo@ing.uniroma2.it

1.1 Introduction

In recent years advances in information technology and telematics have helped transit agencies achieve more efficient public transportation systems while, supporting real-time transit operations control and traveller information (Fig. 1.1).

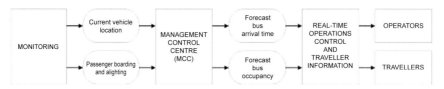

Figure 1.1 Advanced transit operations control and info system.

The introduction of global positioning system (GPS) mobile devices enables not only real-time individual information to be provided *en route*, but also traveller location to be received. This means that travellers can be tracked, representing a milestone in the evolution of real-time information (Fig. 1.2). An example of these new systems is given by App Moovit (Moovit, 2009), which provides real-time pre-trip and *en-route* information based on collecting the real-time position and speed of travellers using the app.

At the same time, bi-directional communication and big data processing allow advances in real-time short-term forecasting, both of transit network conditions and of traveller numbers on board and at stops. Such data can be used in applications of operations control strategies to

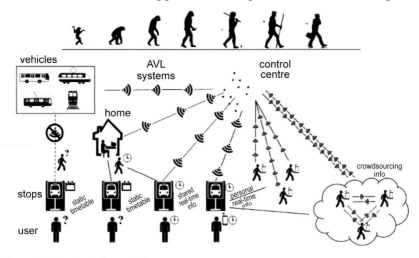

Figure 1.2 Traveller info evolution.

improve transit vehicle trip regularity and mitigate crowding. This type of information can also be provided to travellers, who can choose to skip overloaded runs and wait for less crowded ones, thereby making a trade-off between a longer waiting-time and higher on-board comfort.

The opportunities given by new telematics applications in supporting service operations control are investigated in depth in Chapter 2 (Ceder, 2016), while Chapter 3 (Comi et al., 2016) analyzes the developments of individual real-time traveller info systems.

The large quantity of available data can also be used to investigate the scheduling problem of transit services in multimodal transit networks and to design transit networks with stochastic demand. Indeed, as explored in Chapter 9 (Lam and Li, 2016), the short-term timetables of multimodal transit lines for operations and service planning purposes can be obtained. In Chapter 10 (An and Lo, 2016) the transit network design problem with stochastic demand is studied with reference to two types of services (i.e., rapid transit line and dial-a-ride services).

As reported in this chapter, with the availability of large amounts of data and bi-directional communication, researchers are being encouraged to develop new modelling approaches for real-time transit network simulation and forecasting of vehicle occupancy and crowding degree, with methods which upgrade and update demand flows and path choice model parameters in real time. Further, traveller-tailored models, which take into account personal preferences, can be defined.

This chapter is structured as follows. Section 1.2 focuses on methods and models which can be used to estimate in real time and to forecast vehicle occupancy and crowding degree. Section 1.3 summarizes the latest developments in transit assignment models for real-time applications, while Section 1.4 focuses on advances in traveller path choice modelling, which allows unreliable networks to be simulated and personal preferences to be taken into account. In Section 1.5, the methods which can be used to update and upgrade demand OD flows and model parameters are reconsidered in the new ITS context. Finally, Section 1.6 deals with the main methodological issues and challenges involved in the new research methods.

1.2 On-board Load Forecasting Methodologies

Short-term forecasting of on-board passenger flows is traditionally obtained through data-driven projection methods (Tsai et al., 2009; Chen et al., 2011; Wei and Chen, 2012; Jiuran and Bingfeng, 2013; Moreira-Matias et al., 2015; Sun et al., 2015) similar to those used for short-term road traffic forecasting (Vlahogianni et al., 2014). Such methods present some general limitations, not only in the case of snap disruptions, which bring about major changes in path choices, that cannot be properly taken into consideration, but also in more general conditions. Indeed, when these types of methods are used for public transport, they lead to approximations in both accuracy and time response, as public transport can be accessed only at given points and is available only at given instants. For example, random characteristics of transit services can influence the arrival order of bus lines at stops and hence the number of passengers boarding.

For this reason, some authors have focused on forecasting methods and simulation models that reproduce the traveller's behavior and the way in which travellers respond on the transit network according to real-time demand and transit service configurations. These models, classified as *topological-behavioral*, are based on the *Network Assignment Modelling* approach traditionally used by transportation engineering (Cascetta, 2009; Ortuzar and Willumsen, 2011) and are adapted to transit networks and to short-term forecasting.

As summarized in Fig. 1.3 and Fig. 1.4, and explored in the following chapters of this book, the real-time network assignment model system consists of four main components:

- *network sub-system*, which reproduces real-time supply functioning, using cost functions obtained through historical and real-time data;
- *demand sub-system*, using traditional models and methods in combination with real-time data collecting, this sub-system provides real-time origin-destination flows;
- *assignment sub-system*, which simulates the interaction between demand and supply and provides real-time transit vehicle loads;
- *origin-destination matrix and model parameter up-grading*, which uses historical and real-time data to correct initial values (see Fig. 1.4).

This approach entails that historical and real-time data of passenger flows (and loads) and of network states are used to calibrate network system model components rather than as inputs of projection methods to forecast flows and on-board loads directly.

Recent ITS developments allow several of the main traditional limitations of network assignment modelling to be overcome. Such limitations were found in data collecting, which was usually expensive, not always precise, increasingly complex (right of privacy) and often represented only some aspects of the mobility of specific (limited) temporal periods. Bi-directional communication between travellers and info centres generates data, which can be used not only to estimate improvements in real-time origin-destination matrices and model parameters, but also for travel advice, taking personal preferences into account, as reported in Section 1.4. With transit network big data collecting and processing, a large quantity of data can be obtained at low cost and used to improve the implementation of network supply models, as well as in estimation of origin-destination matrices and model parameters.

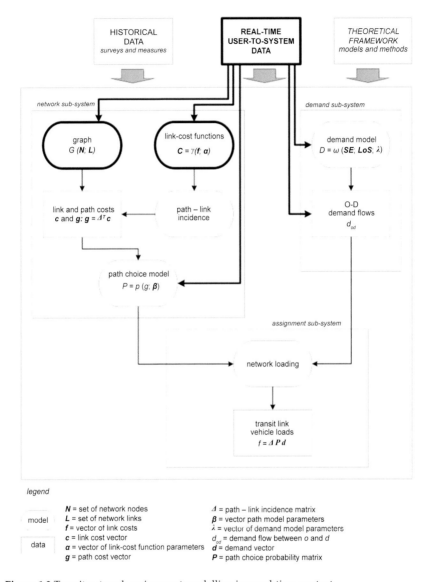

Figure 1.3 Transit network assignment modelling in a real-time context.

An example of logical architecture for a tool which forecasts on-board occupancy, developed within the network assignment approach (Nuzzolo et al., 2013), is depicted in Fig. 1.4. It represents the core of a more general tool under development for the short-term prediction of transit vehicle occupancy which, integrated through simple communication interfaces into an existing real-time info system, can enhance pre-trip and *en-route*

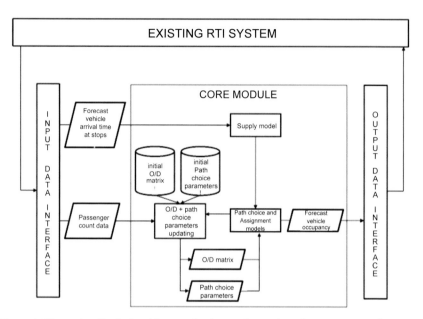

Figure 1.4 Example of logical architecture for forecasting on-board passenger numbers.

information concerning vehicle occupancy. This logical architecture consists of two input data interfaces (enabling the real-time forecast of vehicle arrival times at stops and passenger count data to be received), a core module and an output data interface. The core module allows in real time the on-board passenger number to be forecast for each vehicle at each stop, using network assignment modelling and demand data and model parameters upgrading, explored in the second part of this chapter and in Chapters 4 (Cats, 2016) and 8 (Russo and Vitetta, 2016) of this book. The output data interface provides information on arrival time of transit vehicles at stops and on their occupancy degree to the real-time transit information system (RTIS) and then to travellers (and transit operators).

The recent trends in the development of real-time short-term forecasting and traveller information tools indicate that major changes are required in transit modelling. Such changes, explored in the sections below, especially concern:

- *development of more advanced real-time transit assignment methods*, for example, mesoscopic simulation-based assignment implemented in a real multimodal transit network;
- *path choice models* for unreliable networks which use a strategy-based approach, take into account real-time individual information and implement *personal path choice modelling*, with traveller-tailored model parameters;

- *demand data and model parameters upgrading* methods, which allow use of the considerable availability of passenger data and bi-directional communication in order to improve real-time estimations of origin-destination (O-D) matrices and path choice model parameters.

1.3 Real-time Transit Assignment Modelling

In order to identify the most suitable assignment modelling approach to short-term forecasting of at-stop waiting and on-board passenger numbers, to be used in real-time operations control and information to travellers, a classification of existing transit assignment models (TAMs) is proposed below. Current mesoscopic simulation-based assignment models, which seem to be most useful for real-time applications, are then summarized and appraised. An analysis of mesoscopic models is also carried out in Chapter 4 (Cats, 2016) and Chapter 6 (Comi and Nuzzolo, 2016).

1.3.1 Transit Assignment Model Classification

From the point of view of this chapter, existing TAMs can be classified as:

- *run-oriented*, with results concerning each run (transit trip) of each line, in terms of on-board passengers and vehicle dwelling and running times, for each section of the line route;
- *line-oriented*, with results concerning average passenger flows and average vehicle travel time of each link on a transit line.

The most widely studied and commonly applied line-oriented models are frequency-based transit assignment models (Gentile and Nokel, 2016). These models are mainly used in network strategic planning and, as they are unable to consider single runs, are not further considered in this chapter, where the assignment results for each run are essential.

Run-oriented TAMs can in turn be classified into:

- *analytical models*: model results, such as passenger flows, derive from mathematical equations. As it is not easy to obtain analytical functions which describe all demand and supply processes and connected interactions, the following further TAM classes have been developed;
- *simulation-based models*, which reproduce interactions over time among different agents involved in the transit system;
- *mixed analytical-simulation models*, which use both previous approaches.

Run-oriented analytical TAMs (among the most recent papers, see Sumalee et al., 2009; Khani, 2013; Hamdouch et al., 2014; Gentile and Nokel, 2016; Gentile et al., 2016) simulate transit functioning, taking into account the arrival time of each vehicle at stops, with time-dependent origin-destination flows and traveller choices at the level of the single run.

These models use a space-time representation of services and loading on these networks, which is intrinsically within-day dynamic, is carried out by using pre-trip path choices. Equilibrium or stable configurations can be obtained through traditional assignment algorithms.

Simulation-based or *agent-based* models (Cats, 2013) reproduce the interactions among different agents over time with travellers, transit vehicles and sometimes also other vehicles sharing the right of way. Obviously, these models are intrinsically run-oriented.

In relation to the details of interactions considered, the simulation-based models can be further classified into:

- *microsimulation models*, when interactions among agents are reproduced in detail, moment by moment;
- *mesoscopic models*, when some interactions are omitted or simplified in relation to their application fields.

Mesoscopic models have the advantage of shorter computer time and are therefore most suitable for use in real-time simulations. For this reason, they are analyzed in more detail in Section 1.3.2 below.

In order to overcome the drawback of analytical models in relation to the difficulty of taking account of explicit vehicle capacity constraints and failures to board, *mixed analytical-mesoscopic* models have been proposed. During the network loading step of the assignment procedure (e.g., method of successive averages, MSA), such models simulate competition among travellers at each stop (Sumalee et al., 2009; Khani, 2013; Zhang et al., 2010).

1.3.2 Mesoscopic Simulation-based Models

In the field of mesoscopic assignment models for transit networks in the presence of traveller information, the literature is very limited. Among the few contributions, models proposed by Nuzzolo et al. (2001, 2012, 2016), Wahba and Shalaby (2009) and Cats et al. (2011, 2013) may be mentioned.

Wahba and Shalaby (2009) presented the MILATRAS modelling framework to simulate time-dependent and stochastic transit services in the presence of ITS. Demand is divided into segments characterized by desired arrival time at the destination and the origin and destination locations of single travellers (agents) are randomly obtained using a GIS platform based on residential and employment proportions. Travel choices consider departure time, access stop and boarding run and are based on a decision process founded on modelling utility perception, strategy and experience at the single agent level.

Cats et al. (2011, 2013) proposed a mesoscopic model in BUSMEZZO, a dynamic transit operation and assignment model, developed in the framework of the agent-based approach, which simulates the progress of vehicles and travellers in the transit system and yields temporal and spatial distribution of the latter over the former. Each traveller makes a sequence of travel decisions in reaction to changing environmental conditions (such as a bus arriving at the stop, announcement of a delayed train or consideration of elapsed waiting time), which are simulated using random utility choice models. Such models also consider individual information about waiting and in-vehicle times.

Nuzzolo et al. (2001) presented a run-oriented assignment model for congested networks by using a mesoscopic approach and later extended it over time to consider the presence of vehicle arrival time information and capacity constraints (Nuzzolo et al., 2012). This model is part of DYBUS, a transit simulation framework, oriented to network planning and traveller information assessment. Travellers are characterized by origin-destination and target time and are assumed to have a flexible target time, within a certain range, in order to avoid congestion and mitigate effects of perturbations (irregularity) on the scheduled timetable. Travellers' choices are assumed strictly dependent on anticipated path attributes, which are a function of those experienced and of that supplied by the info system. by the info system (Nuzzolo et al., 2015). Such a mesoscopic assignment model was further developed (Nuzzolo et al., 2016), considering, amongst the information provided, also on-board crowding (Nuzzolo et al., 2015).

Depending on objectives and fields of application (operations control or travel advice), the above-cited models present different modelling features. This is exemplified by the way in which they consider transport demand: Wahba and Shalaby (2009) and Cats et al. (2011) consider a disaggregate representation of demand at the individual passenger level, while Nuzzolo et al. (2001) use groups or packets of passengers with the same origin, destination and desired departure or arrival time. All the models reproduce the traveller's choice as a result of the application of a travel strategy by including transit path alternatives defined as combinations of departure times, boarding stops, lines and walking connections, although Cats et al. (2011) exclude departure times from the above combinations. Finally, in relation to the formalization of path choice modelling, different approaches are adopted: Wahba and Shalaby (2009) use bounded rationality choice models, while Nuzzolo et al. (2001, 2012, 2016) and Cats et al. (2011, 2013) specify random utility choice models. Other differences concern segmentation of demand over time; approach in the representation of the transit network; behavioral assumptions in traveller path choice according to type and contents of information provided and to the learning process of choice attributes; choice dimensions in path choice; choice set generation; model specification (e.g., functional form) and so on.

1.3.3 General Requirements of Real-time Mesoscopic TAMs

To date, the mesoscopic models reported above have been implemented and used mainly in off-line applications for planning or design assessments, seeking to simulate. To the authors' knowledge, real-time application of transit assignment is still a research issue. The main components of TAM systems have to be improved in order to highlight the requirements of such systems for real-time simulations, given that real-time assignment for short-term forecasting requires computing times compatible with the temporal step of simulation. Further, maximum precision of results is required, given that forecasted values are compared by travellers with actual ones, with possible negative effects on information system compliance. In terms of result precision, one of the main success factors is the way demand and supply uncertainty are dealt with. In the real-time transit assignment procedure, the stochastic characteristics of demand can be considered through real-time upgraded origin-destination matrices and real-time upgraded path choice model parameters, as indicated in Section 1.5. Stochastic supply can be considered using strategy-based path choice models for unreliable networks, although the methodologies for hyperpath choice set generation and choice are still an open field of research. Further, individual path utility function parameters can be applied, such as those considered in Section 1.4 below.

In terms of computing time, relationships between simulation times and network dimensions have to be analyzed, trying to optimize the component procedures of simulation codes (see for example, Chapter 7, Gentile, 2016).

1.4 Advanced Path Choice Modelling

Two aspects of path choice modelling advancements are presented below—the first entails simulating traveller behavior on unreliable networks and the second concerns models with individual path utility function parameters.

1.4.1 Path Choice Modelling for Unreliable Networks

Path choice decision-making is influenced by several factors. One of the more complex situations occurs when travellers move on an unreliable network, in which there are certain nodes (*diversion nodes*) where travel decisions have to be carried out according to random transit service occurrences. Even if a system of predictive information on the network states is available to travellers, due to the uncertainty of such forecasting, the choice of one specific path until destination may not be the best decision. Rather, a path strategy should be used, with a choice rule among diversion links according to the phenomenon that occurs. In the context of strategy-based path choice behavior, normative and descriptive strategy-based path choice models can be applied. In this regard, a normative approach is presented in Chapter 5 (Nuzzolo and Comi, 2016a) and can be applied in a trip planner path device, as indicated in Chapter 3 (Comi et al., 2016).

Of course, due to cognitive processes, the behavior of travellers is not exactly normative. Hence, in order to reproduce traveller behavior, a descriptive model has to be used (see Chapter 6, Comi and Nuzzolo, 2016). Descriptive strategy-based path choice behaviors and models for unreliable networks with individual predictive information and suitable for real-time simulation-based assignment modelling, have been presented by Wahba and Shalaby (2009), Cats (2013) and Nuzzolo et al. (2016).

In strategy-based path choice modelling, there are several issues which have to be explored in order to improve simulation procedures. Such issues include: how to combine experience and info supplied, dynamic real-time generation of the path choice set and dynamic real-time expected utility computation for each strategy (pointing group or individual preferences). In the case of a descriptive strategy, modelling strategy-based choice set generation (as explored in Chapter 6, Comi and Nuzzolo, 2016), strategy choice (see, for example, Schmocker et al., 2013) and travel option overlapping (see, for example, Russo and Vitetta, 2003) poses a major research task, such as the development of non-expected utility approaches, e.g., using prospect theory and regret theory (de Palma et al., 2008; Ramos et al., 2014).

1.4.2 Individual Path Choice Modelling

It is easier to have many revealed path choices from several decision makers rather than several revealed choices from the same user. Therefore, contrasting with individual models, path choice models have traditionally been developed with group-level parameters obtained by aggregating several sampled users. Although various types of group models have been proposed, their performance seems limited if applied to simulate single decision making due to variations in taste or preferences among users (Dumont et al., 2015; Nuzzolo and Comi, 2016b), as shown below.

Therefore, in the context of ITS, where bi-directional communication allows individual travellers to be traced, a new challenge involves developing path choice models that take personal preferences and attitudes into account, using choice samples of the single decision maker. In individual modelling, one of the main issues concerns parameter estimation with single user repeated observations, because, as widely detailed in the literature (Lancsar and Louvier, 2008; Frischknecht et al., 2011), it could cause an obvious correlation of disturbances.

Further, heterogeneity could be present, as preferences can also vary for the same user over time. Therefore, estimates of model coefficients will be biased if heterogeneity and correlation are not properly treated.

The Mixed Logit (ML) model offers much in terms of effects due to repeated observations (Ortuzar et al., 2000; Ortuzar and Willumsen, 2011). Revelt and Train (1998) proposed a mixed logit framework which accommodates inter-respondent heterogeneity but assumes intra-respondent homogeneity in tastes (i.e., it includes the effect of repeated choices by assuming that tastes vary across respondents but stay constant across observations for the same respondent). Hess and Rose (2009) relaxed the assumption of intra-respondent homogeneity of tastes.

Besides, other issues refer to data collected. Repeated observations in a short-survey panel can increase the number of observations but might reduce data variability, because observations which are identical do not bring new information about attribute trade-offs.

Tests were carried out by Nuzzolo and Comi (2016b) to identify the best specification of the individual path choice model and compare the performance of individual and aggregate modelling. The tests were carried out in a transit corridor of the metropolitan area of Rome served by a multiservice transit network operated by different companies, where urban bus, tram and metro regional railway lines and regional bus lines with an integrated fare policy are available. As regards path choices at origin, different model specifications were tested, starting from the simplest multinomial logit model (MNL) to the mixed-logit (MXL) and nested-logit (NSL). Mixed or nested logit did not provide significant improvements in parameter estimations, suggesting the use of MNL models for their well-known easy-to-apply advantages. On the other hand, major differences among the travellers' preferences were noted, with different attributes and parameters present in the utility functions of different travellers.

Individual modelling, in general, should, therefore, be preferred to group-level modelling in travel information tools, such as trip planners (TP) as reported in Chapter 3 (Comi et al., 2016), and in simulation-based assignment with the simulation of each traveller's behavior.

1.5 Real-time Upgrading of the O-D Matrix and Model Parameters

Given the quantity and types of data currently available, including bi-directional communication, a new research challenge consists in real-time updating and upgrading of origin-destination matrices, link cost function parameters and path choice model parameters, starting from passenger counts and other data collected from transit systems.

In general, up to now, transit model upgrading has been carried out with an off-line approach and separately for cost functions for OD flows and for model parameters. Among the studies dealing with transit link cost functions, especially transit link travel time as a function of passenger flows and other variables, see Parveen et al. (2007), Moreira-Matias et al. (2015) and Mori et al. (2015). The problem of estimating time-varying OD flows using traffic counts for transit networks was studied in Wong and Tong (1998), Yuxiong et al. (2015), Ji et al. (2015). While several studies exist for car networks (Siripirote et al., 2014), the number of papers dealing with estimation of transit path choice model parameters from traffic counts, is very small.

Conjoint estimation of O-D flows and model parameters can be performed, as reported for car networks in Nguyen et al. (1988) and Cascetta and Russo (1997). A conjoint approach for transit network has been used (Nuzzolo et al., 2013) in a real test application of a short-term forecasting procedure, under development in Santander (Spain). This city has a population of about 180,000 and is served by a bus transit network of 43 lines and 430 stops. The problem is solved by minimising the distances between simulated and measured variables, as follows:

$$\left(\underline{d}^*, \underline{\beta}^*\right) = \underset{\underline{d} \in S_d, \underline{\beta} \in S_\beta, \underline{f}^* \in S_f}{\arg \min} \left[z_1(\underline{d}, \tilde{\underline{d}}) + z_2(\underline{\beta}, \tilde{\underline{\beta}}) + z_3(\underline{f}^*, \hat{\underline{f}}) \right] \tag{1.1}$$

where:

- $\underline{d}^*$ is the updated demand vector;
- $\underline{\beta}^*$ represents the updated path choice model parameters;
- z_1, z_2, z_3 are distance functions;
- S_d is the feasibility domain of demand flows;
- S_f is the feasibility domain of link flows;
- S_β is the feasibility domain of path choice model parameters;
- $\underline{d}$ is the demand vector;
- $\tilde{\underline{d}}$ is the *a-priori* demand vector;
- $\underline{f}^*$ is the vector of link flows;
- $\hat{\underline{f}}$ is the vector of traffic counts;
- $\underline{\beta}$ are the path choice model parameters;
- $\tilde{\underline{\beta}}$ are the *a-priori* path choice model parameters.

The choice of the functional form for z_1, z_2, z_3 depends on the type of available information and on the weights which can be introduced in each term to give a different sensitivity to each calibrated category. In Eq. 1.1, some constraints are necessary to describe transit flows being linked to transit demand and path choice parameters through the function $\underline{f} = v(\underline{d}, \underline{\beta})$ where relation v represents the transit assignment matrix.

To solve the problem, a bi-level approach (Fig. 1.5) was used: the algorithm starts from a given transit demand vector $\underline{d}$ and path choice model parameter β configurations, and calculates distance functions $z_1 = (\underline{d}, \tilde{d})$ and $z_2 = (\beta, \tilde{\beta})$. At the same time, the first level of the optimisation algorithm computes flow vector $\underline{f}^*$ using vectors $\underline{d}$ and β through a mesoscopic run-oriented assignment model.

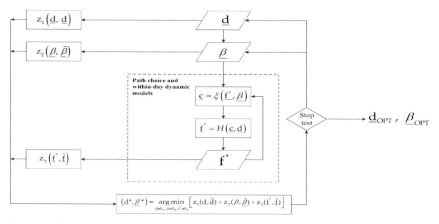

Figure 1.5 Demand and path choice model parameter updating.

The initial stop-to-stop O-D matrix was estimated using classic four-stage travel demand models. The path choice model parameters were adapted starting from those proposed in Nuzzolo et al. (2001).

In order to evaluate the performance of short-term forecasting methods, the boarding flows at stops for each transit ride were predicted and then compared with the real boarding flows revealed by boarding sensors. If only significant boarding flows (e.g., those greater than five travellers) are considered, the MSE is about 7 and the RMSE percentage is about 32 per cent. These results are quite satisfactory given that path choice model parameters were adapted from other studies. Aiming to improve the accuracy and efficiency of real-time predictions, further developments are in progress. Such developments mainly concern the improvement in the methods for real-time joint upgrading of the demand and the path choice model parameters and the use of a specific path choice model calibrated by using ad-hoc surveys and counts. The open research perspectives also concern the investigation of existence and uniqueness of a problem solution and new algorithms for efficient problem solving.

According to Russo and Vitetta (2012), a unified formulation can be used to obtain at the same time the parameters of link performance functions, the origin-destination demand values and the values of demand model parameters, namely the *reverse assignment problem*. This approach is extended to real-time application in Chapter 8 (Russo and Vitetta, 2016).

1.6 Concluding Remarks

Bi-directional communication between travellers and info centres, and transit network big data collecting and processing appear to be two new factors which can improve the tools for real-time transit network simulation and traveller info, which in turn will improve service efficiency and quality. Several methodologies are under development to capture the new opportunities offered by ITS developments and to forecast on-board occupancy as well as support travellers, providing personalised pre-trip and *en-route* information.

For short-term vehicle occupancy forecasting, useful both for transit operations control and traveller advice, the network assignment approach rather than data-driven projection methods, seems to allow more precise forecasts of on-board loads and numbers of users waiting at stops.

Further, as the performance of group parameter path choice models appears limited due to dispersion among travellers and their variations in taste/preferences, estimations of individual path choice model parameters have to be obtained from a sample of observations of single travellers.

However, several theoretical issues remain a matter for in-depth research, namely, better understanding of traveller path choice decisional process and information effects; specification of more advanced real-time strategy-based path choice models using individual path utility parameters and optimization of relative parameter learning processes; development of single traveller dynamic real-time network loading when individual path choice models are used; investigation of theoretical proprieties of the demand-supply interaction process and development of more efficient and effective methodologies to update O-D matrices and model parameters.

Keywords: dynamic transit assignment; advanced traveller information systems; personalized multimodal advisor; real-time transit network modelling; transit dynamic reverse assignment.

References

An, K. and H. K. Lo. 2016. Transit network design with stochastic demand. *In*: A. Nuzzolo and W. H. K. Lam (eds.). Modelling Intelligent Multi-modal Transit Systems, CRC Press.

Cascetta, E. 2009. Transportation Systems Analysis: Models and Applications. Springer.

Cascetta, E. and F. Russo. 1997. Calibrating aggregate travel demand models with traffic counts: Estimators and statistical performance. *In*: Transportation 24: 271–293.

Cats, O., H. N. Kousopoulos, W. Burghout and T. Toledo. 2011. Effect of real-time transit information on dynamic path choice of passengers. *In*: Transportation Research Record 2217: 46–54.

Cats, O. 2013. Multi-agent transit operations and assignment model. Procedia of Computer Science 19: 809–814.

Cats, O. 2016. Real-time operations management decision support systems. *In*: A. Nuzzolo and W. H. K. Lam (eds.). Modelling Intelligent Multi-modal Transit Systems, CRC Press.

Cats, O., H. N. Koutsopoulos, W. Burghout and T. Toledo. 2013. Effect of real-time transit information on dynamic passenger path choice. CTS Working Paper 2013: 28.

Ceder, A. 2016. New applications of ITS to real-time transit operations. *In*: A. Nuzzolo and W. H. K. Lam (eds.). Modelling Intelligent Multi-modal Transit Systems, CRC Press.

Chen, Q., W. Li and J. Zhao. 2011. The use of LS-SVM for short-term passenger flow prediction. *In*: Transport 26(1), Taylor & Francis.

Comi, A. and A. Nuzzolo. 2016. A dynamic strategy-based path choice modelling in real-time transit simulation. *In*: A. Nuzzolo and W. H. K. Lam (eds.). Modelling Intelligent Multi-modal Transit Systems, CRC Press.

Comi, A., A. Nuzzolo, U. Crisalli and L. Rosati. 2016. A new generation of individual real-time transit information systems. *In*: A. Nuzzolo and W. H. K. Lam (eds.). Modelling Intelligent Multi-modal Transit Systems, CRC Press.

de Palma, A., M. Ben-Akiva, D. Brownstone, C. Holt, T. Magnac, D. McFadden, P. Moffatt, N. Picard, K. Train, P. Wakker and J. Walker. 2008. Risk, uncertainty and discrete choice models. *In*: Market Letters 19: 269–285.

Dumont, J., M. Giergiczny and S. Hess. 2015. Individual level models vs. sample level models: Contrasts and mutual benefits. pp. 465–483. *In*: Transportmetrica A: Transport Science 11(6), Taylor & Francis.

Frischknecht, B., C. Eckert and J. Louviere. 2011. Simple ways to estimate choice models for single consumers. Centre for the Study of Choice (CenSoC), Working Paper Series, No. 11-006, University of Technology of Sydney.

Gentile, G. 2016. Dynamic routing on transit networks. *In*: A. Nuzzolo and W. H. K. Lam (eds.). Modelling Intelligent Multi-modal Transit Systems, CRC Press.

Gentile, G. and K. Noekel (eds.). 2016. Modelling Public Transport Passenger Flows in the Era of Intelligent Transport Systems. Springer International Publishing.

Gentile, G., M. Florian, Y. Hamdouch, O. Cats and A. Nuzzolo. 2016. The theory of transit assignment: Basic modelling frameworks. pp. 287–386. *In*: G. Gentile and K. Noekel (eds.). Modelling Public Transport Passenger Flows in the Era of Intelligent Transport Systems: COST Action TU1004 (TransITS), DOI: 10.1007/978-3-319-25082-3_6, Springer Tracts on Transportation and Traffic 10, Springer, Switzerland.

Hamdouch, Y., W. Y. Szeto and Y. Jiang. 2014. A new schedule-based transit assignment model with travel strategies and supply uncertainties. *In*: Transportation Research Part B: Methodological 67: 35–67.

Hess, S. and J. M. Rose. 2009. Allowing for intra-respondent variations in coefficients estimated on repeated choice data. Transportation Research 43B: 708–719.

Jiuran, H. and S. Bingfeng. 2013. The application of ARIMA-RBF model in urban rail traffic volume forecast. Proceedings of the 2nd International Conference on Computer Science and Electronics Engineering (ICCSEE 2013).

Ji, Y., R. G. Mishalani and M. R. McCord. 2015. Transit passenger origin–destination flow estimation: Efficiently combining on-board survey and large automatic passenger count datasets. *In*: Transportation Research Part C: Emerging Technologies 58: 178–192.

Khani, A. 2013. Models and Solution Algorithms for Transit and Intermodal Passenger Assignment (Development of Fast Trips Model). Ph.D. Thesis, University of Arizona, USA.

Lam, W. H. K. and Z. C. Li. 2016. Optimal schedules for multimodal transit services: An activity-based approach. *In*: A. Nuzzolo and W. H. K. Lam (eds.). Modelling Intelligent Multi-modal Transit Systems, CRC Press.

Lancsar, E. and J. Louviere. 2008. Estimating individual level discrete choice models and welfare measures using best worst choice experiments and sequential best worst MNL. http://www.censoc.uts.edu.au/.

Moovit. 2009. Moovit – Public Transport. www.moovitapp.com.

Moreira-Matias, L., J. Mendes-Moreira, J.F. de Sousa and J. Gama. 2015. Improving mass transit operations by using AVL-based systems: A survey. *In*: IEEE Transactions on Intelligent Transportation System, DOI 10.1109/TITS.2014.2376772.

Mori, U., A. Mendiburu, M. Álvarez and J. A. Lozano. 2015. A review of travel time estimation and forecasting for advanced traveller information systems. *In*: Transportmetrica A: Transport Science 11(2): 119–157.

Nguyen, S., E. Morello and S. Pallottino. 1988. Discrete time dynamic estimation model for passenger origin/destination matrices on transit networks.

Nuzzolo, A. and A. Comi. 2016a. Real-time modelling of normative travel strategies on unreliable dynamic transit networks: A framework analysis. *In*: A. Nuzzolo and W. H. K. Lam (eds.). Modelling Intelligent Multi-modal Transit Systems, CRC Press.

Nuzzolo, A. and A. Comi. 2016b. Individual utility-based path suggestions in transit trip planners. *In*: IET Intelligent Transport System, DOI: 10.1049/iet-its.2015.0138, The Institution of Engineering and Technology.

Nuzzolo, A., F. Russo and U. Crisalli. 2001. A doubly dynamic schedule-based assignment model for transit networks. Transportation Science 35: 268–285.

Nuzzolo, A., U. Crisalli and L. Rosati. 2012. A schedule-based assignment model with explicit capacity constraints for congested transit networks. *In*: Transportation Research Part C: Emerging Technologies 20(1): 16–33.

Nuzzolo, A., U. Crisalli, A. Comi and L. Rosati. 2016. A mesoscopic transit assignment model including real-time predictive information on crowding. *In*: Journal of Intelligent Transportation Systems: Technology, Planning and Operations, DOI: 10.1080/15472450.2016.1164047, Taylor & Francis.

Nuzzolo, A., U. Crisalli, L. Rosati and A. Comi. 2015. DYBUS2: A real-time mesoscopic transit modelling framework. pp. 303–308. *In*: 18th IEEE International Conference on Intelligent Transportation Systems (ITSC 2015), DOI: 10.1109/ITSC.2015.59.

Nuzzolo, A., U. Crisalli, L. Rosati and A. Ibeas. 2013. STOP: A short-term transit occupancy prediction tool for APTIS and real time transit management systems. IEEE Proceedings of the 16th International IEEE Conference on Intelligent Transportation Systems: Intelligent Transportation Systems for All Modes, ITSC, Art. No. 6728505, DOI: 10.1109/ITSC.2013.6728505, 1894–1899.

Ortuzar, J. de D. and L. G. Willumsen. 2011. Modelling Transport. John Wiley & Sons, Ltd.

Ortuzar, J. de D., D. A. Roncagliolo and U. C. Velarde. 2000. Interactions and independence in stated preference modelling. *In*: J. de D. Ortuzar (ed.). Stated Preference Modelling Techniques. Perspectives 4, PTRC, London.

Parveen, M., A. Shalaby and M. Wahba. 2007. G-EMME/2: Automatic calibration tool of the EMME/2 transit assignment using genetic algorithms. *In*: J. Transp. Eng. 133(10): 549–555.

Ramos, G. M., W. Daamen and S. Hoogendoorn. 2014. A state-of-the-art review: Developments in utility theory, prospect theory and regret theory to investigate travellers' behaviour in situations involving travel time uncertainty. *In*: Transport Reviews: A Transnational Transdisciplinary Journal 34(1): 46–67, DOI: 10.1080/01441647.2013.856356.

Revelt, D. and K. Train. 1998. Mixed logit with repeated choices: households' choices of appliance efficiency level. pp. 647–657. *In*: Review of Economics and Statistics.

Russo, F. and A. Vitetta. 2003. An assignment model with modified Logit, which obviates enumeration and overlapping problems. *In*: Transportation 30, Springer.

Russo, F. and A. Vitetta. 2012. Reverse assignment: Calibrating link cost functions and updating demand from traffic counts and time measurements. *In*: Inverse Problems in Science and Engineering 19(7): 921–950.

Russo, F. and A. Vitetta. 2016. Real-time reverse dynamic assignment for multiservice transit systems. *In*: A. Nuzzolo and W. H. K. Lam (eds.). Modelling Intelligent Multi-modal Transit Systems, CRC Press.

Schmocker, J. D., H. Shimamoto and F. Kurauchi. 2013. Generation and calibration of transit hyper-paths. Transportation Research C 36: 406–418.

Siripirote, T., A. Sumalee, D. P. Watling and H. Shao. 2014. Updating of travel behaviour parameters and estimation of vehicle trip-chain data based on plate scanning. *In*: Journal of Intelligent Transportation Systems: Technology, Planning and Operations 18(4): 393–409.

Sumalee, A., Z. J. Tanm and W. H. K. Lam. 2009. Dynamic stochastic transit assignment with explicit seat allocation model. Transportation Research Part B 43, Elsevier, 895–912.

Sun, Y., B. Leng and W. Guan. 2015. A novel wavelet-SVM short-time passenger flow prediction in Beijing. pp. 109–121. *In*: Neurocomputing 166, Elsevier.

Tsai, T. H., C. K. Lee and C. H. Wei. 2009. Neural network based temporal feature models for short-term railway passenger demand forecasting. *In*: Expert Systems with Applications 36(2): 3728–3736.

Vlahogianni, E. I., J. C. Golia and G. Karlaftis, M. 2014. Short-term traffic forecasting: overview of objectives and methods. Transport Reviews 24(5): 533–557.

Wahba, M. and A. Shalaby. 2009. Learning-based departure time and path choice modelling for transit information under information provision: A theoretical framework. Proceedings of the 88th Transportation Research Board Annual Meeting, Washington DC, USA.

Wei, Y. and M. Chen. 2012. Forecasting the short-term metro passenger flow with empiric mode decomposition and neural networks. *In*: Transportation Research 21C: 148–162.

Wong, S. C. and T. O. Ton. 1998. Estimation of time-dependent origin-destination matrices for transit networks. *In*: Transportation Research B 32(1): 35–48.

Yuxiong, J., R. G. Mishalani and M. R. McCord. 2015. Transit passenger origin–destination flow estimation: Efficiently combining on board survey and large automatic passenger count datasets. *In*: Transportation Research Part C 58: 178–192.

Zhang, Y. Q., W. H. K. Lam, A. Sumalee, H. K. Lo and C. O. Tong. 2010. The multi-class schedule-based transit assignment model under network uncertainties. Public Transport 2: 69–86.

2

New Applications of ITS to Real-time Transit Operations

Avishai (Avi) Ceder

ABSTRACT

This chapter is based on a research conducted in the last six years, and part of which was presented as a keynote lecture at the COST-TU1004 TransITS conference, Paris, 11–12 May 2015. The major goal of this chapter is to introduce new and improved ideas and methods of transit-related intelligent transportation systems (ITS). It is not only a matter of illustrating new ITS applications in transit planning and operation, but also to find how far the chapter can inspire the reader's imagination to think further. The chapter consists of six main sections: (i) introduction, (ii) multi-agent transit system (MATS), (iii) synchronized transfers, (iv) real-time operational tactics, (v) customized bus, and (vi) vehicle-to-vehicle communication and predictive control. Firstly an introduction on ITS-related updates in transit is presented; secondly the remaining five main sections are introduced, including a literature review.

Department of Civil and Environmental Engineering at both the University of Auckland, New Zealand, and Technion – Israel Institute of Technology, Israel.
Address: 10 HaLilach street, Apt 2, Haifa, Israel.
E-mail: a.ceder@auckland.ac.nz; ceder@technion.ac.il

2.1 Introduction

The definition of ITS is provided in the EU Directive 2010/40/EU in Article 4 (Directive, 2010) describing ITS as systems in which information and communication technologies are applied. Two recent ITS transit-related studies are henceforth briefly illustrated. Schweiger (2015) describes the availability and opportunity of open transit data to the public. In this TRCP (of USA) report, Schweiger (2015) shows the practice and policies in use of open data for improved transit operations planning, transit reliability and customer information. In addition this report depicts the implications of open data and open documentation policies while focusing on successful practices. For instance, the state of the open data for transit timetables across a large number of countries is shown online (Timetables, 2015). This online graphical description contains icons to indicate data accessibility and availability. In another recent study, Pessaro (2015) provides an overview of the state of automated vehicle technology in transit operations. He describes the involvement of the international association of public transport (UITP) in two projects: the European bus system of the future and the follow-on 3iBS project. However, these two projects mainly focus on improving vehicle aesthetics and not so much on automated vehicle technology. In addition Pessaro (2015) found that in the USA, there are only two transit-related automated vehicle technology studies: (i) the University of Minnesota developed a GPS-based driver assist system to improve safety during bus shoulder operations, and (ii) the PATH program of the University of California at Berkeley developed a magnetic guidance system that is used for precision docking by the EmX bus rapid transit (BRT) system at three stations.

2.2 Multi-Agent Transit System (MATS)

This section refers to what is known as multi-agent systems (MAS) and starts with a brief background. Van Dyke Parunak (1997) defined MAS as collections of autonomous agents within an environment that interact with each other for achieving internal and/or global objectives. Minsky (1986) argued that an intelligent system could emerge from non-intelligent parts. His definition of the 'Society of Mind' makes use of small, simple processes, called agents, each of which performs some simple action, while the combination of all these agents forms an intelligent society (system). Bradshaw (1997) classified agents by three attributes: autonomy, cooperation and learning. He defined four agent types from these attributes: collaborative, collaborative learning, interface and smart. The most interesting type so far as the public transit system is concerned is the collaborative agent. Such an agent is simple and can perform tasks independently, but can collaborate with other agents if necessary in order to achieve a better solution. Zhao et al. (2003) developed MAS for a bus-holding algorithm. The authors treat each bus-stop as an agent; the agents negotiate with each other, based on marginal cost calculations, to devise minimal passenger waiting-time costs.

The multi-agent transportation system (MATS) described in this section is based on Hadas and Ceder (2008b). The MATS will be composed of the following agents: public transit vehicles, passengers, road segments, transit agencies and transit authorities. In order to construct the system as a whole, it is necessary to define each agent and the interrelationship of agents. Since the system is complex and cumbersome, each agent will be explored separately. This will also make it easier to define the system's elements. Figure 2.1 presents the main activities and interrelationships of the proposed system.

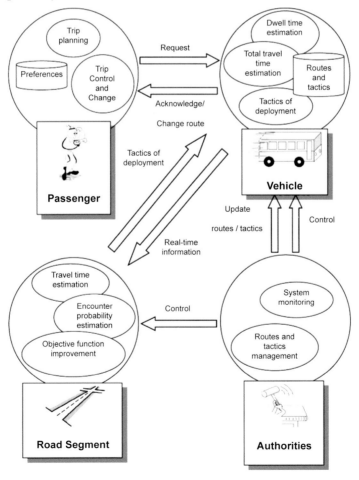

Figure 2.1 Agents' activities and relationships (Source: Hadas and Ceder, 2008b).

Passenger agent: The passenger agent plans the trip according to passenger preferences and based on the real-time information available from road-segment agents (travel time) and vehicle agents (routes and dwell times). Passengers will input to their cell phones or personal digital assistant (PDA) every trip desired from point A to point B. The agent, referring to each passenger preference, will search for the best trip based on the public transit data available at that time. The passenger will choose among the possible options and book a trip in the system. Then the passenger will be notified by SMS or a different means of communication about path changes owing to, for example, traffic congestion or a system-wide optimization deploying tactics that change the planned schedule of route legs. In response to the route changes, the passenger agent will try to find a better route, if possible by changing the existing planned path. The agent himself/herself can be a small software program running on a cell phone or PDA.

Road-segment agent: The road-segment agent can reside physically, as part of the road infrastructure (e.g., at a traffic light control), or virtually, as part of a multi-agent software system. The agent is responsible for the following activities: travel-time estimation, encounter probability estimation, improving the system's objective function and instructing vehicles on tactical deployment. Each road-segment agent continuously collects local traffic-flow information and estimates the travel time. The agent evaluates the encounter probability (for vehicles transferring passengers), based on the adjacent road-segment travel-time estimations and vehicle locations. This probability is described in the next section. Using dynamic programming, each agent or group of agents calculates the optimal tactical deployment that will optimize the system's objective function, which is the total expected travel time.

Vehicle agent: The vehicle agent can be part of the on-board automatic vehicle location (AVL) system or, virtually, part of a multi-agent software system. The agent estimates the vehicle's dwell time according to booked trips and demand forecasts, using travel time estimations (from the road-segment agents) and estimated deviations from the planned timetable schedule. The vehicle agent receives instructions for the deployment of tactics from the road-segment agent in order to improve the system's objective function.

Agency agent: This agent is responsible for designing and managing the transit network for updating timetables and for configuring the possible operational tactics available on each road-segment for each route.

Authority agent: The authority (local authority or federal government) is responsible for monitoring system performance according to the determined/decided indicators.

The MATS offers the following benefits, which are inherent in the multi-agent approach:

Extensibility: Easily allows the system's growth and adding of new resources. Each vehicle has computation power to contribute to the entire network. Adding a new vehicle is similar to plugging in a new computer to a local area network (LAN).

Fault tolerance: The proposed system will handle failures. Critical operations that are heavily dependent on computation and that are built on a standard central computing architecture must have a redundancy system in order to maintain a certain level of service. Redundant systems are expensive and cumbersome; however, MAS agents are distributed. Consequently, if some agents are down, the others will continue to perform because of their autonomous capability. Table 2.1 presents the mode of operation and outcome for different communication scenarios in case of communications disturbances.

Table 2.1 Communication Scenarios (Source: Hadas and Ceder, 2008b).

Scenario	Mode of Operation	Outcome
Full communication	On-line collaboration	Synchronized transfers; Total travel-time reduction
Partial communication	⇕	⇕
No communication	Autonomous; According to timetables	Ordinary transit system with reliability problems

Scalability: Distributed systems can theoretically grow without limit (e.g., the Internet).

Adaptability: Changing rules and transmitting data to the agent is quick and simple, similar to the spreading of a virus.

Efficiency: Negotiations between agents can reach an optimal or a near-optimal solution efficiently (e.g., see Raiffa, 1982).

Distributed problem solving: Cooperating agents can distribute sub-tasks to other agents that are idle and can contribute their computer time to solve these sub-tasks (e.g., see Smith, 1980; Davis and Smith, 1983).

Stability: The use of a closed set of operations tactics for each road-segment in order to eliminate solution sets that are not stable.

2.3 Synchronized Transfers

A brief review is introduced for describing synchronized transfers. The availability of real-time information on bus locations, estimated arrival times, number of passengers and their destinations enabled Dessouky et al. (1999) to develop an algorithm of the bus-dispatching process at timed-transfer points. Such an algorithm can intelligently decide whether or not a bus should be held back in order to achieve a transfer with a late bus. In another study, based on automatic vehicle location (AVL) technology, Dessouky et al. (2003) present a method to forecast accurately the buses' estimated arrival times and to use bus-holding strategies to coordinate transfers. The use of advanced public transit systems on a fixed route and with para-transit operations was found to be important for improving departure times and transfers by Levine et al. (2000). In order to overcome the complexity of trip planning, Horn (2004) suggests an algorithm for the planning of a multi-legged trip. Its objective is to construct a journey while minimizing the travel time, subject to time-window constraints.

Figure 2.2 illustrates the process of improving the total travel time (TTT) which is the sum of the trip legs and waiting time of all passengers of a transit system. Use of operational tactics can help in performing transfers at point X by changing the arrival time of a vehicle (or vehicles) to the transfer point; this will decrease TTT (because of a shorter waiting time). On the other hand, the deployment of tactics may increase the travel time of passengers on-board the vehicles (i.e., using hold tactic) or increase the waiting time of passengers at the bus stops (i.e., hold and skip-stop tactics).

The illustration of Fig. 2.1 shows that the time change can be reached by deploying tactics along the upstream road segments of X (X-1, X-2, etc.) with the effect of the change on all road segments downstream to

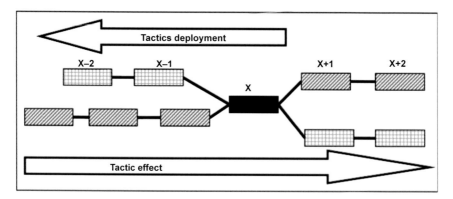

Figure 2.2 Routes of two vehicles scheduled to transfer passengers at road segment X (Source: Hadas and Ceder, 2008a).

the furthest road segment in which tactics are deployed (X-2, X-1, X, X+1, X+2, etc.). The optimization process needs to balance travel time changes in order to minimize TTT. Clearly this process fits into the description of a dynamic programming problem which is formulated as an allocation problem with G (the estimated time gap of arrival to the transfer point between the two vehicles) as a resource, TTT as the objective function and N road segments as stages.

Because of incorporating n stages in the model (each state is a road-segment, as described in Fig. 2.2), m tactics (each tactic is an optional tactic that can be deployed at a road-segment) and s transfer points (state variables representing the scheduled transfers based on the synchronized time tables), its complexity will rise if the number of simultaneous transfers is high. In order to reduce the level of complexity, a distributed dynamic programming model was developed by Hadas and Ceder (2008a) breaking down the global problem into s sub-problems, each with a single state variable. Such a model can be executed in a reasonable time because the complexity of each sub-problem is $O(n \cdot m)$. This optimization model integrates the public transit network (timetables, synchronized transfer points, operational tactics) with on-line data (such as public transit vehicle locations, travel time estimation and demand forecasts) for the reduction of TTT and increase in travel comfort. For example, if the estimated arrival time of a bus to the transfer point is delayed (because of traffic congestion or increased dwell time) and as a result the estimated encounter probability decreases significantly, the system optimization model will recommend the deployment of tactics to increase the encounter probability and at the same time reduce the total travel time.

2.3.1 Network Simulation

Hadas and Ceder (2008a) also developed a simulation model to especially validate the results of the optimization model for transit transfer synchronization. This simulation model is constructed for (a) validating, evaluating and analyzing the benefits of the optimization and (b) comparing the performance of a highly complex global-optimization problem to a low complexity sub-optimal local optimization. This simulation model is composed of two components: (a) simulation model of a public transit network, and (b) dynamic programming optimization model.

Simulation principles

The simulation model is discrete. In each step buses are moved to the next road segment in which the arrival time for each bus to each road segment is known, but all activities within the road segment are performed in a single simulation step. The simulation model can be executed in three modes:

- Not optimized, in which the operation of the buses is not altered during the simulation run; this is the basis for evaluating the optimization.
- Global optimization, in which the optimization is carried out for the whole network.
- Local optimization, in which the optimization is performed locally for each road segment.

For the three optimization modes, two parameters are in use: time horizon and space horizon. The time-horizon parameter determines the forecast range of optimization. Each time optimization is carried out, all future arrivals of buses to the road segment within the time horizon are treated. This parameter is relevant for global and local optimization processes. The space-horizon parameter is relevant for local optimization only. This parameter effects the collaboration of neighbor road segments. If two or more road segments are within the range of the space horizon parameter, then the optimization process is performed jointly; otherwise the optimization is carried locally.

Figure 2.3 illustrates the behavior of the space horizon parameter on a grid-shape transit network. The straight lines are the transit network, the black dots are two road segments with possible transfers between transit routes and the circles are the space horizon. In part (a) of Fig. 2.3, calculations are done jointly as in global optimization because of the intersection of the space horizons; in Fig. 2.3 (b) there are two separate calculations, without interactions, because both space horizons do not intersect.

The total travel time is composed of riding time and waiting time according to the total travel time objective function. Riding time is estimated from a riding-time estimation matrix and waiting time is estimated as follows: (i) calculate the estimated arrival time of both buses to the transfer road segment, (ii) compute the direct transfer probability, and (iii) compute the waiting time.

An example, shown in Fig. 2.4, represents a public transit network with the main line connecting the railway and two feeding lines. It is difficult to solve such a network in a real-life situation because of the three transfers locations associated with the main line sharing all transfer areas.

The example network in Fig. 2.4 illustrates 14 road segments, three bus routes and a train line. This network is used for evaluation and validation of the model. The bus routes and the train line characteristics are described in Table 2.2. Planned transfers are along $S2$ (for Routes 1 and 2) and along $S4$ (for Routes 1 and 3). Transferring passengers to the train line is from Route 1 at a fictitious road segment ($S13$). The transfer road segments are marked with circles in Fig. 2.4. The demand between each pair of road

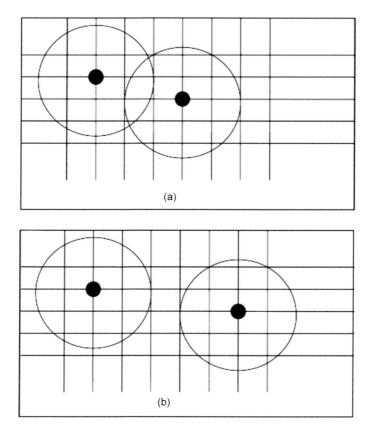

Figure 2.3 Space horizon parameters in a grid-shape transit network: (a) joint calculation and (b) two separate calculations (Source: Hadas and Ceder, 2008a).

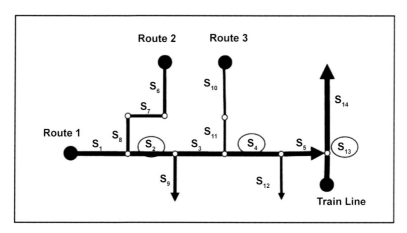

Figure 2.4 Network example (Source: Hadas and Ceder, 2008a).

segments appears in Table 2.3. Passenger arrival at each road segment is shown in Table 2.4. The arrival time distribution reflects the case in which most of the passengers are aware of the published timetables and arrive before the planned departure.

Table 2.2 Routes Characteristics (Source: Hadas and Ceder, 2008a).

Route No.	First Departure	Headway (min.)	Route Layout
1	6:00	20	S_1-S_2-S_3-S_4-S_5- S_{13}
2	6:25	20	S_6-S_7-S_8-S_2-S_9
3	6:40	15	S_{10}-S_{11}-S_4-S_{12}
Train Line	7:00	20	S_{13}- S_{14}

Table 2.3 OD Matrix in Passengers per hour (Source: Hadas and Ceder, 2008a).

From \ To	S_9	S_{12}	S_{14}
S_1	300	300	100
S_2	100	100	50
S_3	0	100	100
S_4	0	100	50
S_5	0	0	10
S_6	200	100	100
S_7	150	150	50
S_8	150	150	100
S_9	0	0	0
S_{10}	0	150	50
S_{11}	0	150	50
S_{12}	0	0	0
S_{13}	0	0	0
S_{14}	0	0	0

Riding time along each segment is distributed lognormal with the parameters LN (5, 2) in minutes, except for the train road segment, which is a constant of 30 minutes. Because the simulation model is only at the road segment level, not dealing with intersegment characteristics (such as bus stops), a probability table for a direct transfer was constructed based on the time gap between two buses entering a road segment—that is, for a time gap of one minute or less, the encounter probability is 0.95, whereas for 2, 3, 4, and 5 and above the probabilities are 0.80, 0.60, 0.30 and 0.00, respectively.

Table 2.4 Arrival Time Distribution (Source: Hadas and Ceder, 2008a).

Arrival Time	Percentage
(Planned Departure) – (Headway)	10
(Planned Departure) – (Headway/2)	20
Planned Departure	55
(Planned Departure) + (Headway/2)	10
(Planned Departure) + (Headway)	5

For each pair of buses (or bus and train) and according to the estimated time gap, an encounter probability will be selected directly from that table. A number between 0 and 1 will be drawn from a uniform distribution and if the result is less or equal to the encounter probability, then a direct encounter has occurred. In the case of a direct encounter, transferred passengers will not experience any delay; else, the transferred passengers will be dropped at the road segment to wait for the next bus or train. For comparison of different optimization types, the seed for the pseudo-random number generator can be selected, hence generating the same set of data. The following two discrete tactics are available: (a) hold for 1, 2, 3, 4 or 5 minutes; and (b) skip a segment which results with travel time reduction of 4 minutes.

Output: For each passenger group departing from the same origin, heading towards the same destination and arriving at the bus stop at the same time, the information of group number, origin, destination, number of passengers, start and end of trip time will be accumulated and saved. For each trip leg, leg number, route number, bus number, transfer point, transfer type (1 = direct, 2 = arrive before second bus, 3 = arrive after second bus), boarding and alighting time will be collected.

Scenarios: Eleven scenarios were constructed for the simulation on the basis of the parameters summarized in Table 2.5. These scenarios present the combined effect of major attributes of public transit networks on the model: (a) headway—routes with long versus short headways; (b) riding time variance—low variance (fewer disruptions in the traffic flow) versus high variance; and (c) synchronized transfers—whether or not the transit network incorporates timed-transfers (planned synchronization). The time horizon attribute is relevant for the performance of the model and was included for comparison purposes.

Scenarios 1 and 2 present synchronized timetables; Scenario 3, in comparison, does not have synchronized timetables. The purpose of the simulation, in this case, is to provide insights on the effect of the model in synchronized and non-synchronized networks. Scenario 4 provides a higher load to the train (which, due to fixed timetables, cannot deploy

Table 2.5 List of Scenarios (Source: Hadas and Ceder, 2008a).

Scenario	Synchronized Timetables	Time Horizon [min.]	Headway	Variance
1	yes	10	Short	base
2	yes	15	Short	base
3	no	10	Short	base
4	yes + high load to train	10	Short	base
5	yes	10	Long	base
6	yes	20	Long	base
7	yes	30	Long	base
8	no	10	Long	base
9	yes	10	Short	high
10	yes	15	Short	high
11	yes	10	Short	low

operational tactics). Scenarios 5 through 8 are characterized by long headways and Scenarios 9 through 11 are characterized by high and low travel time variance. These scenarios represent typical examples of transit networks. One of the main assumptions for the model is that the transit network of routes is synchronized (at the planning stage). The objective of the optimization process is to approach better synchronization considering the changes of travel times that may be the result of traffic congestion or change in passenger demand. Hence, the simulation model was executed with synchronized timetables (Table 2.2) and non-synchronized timetables (Table 2.6). Few options of increased time horizon were checked. The larger the time horizon the higher the complexity; however, it increases the ability to react earlier to more events. The space horizon for the example was set to 1 to force local optimization.

Results: Each scenario was executed for a time period of six hours, once per optimization type. The full list of results appears in Hadas and Ceder

Table 2.6 Headways for Non-synchronized Scenarios (Source: Hadas and Ceder, 2008a).

		Scenario 3	Scenario 8
Route	First Departure	Headway (min.)	Headway (min.)
1	6:00	21	32
2	6:25	18	42
3	6:40	12	27
Train Line	7:00	20	20

(2008a). The main conclusions reached in this study were: (i) by using tactics it is possible to attain a reduction in the total travel time between 3 per cent and 17 per cent; (ii) global optimization tends to yield better results than local optimization; (iii) because local optimization has only one state variable and the execution of each optimization problem is carried out in parallel, the methodology is suitable for online optimization problems; (iv) aside from the reduction in travel time, the number of direct (without wait) transfers increased significantly, resulting in a significant effect on the comfort of the ride and ease of transfer; and (v) the simulation results, which depend on the scenario's characteristics as well as the optimization parameters, call for the development of a full-scale simulation system for the analysis of real-world networks and the calibration of these parameters.

2.4 Real-time Operational Tactics

Another way of improving synchronization of transfers is by using specific operational tactics in real time (Eberlein et al., 2001; Hickman, 2001; Ceder, 2007, 2016; Nesheli and Ceder, 2015a; Nesheli et al., 2015; Ibarra-Rojas et al., 2015). The specific tactics evaluated by Hadas and Ceder (2008a) included stalling of buses at stops in anticipation of connecting buses and instructing skip-stops and shortcuts of routes to meet subsequent connections. Another study by Ceder et al. (2013) investigates how to use selected operational tactics in transit networks for increasing the actual occurrence of scheduled transfers. The model presented determines the impact of instructing vehicles to either hold on at or skip certain stops on the total passenger travel time and the number of simultaneous transfers. A recent study by Nesheli and Ceder (2014) introduces the possibility of skip-segment in addition to skip only an individual stop and for real-time operational control; the refinement takes place in the optimization formulation. The objectives are to create simulation and optimization frameworks for optimally use the three scenarios and compare the scenarios using a case study. Finally, Nesheli and Ceder (2015b) investigated the effect of each operational tactic on transit-performance measures using *system reliability theory*. This section follows the work of Nesheli and Ceder (2014, 2015a).

2.4.1 Holding and Skip-stop/Segment Tactics for Transfer Synchronization

Formulation and modelling framework

The methodology of work commences with the use of TransModeler simulation tool (Caliper, 2013) to represent a real-life example and to generate random input data for the proposed optimization model. Then standard optimization software, ILOG (IBM, 2012), was used to solve optimally a range of different scenarios determined by the simulation runs. Finally, more simulation runs, containing the tactics determined by the optimization program are made, so as to validate the results attained by the model.

Model description

The model developed (Nesheli and Ceder, 2014) considers transit networks consisting of main and feeder routes. The transfers occur at separate transfer points on each route. The formulation contains all the implemented tactics using a deterministic modelling. Analytically the model seeks to attain minimum total passenger travel time and to increase, in this way, the total number of direct transfers. The model formulates the tactics of stalling vehicles, skipping individual stops and skipping segments, as well as indication of missing or making a direct transfer. Thus the components of the model are: (i) the effect on total passenger travel time due to stalling a vehicle, (ii) the effect on total passenger travel time of skipping a stop/segment, and (iii) the effect on total passenger travel time of the vehicle being late or not, to a transfer.

State variables

N	Set of all bus stops, in which $n \in N$
R	Set of all bus routes in which $\{r, r'\} \in R$
TF	Set of all transfer points in which $tr \in TF$
Q_r^{max}	Passenger capacity of bus of route r
l_r^n	Passengers' load of route r at stop n
b_r^n	The number of boarding passengers of route r at stop n
a_r^n	The number of alighting passengers of route r at stop n
$p_{rr'}^n$	The number of transferring passengers of route r to route r' at stop n
d_r^n	Bus dwell time of route r at stop n *(in seconds)*
h_r	Bus headway of route r
c_r^n	Bus running time of route r at stop n from the previous stop
A_r^n	Bus arrival time of route r at stop n
D_r^n	Bus departure time of route r at stop n
$\Omega(t)_r^n$	Time penalty function of route r at stop n

Γ_r^n	Time to reach a desired stop skipped of route r at stop n
T_r	Bus schedule deviation of route r
TP_r^{tr}	Transfer stop of route r at transfer point of tr
E_r	Bus elapsed time of route r from the previous stop to the current position
m_r	Maximum total number of stops of route r
k_r	Positional stop of route r for a snapshot
ω	Ratio between the average speed of a bus and the average walking speed of pedestrian (same ratio for all routes and stops).

Parameters

θ_r^n	The number of passengers of route r for a bus departing stop n
β_r^n	The number of passengers waiting at stops further along the routes with respect to route r and stop n (future passengers)
γ_r^n	The number of passengers who wish to have transfers at transfer points with respect to route r and stop n
λ_r^n	The waiting time per passenger at previous stops due to applied tactics.

Decision variables

HO_r^n	Bus holding time of route r at stop n
S_r^n	Bus skipping stop of route r at stop n; if stop skipped = 1, otherwise = 0
$Y_{rr'}^n$	Possible transferring from route r to route r' at transfer stop n, pre-tactics; if a possible transfer occurs = 0, otherwise = 1
$Z_{rr'}^n$	Possible transferring from route r to route r' at transfer stop n, post-tactics; if a possible transfer occurs = 0, otherwise = 1.

Assumptions: The model is designed deterministically. Therefore the following assumptions are made: (i) there is foreknowledge of the route information, including average travel times, average passenger demand, average number of transferring passengers and average dwell times, (ii) passenger demand is independent of bus arrival time, (iii) vehicles are operated in FIFO manner with an evenly scheduled headway, (iv) passengers will wait at their stop until a bus arrives (none leaves the system without taking the first arrived bus), (v) the bus arriving subsequently to a bus that skipped stop cannot use any of the two tactics considered, (vi) passengers onboard a bus that will skip segment will be informed of this action at the time of the decision so as they can alight before or after the skipped segment. It is to be noted that the formulation of optimization minimizes these types of passengers and in most cases tested, it is nil, and (vii) stops where passengers want to transfer cannot be skipped.

Formulation and properties of holding tactic

Holding or stalling a vehicle is a tactic considered operationally for regulating undesired scheduled deviations, reducing bunching and approaching a direct transfer at transfer points. However, use of the holding tactic has some drawbacks on the total travel time of the passengers. That is, the holding tactic would affect three groups of passengers: (a) those on-board the bus defined as θ_r^n, (b) those waiting for the bus further along the route defined as β_r^n, and (c) those who wish to have transfers defined as γ_r^n. The following formulation can now takes place. It is to be noted that the term 'route' describes a transit service that serves a series of stops. A route is made up of a collection of 'trips'—each trip represents a single run, based on a certain departure time, along the series of stops on the route. For instance, a_r^n describes the number of alighting passengers at stop n of a bus serving route r.

$$\theta_r^n = l_r^n + b_r^n - a_r^n \tag{2.1}$$

$$\beta_r^n = \sum_{i=n+1}^{m_r} \left[b_r^i + \sum_{r' \in R} (1 - Z_{r'r}^i) p_{r'r}^i \right] \tag{2.2}$$

$$\gamma_r^n = \sum_{r' \in R} \left(\left(1 - Y_{r'r}^n\right) p_{r',}^n - p_{rr'}^n \right) \tag{2.3}$$

The effect of the holding tactic on the change in total passenger travel time with respect to route r and stop n is:

$$\Delta TPTT (\text{Holding})_r^n = HO_r^n [\theta_r^n + \beta_r^n + \gamma_r^n] \tag{2.4}$$

Formulation and properties of skip-stop and skip-segment tactics

Skip-stop: When a major disruption occurs, holding a vehicle as the only tactic available cannot guarantee to obviate the headway variation from the schedule. This is true even when holding the following vehicles because these actions may lead to greater schedule deviations (Sun and Hickman, 2005). Skip-stop is another tactic that can be implemented to decrease the irregularity of service and increase the number of direct transfers. The advantage of skip stop is for passengers who already are onboard the bus and those to board downstream. On the other hand, it has an adverse effect on passengers who want to alight or board at the skipped stop.

Skip-segment: The skip of individual stop tactic has the limit imposed— that no more than one stop can be skipped in a row. However if, for instance, some stops are close to each other and only a very few passengers will be impacted by skipping those stops, another tactic can take place as is illustrated by Fig. 2.5. That is, consideration of a skip-segment tactic, where a segment is defined as a group (one or more) of adjacent stops.

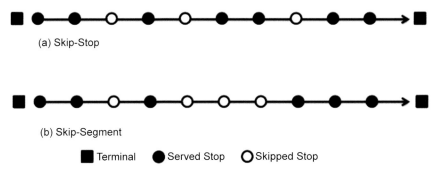

(a) Skip-Stop

(b) Skip-Segment

■ Terminal   ● Served Stop   ○ Skipped Stop

Figure 2.5 Skip-stop and skip-segment tactics (Source: Nesheli and Ceder, 2014).

One of the assumptions made is that passengers onboard a bus that will skip segment will be informed immediately of this action at the time of the decision, so that they can alight before or after the skipped segment. This assumption is considered in the optimization formulation and in most of the cases simulated, the amount of this type of passengers approached zero. It is because of minimizing the total travel time. In other words, if the amount of passengers of this type is large, this tactic of skip segment won't take place. The formulation of this type of passengers is explicated as follows:

Let us consider the end and start-again service stops of the skipped segment; that is, the end stop is the last stop served before the skipped segment and the start-again stop is the first stop served after the skipped stop. Passengers who want to alight in the skipped segment and alight at the end-service stop will have extra time to reach their destination (within the skipped segment) to be termed '*forward time*', Γ (*forward*). However to determine the actual additional travel time, the running time of the bus to the desired skipped stop has to be subtracted and thus the following formulation is used.

$$\Gamma^n_{r(forward)} = (\omega - 1) \sum_{i=1}^{n} c_r^i \left(\prod_{i=q}^{n} S_r^i \right) \qquad \forall (n, q \in N)\{1 \leq q < n\} \tag{2.5}$$

At the same time passengers alighting at the start-again service stop will need to go back to their destination stop, and their time is termed *'backward time'*, Γ *(backward)*. In this case additional bus running is added, and the passengers, to use this way to reach their destination, will save the dwell times of the skipped stops. These considerations yield:

$$\Gamma^n_{r(backward)} = (\omega+1) \sum_{i=n+1}^{m_r} c^i_r \left(\prod_{i=n}^{q} S^i_r\right) - \sum_{i=1}^{n} S^i_r d^i_r \ \forall (n,q \in N)\{1 \le n < q\} \qquad (2.6)$$

It is possible that these passengers will decide to walk; thus a *'walking time'* penalty function is described as:

$$\Omega(t)^n_{r(walking)} = min(\Gamma^n_{r(forward)}, \Gamma^n_{r(backward)}) \qquad (2.7)$$

The alternative of walking is waiting for the next bus to bring these passengers to their destination. The waiting time associated with upstream stops is designated λ^n_r to be:

$$\lambda^n_r = \sum_{i=k_r}^{n-1} (S^i_r d^i_r - HO^i_r) \qquad (2.8)$$

The *'waiting time'* penalty function is then determined by the following equation with consideration of the schedule deviation:

$$T_r : \Omega(t)^n_{r(waiting)} = (h_r - T_r + \lambda^n_r) \qquad (2.9)$$

Two penalty functions were established: walking time and waiting time penalties, with a new definition of *'total time'* penalty being the minimum of the two as follows:

$$\Omega(t)^n_{r(total)} = min\{\Omega(t)^n_{r(walking)}, \Omega(t)^n_{r(waiting)}\} \qquad (2.10)$$

Consequently the effect of the skip-segment tactic on the change of the total travel time for route r and stop n is:

$$\Delta TPTT(\text{Skip} - \text{Segment})^n_r = S^n_r [a^n_r \Omega(t)^n_{r(total)} + b^n_r (h_r - T_r + \lambda^n_r) - d^n_r (l^n_r + \beta^n_r)] \qquad (2.11)$$

Formulation and properties of transfers

Synchronized transfers mean that two or more buses, or other transit vehicles, meet at the same time at a transfer point. As told, this can be improved by the use of real-time operational tactics. For missed transfers, passengers have to tolerate a waiting time of $(h_r - T_r)$ where h_r is the headway at route r. Therefore, a direct transfer occurs if the following holds true:

$$Y^n_{rr'} + Y^n_{r'r} = 0 \qquad (2.12)$$

$$Z^n_{rr'} + Z^n_{r'r} = 0 \qquad (2.13)$$

In this model a possible transfer $Y_{rr'}^n$ (before implementing tactics) occurs if departure time of the bus on route r' is after the bus arrival time on route r and vice versa for $Y_{r'r}^n$. Same arguments apply for $Z_{rr'}^n$ and $Z_{r'r}^n$ after utilizing tactics. If the following conditions hold, direct transfers will be possible and none of the buses will be late at a transfer point:

$$A_r^n = \sum_{i=k_r}^n c_r^i - E_r \tag{2.14}$$

$$D_{r'}^n = \sum_{i=k_{r'}}^n c_{r'}^i - E_{r'} + d_{r'}^n \tag{2.15}$$

$$\text{if } D_{r'}^n \le A_r^n, \text{ then } Y_{rr'}^n = 1 \quad \forall (p_{rr'}^n + p_{r'r}^n \ge 1) \tag{2.16}$$

Subsequently the effect of transfers on the change of total travel time for route r and stop n is:

$$\Delta TPTT(\text{Transfer})_r^n = \sum_{r' \in R} [p_{rr'}^n \cdot (Z_{rr'}^n (h_{r'} - T_r + \lambda_r^n) - Y_{rr'}^n (h_{r'} - T_r))] \tag{2.17}$$

Objective function

According to the formulations derived of the total passenger travel time, an objective function for the proposed model can be written as:

$$min \sum_{r \in R} \sum_{n \in N} \Delta TPTT \{(\text{Holding})_r^n + (\text{Skip} - \text{Segment})_r^n + (\text{Transfer})_r^n\} \tag{2.18}$$

This objective function, of minimum total travel time, is subject to the following constraints:

Constraints

Transfer points cannot be skipped and is stated as:

$$S_r^n \left[\sum_{r' \in R} (p_{rr'}^n + p_{r'r}^n) \right] = 0 \tag{2.19}$$

Direct transferring would occur only if the following exists:

$$A_r^n - \lambda_r^n - D_{r'}^n - HO_{r'}^n + \lambda_{r'}^n \le M * Z_{rr'}^n \quad \forall (p_{rr'}^n + p_{r'r}^n \ge 1) \tag{2.20}$$

where M is a large number.

The following is to state that tactics can't be applied at not-served stops and stops already passed:

$$HO_r^n = 0 \qquad \text{when } (n < k_r) \tag{2.21}$$

$$S_r^n = 0 \qquad \text{when } (n < k_r) \tag{2.22}$$

It is not allowed to skip the first stop:

$$S_r^1 = 0, \tag{2.23}$$

Moreover, no skipping and holding at the same stop are allowed:

$$\{S_r^n * HO_r^n = 0\} \tag{2.24}$$

This constraint can be simplified and reformulated. Let M denote a large number; thus Constraint (24) is exchanged with the following constraints:

$$\text{if } S_r^n = 1, \text{ then } HO_r^n \le M * (1 - S_r^n) \tag{2.25}$$

$$\text{if } HO_r^n > 0, \text{ then } M * S_r^n \le HO_r^n \tag{2.26}$$

The maximum holding time is not more than half the headway of the route:

$$HO_r^n \le 1/2 h_r \tag{2.27}$$

If a transfer occurs at pre-tactics situation, the same applies to with-tactics situation:

$$Z_{rr'}^n \le Y_{rr'}^n \tag{2.28}$$

If direct transferring is possible without the use of any tactic, there is no need to interfere. This constraint is expressed as follows where M is a large number:

$$\sum_{i=TP_r^{tr-1}}^{TP_r^{tr}} S_r^i \le \sum_{tr \in TF} M * Y_{rr'}^{tr} \quad \forall (TP_r^{tr} > TP_r^{tr-1} > k_r) \tag{2.29}$$

$$\sum_{i=k_r}^{TP_r^{tr}} S_r^i \le \sum_{tr \in TF} M * Y_{rr'}^{tr} \quad \forall (TP_r^{tr} > k_r > TP_r^{tr-1}) \tag{2.30}$$

$$\sum_{i=k_r}^{TP_r^1} S_r^i \le M * Y_{rr'}^1 \quad \forall (TP_r^1 > k_r) \tag{2.31}$$

If the use of tactics does not result in a transfer, no tactics are applied:

$$\sum_{i=TP_r^{tr-1}}^{TP_r^{tr}} S_r^i * \prod_{tr \in TF} Z_{rr'}^{tr} < 1 \quad \forall (TP_r^{tr} > TP_r^{tr-1} > k_r) \tag{2.32}$$

$$\sum_{i=k_r}^{TP_r^{tr}} S_r^i * \prod_{tr \in TF} Z_{rr'}^{tr} < 1 \quad \forall (TP_r^{tr} > k_r > TP_r^{tr-1}) \tag{2.33}$$

$$\sum_{i=k_r}^{TP_r^1} S_r^i * Z_{rr'}^1 < 1 \quad \forall (TP_r^1 > k_r) \tag{2.34}$$

Constraints (32–34) can be simplified. Let us denote Im as:

$$Im = \sum_{tr \in TF} Z_{rr'}^{tr} \tag{2.35}$$

The maximum value of Im implies that the use of tactic does not result in a transfer and $Z = 1$ for all routes at all transfer points. Thus the following constraint ensures the use of tactics when a transfer occurs:

$$\text{if } Im = size \text{ of } Z_{rr'}^{tr}, \text{then} \sum_{i=TP_r^{tr-1}}^{TP_r^{tr}} S_r^i = 0 \quad \forall (TP_r^{tr} > TP_r^{tr-1} > k_r) \tag{2.36}$$

The same applies for Constraints (32) and (33). For using skip-stop (not segment) tactic, the following constraints is used for skipping only one stop in a row:

$$S_r^n + S_r^{n+1} \leq 1 \tag{2.37}$$

To restrict the number of passengers onboard when the bus departs from the stop, the following constraint applies:

$$\theta_r^n \leq Q_r^{max} \tag{2.38}$$

Model optimization

The formulation of the problem with the three scenarios (no-tactics, holding & skip-stop, holding & skip-segment) can be solved using constrained programming (CP) technique. The CP is an efficient approach to solve and optimize problems that are too irregular for mathematical optimization. The reasons for these irregularities as per IBM (2012) are: (i) the constraints are nonlinear in nature, (ii) a non-convex solution space exists with many local-optimal solutions, and (iii) multiple disjunctions exist, resulting in poor returned information by a linear relaxation of the problem. A CP engine makes decisions on variables and values and, after each decision, performs a set of logical inferences to reduce the available options for the remaining variables' domains. This is in comparison to a mathematical programming engine that uses a combination of relaxations (strengthened by cutting-planes) and 'branch and bound'.

2.4.2 Case Study of Real-time Tactics Implementation

Examination of the model developed (Nesheli and Ceder, 2014) took place with a real-life case study. The data of the case study was collected in Auckland, New Zealand. It is to note that all buses of the Auckland transport network are equipped with AVL systems. The primary benefits of the AVL system are in the communication and processing of data for service monitoring, fleet management and traveller information. In addition, AVL data are used to analyze the bus travel times and schedule adherence (to implement holding tactics). This means that real-time information on bus locations can be used to predict bus arrivals at the transfer points and to determine the bus schedule deviations.

The transit network of the case study consists of three bus routes and two transfer points. The first route is known as Northern Express with a dedicated lane that runs from the suburb across the Auckland CBD area and has quite a high number of passengers during peak hours. The second route, Route 858, runs north-south (east of the first route), and the third, Route 880, is a loop that serves as a feeder route. Figure 2.6 illustrates the three routes and the two transfer points used.

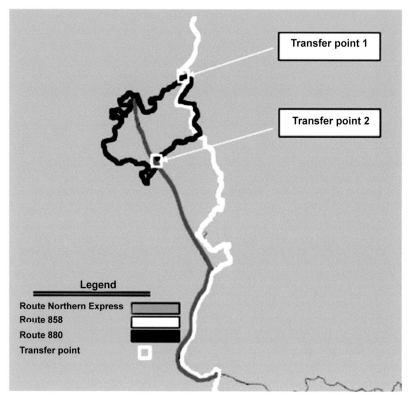

Figure 2.6 Bus system and the study routes, Auckland, New Zealand (Source: Nesheli and Ceder, 2014).

Data: The data was recorded and available are the route characteristics-based; that is, stop ID, longitude, latitude, stop sequence, stop flag, stop code, stop name, route ID, user ID, point ID and route number. Bus capacity of 60 passengers is assumed with 40 seated passengers and 20 standing. The simulation package used TransModeler 3.0 (Caliper, 2013), to consider passenger crowding for the calculation of the dwell time; that is, it takes longer to board or alight a bus if there are more standing passengers. In addition, bus dwell times were taken from Dueker et al. (2004) for simulation use. It affects whether or not the stop is skipped. The dead time (time spent at the stop without boarding or alighting) is set to 4 seconds as the default value of TransModeler. Passenger demand in terms of origin-destination (OD) matrix is estimated, based on the average number of passengers boarding and alighting at each stop from the first stop of the route to a transfer point; that is, OD is estimated based on proportions of the number of boarding, alighting and onboard passengers. Average headways and dispatch times were assumed to be known at both Transfer-point 1 and Transfer-point 2. In addition, vehicle headways considered are 5, 10, 15 and 20 min. (same for all routes). The case study was simulated for analyzing and validating the performance of the model before and after implementation of the tactics. Moreover, the transit network was simulated for handling the concept of synchronized transfers. Overall, the analysis included simulation (Caliper, 2013) and optimization (IBM, 2012).

Results

Three scenarios were designated 'No-Tactics', H-S (holding and skip-stop) and H-SS (holding and skip-segment). Note the difference between *skip-stop* and *skip-segment*. Figure 2.7 demonstrates the significantly better results for the holding and skip-segment combined tactic than for the holding and skip-stop combined tactic, especially for short headway cases. It illustrates the effect of the model on the total passenger travel time for Transfer-point 1.

In Fig. 2.7 the results of short headways, in Part (a), are completely different then the results of the long headways in Part (b) in terms of the shape of the trend before and after the No-Tactics zone. The figure shows that by using combined tactics, compared with the No-Tactics scenario, a considerable reduction in total travel time is attained. Parts (a) and (b) show that when the schedule deviation tends to be zero, the maximum saving of total travel time occurs without the use of any tactics. This max travel time saving coincides with max numbers of direct transfers. It is also observed that the schedule-deviation interval, in which no tactic is used, is larger for long than for short headways. The different shapes of trend in Parts (a) and (b) of Fig. 2.7 deserve explication. Figure 2.7(a) illustrates that a larger reduction of total passenger travel time is possible when the bus is behind schedule. This suggests that passengers waiting for a late bus prefer, in the short headway cases, to continue to wait than to find an alternative solution (walking or use of another travel mode). In this case, the combined tactics are worth applying. Figure 2.7(b) demonstrates an entirely different pattern; it shows that, in the long headway case, travel time is not saved much for large schedule deviations, that is, unreliable service for long headway cases cannot be compensated by the use of tactics and passengers will tend to find, for such a service, alternative solutions. However, if the deviations are reasonable, the use of tactics can save travel time and increase the number of direct transfers. It is to be noted that the deviations (in seconds) are based on a few assumptions and, thus cannot be translated to exact figures of being behind or ahead of the planned schedule. Instead their trends provide a new insight into when to use the combined tactics proposed. More on this concept of evaluation and decision of best operational tactics in real-time scenarios appears in Nesheli and Ceder (2014).

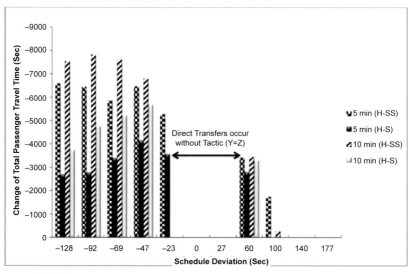

(a) ΔTPTT for Short Headway (5 min, 10 min)

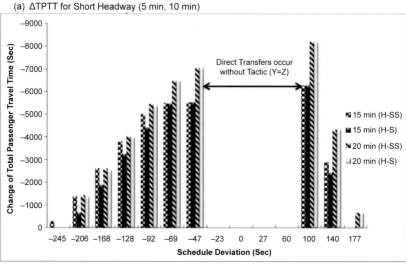

(b) ΔTPTT for Long Headway (15 min, 20 min)

Figure 2.7 ΔTPTT with the two combined tactics of holding and skip-stop/segment at Transfer-point 1 (Source: Nesheli and Ceder, 2014).

2.4.3 A Robust, Tactic-based, Real-time Framework for Transfer Synchronization

A new formulation for the problem of transfer synchronization at the operational level with real-time information was developed by Nesheli and Ceder (2015a). This model allows us to improve transit service performance by optimally increasing the number of direct transfers and reducing the total passenger travel time. The authors developed a hybrid model that uses mixed integer programming (MIP) and constraint programming (CP) techniques to solve the problem. It is a combinatorial problem in which the decision variables are a finite and discrete set. Certainly in order to deal with real-time issues, special attention must be paid to the running time of the algorithm developed. However, because of the problem being NP hard (high complexity), the proposed method is to solve the problem in polynomial time to make it tractable. A rolling horizon approach is utilized to solve this hybrid model. An agent-based, transport-simulation framework is also used to represent a real-life example and to generate random input data for the proposed optimization model. Based on the proposed framework, both the uncertainty of the transit service and the total passenger travel time are reduced by increasing the number of direct transfers.

Algorithm for combined tactic-based problem (CTP)

The following algorithm for the combined tactic-based problem (CTP) generates an optimal combination of tactics for all policies $\Phi = \{\varphi_1, \varphi_2,...,\varphi_n\}$ and different scenarios $= \{\psi_1, \psi_2,...,\psi_n\}$. The service area is divided into $I1$, $I2,...,Im$ segments (one or more consecutive stops), with $n \in N$ vehicle stops and $\{r, x\} \in R$ routes. Let LT be the library of tactics, consisting of holding (HO), skip stop/segment (S) and short turning (SH). To consider the problem of transferring passengers, passenger transfer time (T) is defined. The binary variables Y and Z are used to determine if a direct transfer is made before and after implementation of tactics, respectively.

Step 0. Initial setting: Running the simulation to generate random input data for the proposed optimization model.

Step 1. Planned direct transfer (pre-tactics) $Y_{rx}^n = 0$ schedule deviation $\theta_r^1 \cong 0$ and initial control action $(LT)_r^{I,n} = 0$, where

Step 2. Schedule deviation: $\theta_r^1 \neq 0$, if a direct transfer has not occurred $Y_{rx}^n = 1$; for each $I \in Im$ compute the objective function:

$$\min \sum_{r \in R} \sum_{I \in Im} \sum_{n \in N} \Delta TPTT \left\{ \sum_{\varphi \in \Phi} (LT)_{r,\psi_i}^{I,n} + (T)_{r,\psi_i}^{I,n} \right\}.$$

Step 3. If in post-tactics a direct transfer occurred $Z_{rx}^n = 0$, go to Step 4; otherwise, change the policy φ with a new decision variable and go to Step 2.

Step 4. Verification of the optimum solution by continuing the simulation. If the solution has an acceptable confidence interval (difference between the results of simulation and optimization), stop; otherwise, change the policy φ with new a decision variable and go to Step 2.

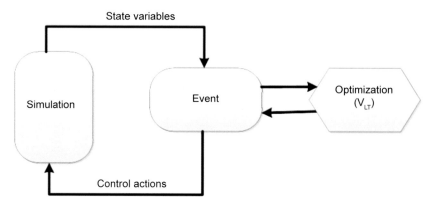

Figure 2.8 CTP framework (Source: Nesheli and Ceder, 2015a).

Based on the CTP algorithm, the value of suggested tactics (V_{LT}) is determined. V_{LT} denotes the optimal value of a tactic for each scenario. This optimal solution will naturally tend to maximize the number of direct transfers and the potential saving in total passenger travel time. This approach is entitled 'optimization- and simulation-based' as opposed to the results of the optimization only entitled as 'optimization-based'. Figure 2.8 illustrates the general analysis of the CTP framework.

2.4.4 Case Study of Different Control Policies

The analysis carried out was based on a case study of Auckland, New Zealand—the country's largest and busiest city, whose transit system has been criticized for not being efficient (Mees, 2010). More recently, the Auckland Transport Regional Public Transport Plan (AT, 2010) was created for the purpose of developing an integrated transport network to provide the people of Auckland with a sustainable transport system in a safe, integrated and affordable manner.

Scenarios: As summarized in Table 2.7, 14 scenarios were tested to evaluate and compare the proposed model under different operational conditions. These scenarios differ in two dimensions: (i) bus schedule-deviation (event) for different bus states; (ii) service headways, which can be high frequency services (i.e., short headways), or low frequency services (i.e., long headways). These scenarios present the combined effect of the major attributes of a transit service on the model.

Control strategies (policy): Four different control policies were tested and compared. The first policy, 'no-tactics', was used for comparison purposes; the second was the combination of 'holding and skip-stops'; the third was a combination of 'holding and short-turning' and the last policy was a combination of the three tactics modelled.

Table 2.7 List of Scenarios (Source: Nesheli and Ceder, 2015a).

Scenario	Event (sec.)	Headway
1	−100	Short
2	−100 ≤ θ < −55	Short
3	−55 ≤ θ < −20	Short
4	−20 ≤ θ < 20	Short
5	20 ≤ θ < 60	Short
6	60 ≤ θ < 110	Short
7	110 ≤ θ	Short
8	−100 > θ	Long
9	−100 ≤ θ < −55	Long
10	−55 ≤ θ < −20	Long
11	−20 ≤ θ < 20	Long
12	20 ≤ θ < 60	Long
13	60 ≤ θ < 110	Long
14	110 ≤ θ	Long

2.4.5 Analysis

The performance of the Auckland system was evaluated by means of the total passenger travel time. The events were extracted from the simulation model for 24 hours. For each trip, the states of the bus with reference to different events at select control stops were computed. Based on optimization, the maximum hold time observed in all scenarios was 186 seconds, subject to the length of headways when the average hold time was 41 seconds. Table 2.8 summarizes the main results of the performance indicators, an analysis of which follows:

Total passenger travel time: The total passenger travel time (travel time and waiting time) was computed for all passengers completing a trip from origin to destination.

Direct transfers: The CTP algorithm was determined for each scenario, so as to minimize the total passenger travel time. This optimal solution will

Table 2.8 Summary of the Results (Source: Nesheli and Ceder, 2015a).

Average Improvement	Headway		Control Policy		
	Short	Long	HO+S	HO+SH	HO+S+SH
TPTT (%)	3.61	2.37	2.14	2.10	4.73
DT (%)	249.55	182.19	101	103	153

Note: HO = holding; S = skip-stops; SH = short-turning.

naturally tend to maximize the number of direct transfers, which were analyzed in terms of the number of transfers.

Improvement: Different policies were compared in terms of total passenger travel time and direct transfers.

From Table 2.8, it is possible to draw an immediate conclusion that applying the library of selected tactics improved the overall system performance. The short headways yielded the best results, as expected. The results revealed that the CTP process improved the system performance considerably by the use of different policies in various scenarios. The combination of all possible tactics leads to the highest improvement in the objective function of the proposed model. CTP shows significant benefits in relation to holding and skip-stops and short-turning in all the scenarios. Interestingly, with this control policy, not only is the expected total passenger travel time reduced (on an average by 4.73 per cent) in comparison with a no-control policy, but it also outperforms other control schemes in terms of reliability in meeting vehicles at transfer points. Thus, combination of all the control tactics leads to an increase in direct transfers of up to 153 per cent on an average for all scenarios.

This can be observed graphically in Fig. 2.9, which illustrates the effect of the model on total passenger travel time and on the number of direct transfers. It also shows that the holding and short-turning policy for a larger schedule deviation results in a better performance than the holding and skip-stops strategy. This indicates that applying such a combination of tactics in worse scenarios leads to considerable progress in transit service. It may also be observed from this figure that with large schedule-deviation intervals, control policies contribute to a higher percentage of direct transfers. More on this concept of evaluation and decision of best operational tactics in real-time scenarios appears in Nesheli and Ceder (2015a, 2016), and Nesheli et al. (2015).

2.5 Customized Bus (CB)

A new and innovative mode of cyber-enabled, demand-interactive transit system, known as customized bus (CB) is introduced in this section. CB provides advanced, attractive and user-oriented minibus service to commuters by aggregating their similar travel-demand pattern and using online information platforms, such as internet, telephone and smartphone apps. Unlike conventional transit service, CB users are actively involved in various interactive operational planning activities, including online demand collection, network route design, timetable development, vehicle and crew scheduling. CB service is more comfortable, convenient and reliable than conventional transit service and more efficient, cost-effective and environment-friendly than private cars. Therefore, CB serves as a

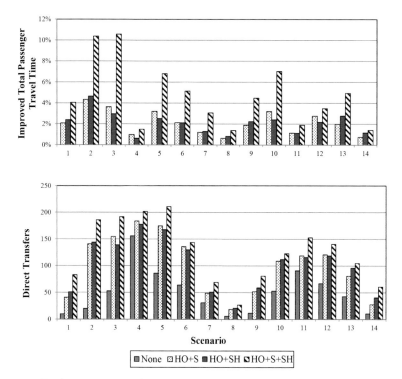

Figure 2.9 Total passenger travel time saved and number of direct transfers per scenario (Source: Nesheli and Ceder, 2015a).

good alternative for reducing urban traffic congestion, improving traffic safety and alleviating energy consumption and greenhouse gas-emission problems.

Because of its various advantages, CB is actively promoted nationally and locally in China and currently operates successfully in 22 cities, with eight more cities having this service under construction and many more municipalities considering launching it. CB is generally regarded as a successful transportation mode that can for the most part meet increasing and diversified mobility needs in China and elsewhere. This section follows the work of Liu and Ceder (2015a).

2.5.1 Demand-based CB Service Design

In transit service design, the fundamental question facing transit agencies is how to design the service well enough so that it can attract private car users to give up their cars. An advanced, attractive and viable transit service needs to be designed to make transit users feel that using transit service is like eating potato chips: once you start, it is hard to stop (Ceder, 2016).

CB is a demand-based transit system that aggregates the travel demand of individual passengers to provide personalized transit service. In transit planning, passenger-demand estimation, i.e., estimating the size, composition and distribution of passenger demand, is usually the first and most vital question facing operators because passenger-demand data are essential input parameters for designing or redesigning any transit service. Traditional demand-estimation methods usually employ mathematical models, such as transit-assignment models or historical data to estimate OD matrices. However, these methods may suffer from the problem of prediction errors because of their failure to accurately describe the complex choice behavior of various passengers. The rapid advances in information and communication technology have now made it possible to conduct real-time online surveys, in which passengers are asked directly about their precise OD, departure times and even personal contact information; thus collecting more accurate passenger-demand data. CB uses online passenger-demand surveys based on interactive, integrated information platforms, such as internet websites and smartphone apps, to collect demand data and provide passengers with customized transit service.

The online demand-based service-design process in these CB systems, as depicted by the flowchart in Fig. 2.10, is a dynamic, interactive and bi-level decision-making problem. Instead of making decisions for all decision variables instantly in one step, both passengers and operators finish their decision-making process in a multi-stage manner, gaining feedback from each other at each decision stage. The whole decision-making process can be divided into four stages as shown in Fig. 2.10: travel survey, call for passengers, seat reservations and seat purchase. Each stage can be formulated as a bi-level programming problem, with passengers as the upper-level decision-maker and operators as the lower-level decision-maker. This path is totally different from traditional transit-planning activities in which operators are treated as the upper-level and passengers as the lower-level decision-maker. During this demand-based service-design process, an interactive, online information platform is built for both passengers and operators to disseminate and collect information to support their decisions. Obviously, CB is user-oriented and designed to cater to different kinds of user demands. The four decision-making stages are explained in detail as follows:

- *Travel survey*: Potential CB users were first asked to register online on an information platform, whether an internet website or a smartphone app. They then can log into the program and submit a travel request, which includes information about travel OD, departure times, whether round trip or one way, etc. Usually, a mobile phone number or an e-mail address is required for receiving verification information from the transit operator after one successfully books a CB service. In the future, passengers can log in, as well, to update their profiles.

- *Call for passengers*: Based on the aggregate travel-demand data provided by potential users, transit operators then design some appropriate initial CB routes. The origin area, destination area, departure and arrival times, boarding and alighting stops of the planned routes are announced on the information platform to recruit users. Potential users then choose the routes that suit them best. If the number of passengers choosing a route is large enough, e.g., more than 50 per cent seats of a vehicle (such as the case in China), then the route will be regarded as effective and put into the final route set with its service scheduled; otherwise, the recruitment of passengers will continue.

- *Seat reservations*: Once final routes are determined and service is scheduled, passengers can see the final scheduled service information and reserve seats through the online information platform. If there are empty seats, passengers who do not participate in the *travel survey* and *call for passenger* stages can also make reservations for seats.

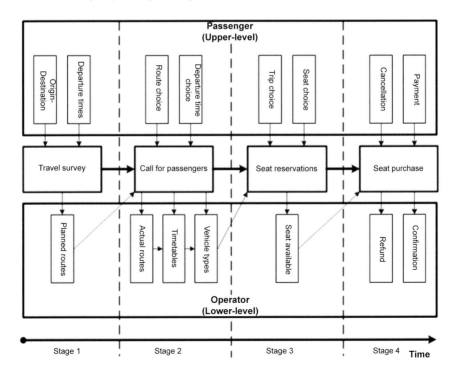

Figure 2.10 Dynamic and interactive demand-based service-design process in CB (Source: Liu and Ceder, 2015a).

CB operators will not stop to announce real-time seat-status information until no seat is available.

- *Seat purchase*: After booking their seats, passengers need to pay for the booked CB service, using online banking, credit card or transit smart cards. If they successfully make a payment, passengers will receive confirmation information, usually in the form of a short message via the mobile phone or by checking the confirmed status directory on the information platform. Passengers who subscribe to the service for a current time-window can directly subscribe to the service for the next time-window without any prior reservation. Passengers can also ask for a refund if the scheduled service does not occur because of some unexpected reason, such as an insufficient number of bookings.

The online travel demand survey provides the foundation for CB service design. Only with accurate travel-demand data can CB operators conduct operation-planning activities, whether route design, timetable development, or vehicle and crew scheduling. This dynamic, interactive, four-stage, demand-based service design process helps transit operators to design a customized travel service that satisfies the various requirements of different passengers.

2.5.2 CB Operations-planning Process

The CB operations-planning process commonly includes five basic activities, usually performed in a sequence: (1) network route design, (2) timetable development, (3) vehicle scheduling, (4) crew scheduling and (5) real-time control. The systematic decision sequence of these five activities is outlined in Fig. 2.11 (Ceder and Wilson, 1986; Ceder, 2016).

Generally speaking, the output of each activity positioned higher in the sequence becomes an important input for lower-level decisions. Occasionally the sequence in Fig. 2.11 is repeated; the required feedback is incorporated over time. Different from traditional transit operation planning process, which is fulfilled in a considerably long planning horizon, e.g., a year, a season or a month, the CB operation-planning activities are completed in a relatively short time. This planning process, furthermore, is conducted in an interactive manner with real-time communication between CB users and operators based on an online information platform. A detailed description of this operation-planning process can be found in Liu and Ceder (2015a).

Advantages of CB

The interpretation of each advantage is as follows:

(a) Personalization

Unlike conventional bus transit systems in which users are passive recipients of a predefined, standardized bus service, CB system users are

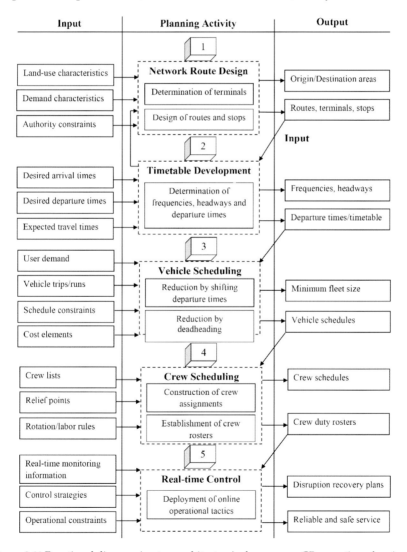

Figure 2.11 Functional diagram (system architecture) of a common CB operation-planning process (Source: Liu and Ceder, 2015a).

actively involved in the various operations planning activities of the whole service-design process. This enables them to develop a personalized transit service by using interactive, integrated information platforms, such as an Internet website, a regular telephone or a smartphone app. CB services are tailored to meet a user's preference and users on their own initiative can affect various operations planning activities, such as network route design, timetable development and vehicle scheduling. The personalized services suit a user's need and improve satisfaction.

(b) Flexibility

Conventional bus systems use fixed routes, fixed stops and fixed timetables, which are not user-oriented and sometimes make users feel clumsy and awkward. In contrast, CB systems make use of flexible route segments, variable stops and adjustable departure times based on real-time user demand and road traffic conditions. CB vehicles are allowed to use dedicated bus lanes, which can significantly reduce travel time, especially at peak hours. There are also no or fewer stops between origin and destination, thus obviating the need for users to wait. Based on real-time traffic information, CB drivers can change route segments to reduce travel time, for example, by bypassing a congested route segment. Furthermore, users can negotiate with one another to achieve a desirable departure time that satisfies everyone. These flexible aspects of CB systems can improve the level of service and reduce operational costs.

(c) Attractiveness

Compared with conventional bus transit, private car and taxi service, CB service is more attractive to users. First, CB service is more comfortable than conventional bus transit service. Every subscriber of a CB service is guaranteed a clean, snug and cosy seat. The overcrowding and dirty seating environment characterizing conventional bus transit systems are successfully avoided. Other amenities include good air-conditioning, free Wi-Fi, TV, radio, newspapers, magazines, drinking water, tissues, air purifier, medical kit, mobile phone chargers and online displays of timetable, time and weather. Second, CB service is more cost-effective than using a private car or a taxi service. The level of service of CB is almost the same as and sometimes better than private cars and taxi, and the cost is much lower. In addition, the cost of construction and operation of a CB system is cheaper than that of metro systems. Some CB transit agencies even buy health and life insurance for users. Given these features of attractiveness, the CB can really 'beat' the car and, thus, successfully attract a significant number of private car users to give up their cars and shift to transit.

(d) Reliability

In CB systems, there are small variances in measures of concern to both users and operators. Because CB routes are designed to include dedicated bus lanes, there are fewer travel delays caused by road traffic disturbances and disruptions. The total travel time is more reliable. Users can receive real-time pre-trip information about the specific stops, such as location and departure time, from an interactive, integrated information platform; drivers, on their part, have information about users, such as name, telephone number, address, etc., which reduce the risk of the users missing their trip. GPS devices are installed in the vehicle and the transit control center can monitor the location of vehicles in real time. Once there are schedule and route deviations, online operational tactics are transmitted to drivers to enable achieving a recovery from errors. Service punctuality and reliability are significantly improved under this advanced, real-time and tactic-based method of operation.

(e) Rapidness

CB vehicles travel much faster than the conventional bus and sometimes even faster than private cars. Usually there are no or few middle stops along CB routes and thus vehicles can move at a stable speed as they do not need to stop to pick up and drop off passengers. What is more, CB vehicles, in addition to the permitted use of dedicated bus lanes, sometimes have signal priority at signalized intersections. Another feature is that CB vehicles can dynamically change route segments in real time according to the current traffic condition in order to avoid traffic congestion. All of these characteristics contribute to improving the rapidness of a CB service.

(f) Smoothness (ease of)

CB service is accessible to a multiclass of users. As described earlier, they can employ the telephone, internet, or smartphone app to subscribe to a service. Based on demand, CB boarding and alighting stops are designed to perfectly match users' origins and destinations; thus, they do not need to walk a long distance to/from the stop. Pictures of spatial positions of the boarding and alighting stops are displayed on the information platforms, so that users can conveniently become familiar with them. Finally, CB vehicles are designed with a low entry chassis and two sets of double doors so that users can board and alight easily and smoothly.

(g) Transfer-free door-to-door

One of the biggest advantages of a CB system is that users do not need to make any transfers. The inconvenience and delay caused by transfers constitutes one of the most important factors deterring citizens from using a transit service. Synchronized transfers in conventional bus transit systems are utilized to reduce inter-route passenger-transfer waiting time and to provide a well-connected service. In practice, however, synchronized transfers do not always materialize because of stochastic and uncertain factors, such as traffic disturbances and disruptions, fluctuations in passenger demand and bus drivers' erroneous behavior. Without transfers, the origin and destination stops along routes are designed directly from users' departure and arrival places, enabling riders to enjoy a no-walking, transfer-free, door-to-door trip chain service.

In addition to advantages enjoyed by users and transit agencies, CB also offers various benefits to the public at large (externalities). First, it provides a promising solution to the problem of traffic congestion in many highly populated cities. It is estimated that a CB vehicle can effectively take the place of at least 30 private cars at peak hours. It can successfully lure people from their cars so as to help relieve the severe traffic problems in many cities with large populations and a high car ownership, such as in Beijing, Shanghai and Guangzhou. Second, by reducing the number of private cars and using electric vehicles, CB contributes to reducing vehicle emissions and improving air quality and, consequently, human health. Statistics from the Beijing Municipal Environmental Protection Bureau (2014) show that vehicle exhaust accounts for 31.1 per cent of PM2.5, which is to blame for causing thick fog, air pollution and human health problems. CB's use of electric vehicles can help to mitigate this serious problem. Third, CB caters to the increasing diversity of travel demand. The rapid development of China's economy and the country's increasing wealth have raised a strong desire for comfortable, convenient and personalized transit service. CB, as a substitute service for private car and a complementary service to conventional transit, serves as a good alternative to meet the diversification of user demand and to accommodate the demand that is not adequately addressed by conventional transit systems. Although we cannot change the direction of the wind (the diversification and evolution of user demand), we can adjust the sails (create advanced and personalized CB service), which eventually will pay off the expense.

2.6 Vehicle-to-Vehicle Communication and Predictive Control

This section considers another aspect related to transfer synchronization which can be used by both conventional transit vehicles and CB; that is, the consideration of a vehicle-to-vehicle (V2V) communication-based scheme to optimize the synchronization of planned transfers in transit networks. By using this scheme, transit drivers can share their real-time information, such as vehicle location, vehicle speed and the number of passengers, with each other through a central communication coordinator. This scheme is especially useful for transit drivers from different routes and driving towards a same transfer point because they can communicate and collaborate with each other, thus dynamically adjusting their running speed on transit route segments or holding time at transit stops. Through this cooperative driving approach, the frequency of simultaneous arrivals of transit vehicles at a transfer point can be improved.

A model predictive control (MPC)-based optimization procedure is introduced to increase the actual occurrence of synchronized transfers in schedule-based transit networks. The procedure aims at reducing the uncertainty of meetings between transit vehicles. The MPC utilizes selected online operational tactics, such as skip-stop, speed change and holding, all of which are based on real-time data. A library of operational tactics is firstly built, as is described previously, to serve as a basis of the sequential receding horizon control process in the MPC. Then, an event-activity network with dynamic moving elements is constructed to represent the logical process of the transit transfer synchronization problem. The MPC procedure for a real-time deployment of operational tactics is explicated. The section follows the work of Liu et al. (2014a, 2014b, 2015) and of Liu and Ceder (2015b, 2016a, 2016b).

2.6.1 Vehicle-to-vehicle Communication

Overview of inter-vehicle communication (IVC)

Inter-vehicle communication (IVC) enables drivers to communicate with other drivers that locate outside the range of line of sight and share real-time traffic information with each other. IVC has the ability to extend the horizon of drivers and on-board devices, and thus has the potential to significantly improve road traffic safety, efficiency and comfort (Luo and Hubaux, 2006). It is recognized as an important component of intelligent transportation systems (ITS) and has attracted significant research attention around the world. Since the 1980s, a number of research projects focusing on IVC have been set up, such as CarTalk 2000, Fleetnet, PATH, Car-to-Car Communication Consortium (C2C-CC), Network on Wheels (NoW) and COMeSafety (Reichardt et al., 2002; Sichitiu and Kihl, 2008). Recent research focuses mainly on vehicular ad hoc networks (VANETs), which are a kind of mobile ad hoc networks offering direct communication between vehicles (Hartenstein and Laberteaux, 2008).

The main application areas of IVC are collision warning, traffic monitoring, traffic coordination, traveller information support, traffic light scheduling, traffic coordination, targeted vehicular communications and car to land communication (Sichitiu and Kihl, 2008). These appealing applications have attracted many researches. However, the majority of current research is focused on cars; not so much research has been done about transit vehicles.

Methodology using IVC

This section describes a semi-decentralized control strategy for transit vehicles transfer synchronization. Consider a directed transit network G = (N, R) with a set N of stops and a set R of routes. The transit routes are divided into a few route segments by transfer points, as is shown in the upper Part (a) of Fig. 2.12. A communication control center is assigned to an intersection point or a shared route segment where passengers make a transfer. The communication control center is responsible for the coordination of vehicles on the route segments leading to it. A route

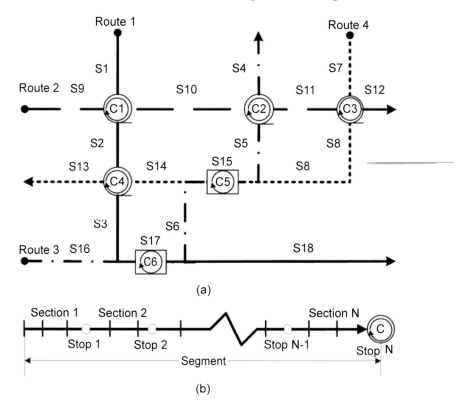

Figure 2.12 Public-transit network example illustrating network-level semi-decentralized control strategy: (a) a basic transit network with communication centers; (b) an illustration of the concept of route sections (Source: Liu et al., 2014a).

segment is divided into a few route sections determined by transit stops along it, as shown in the lower Part (b) of the same figure. The route section is further divided into a few smaller intervals.

An interval is a basic unit controlled by the communication control center which delivers advisory speed information to a transit vehicle when it arrives at each interval. New information is sent to the driver when moving to the next interval. A simple example is shown in Fig. 2.12(a). In this example, there are four transit routes R_i ($i=1,2,...,4$) and the whole network is divided into 18 route segments S_j ($j=1,2,...,18$). The segments S_{15} and S_{17} are shared road-segment encounter transfers while the other segments direct to a fixed single-point encounter transfer. Six communication control centers C_k ($k=1, 2, ..., 6$) are assigned to the corresponding transfers and are responsible for their own route segments. For example, C_1 is in charge of the coordination of vehicles on route segments S_1 and S_9, while C_5 is in charge of S_6, S_8 and S_{15}.

Transit vehicles controlled by a same communication control center are in the same group and can communicate with their peers to share traffic information. The communication coordinator in the control center delivers advice on the real-time optimal running speed. Drivers can adjust their running speed in a cooperative manner in order to achieve a simultaneous arrival. Once a transit vehicle passes the communication control center, it automatically joins another group of vehicles. The communication group is autonomous and self-organized. It is to be noted that from the entire transit network perspective, different communication groups can be performed at the same time, that is, the process is a decentralized and parallel process. However, from a single route segment perspective, the communication activity is severely governed by the communication control center, and thus it is a centralized process. This control strategy is termed semi-decentralized. By doing so, drivers in the same communication group drive in a cooperative manner and the opportunity of simultaneous arrivals improves.

Conceptual illustration

A simple example is used as an explanatory device to show the main components of the proposed IVC system and to outline its potential benefits. The example is shown schematically in Fig. 2.13, in which two transit vehicles are moving towards the same fixed single-point transfer. Both vehicles belong to the same communication group and are controlled by the same communication control center. The center comprises mainly a central server and a communication coordinator.

An on-board device (OBD) is installed on the transit vehicle to receive signals from GPS satellites. The OBD can record information about vehicle ID, vehicle location, vehicle speed, time, route direction and route ID. A SIM

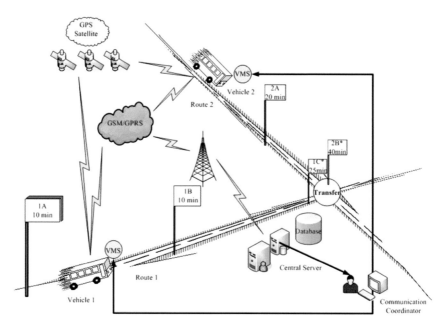

Figure 2.13 Graphical illustration of transfer synchronization of two public transit routes based on the IVC systems (Source: Liu et al. 2014a).

card is embedded in the OBD. The recorded data then can be transmitted to the database in the communication control center through GSM/GPRS networks. These data are visualized in GIS maps. The communication coordinator has knowledge about the relatively accurate location and speed information of the vehicles of the same group. Based on the knowledge, advisory optimal speed information is disseminated to the drivers in the group, so as to meet simultaneously or within a given time-window at the same transfer point. The advisory information can be displayed online to the driver on the on-board variable message sign (VMS) installed in the vehicle. This will allow for a peer-cooperative communication. It is to be noted that the basic assumption of the IVC system is that drivers will comply with the recommended (feasible) speed and holding time so as to materialize the direct transfers of passengers without waiting time. The control center will have a record of this compliance to help minimizing issues associated with driver behavior.

Main features

The main features of the CBVC-B are related to the whole transit network communications comprised of different decentralized and parallel groups. The communication-based control process can be performed at the same time between different communication groups. However, technically bus drivers of the same communication group are not exactly communicated in a direct peer-to-peer (P2P) manner; it is more of a client-server (CS) way. Therefore, the whole control process is termed semi-decentralized group communication. In a communication group, bus drivers leading to the same transfer point serve as clients and the communication control center serves as the central server. Bus drivers through the central server share vehicle and passenger information with their peers. The central communication center is responsible for the communication coordination of vehicles on the route segments and delivers advice on real-time vehicle control tactics to bus drivers. Once a bus passes the communication control center, it automatically joins another group of vehicles. The communication group is self-organized.

A small transit network is used in Fig. 2.14(a) to illustrate the basic concepts. This network is from Ceder et al. (2001). The network is divided into four communication groups by four transfer points shown in Fig. 2.14(b). Each group has two routes leading to the transfer point, and a central server provides communication between bus drivers.

Model

The model developed by Liu et al. (2014a) assumes given GPS information of transit vehicle speed and location. It is a distance-based dynamic speed-adjustment model based on IVC systems for vehicle running speed update under both fixed single-point encounter and flexible road-segment encounter scenarios. In this model drivers of different routes can share their real-time travel speed and location information through the IVC

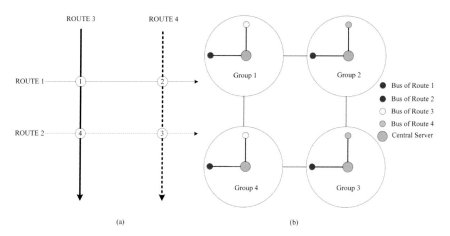

Figure 2.14 Cooperative group communication between bus drivers in a transit network: (a) an example bus transit network; (b) semi-decentralized group communication between bus drivers of the example (Source: Liu et al. 2014b).

communication control center. Drivers can adjust their travel speed or holding time at transfer points dynamically and cooperatively in order to achieve a pre-planned direct transfer. The main objective of the model is to increase the number of simultaneous arrivals at all transfer points. This approach analyzes the impact of the changes on the total passenger travel time, in-vehicle travel time, out-of-vehicle waiting time at stops and transfer waiting time. These impacts are studied on a network-wide scale, using the optimization model presented below:

Notations

R	set of all PT routes $r \in R$
S	set of all segments $s \in S$ of all routes $r \in R$
N^{sr}	set of all stops on segment s of route r, $n \in N^{sr}$
$\lvert N^{sr} \rvert$	total number of stops on segment s of route r
PI_n^{sr}	positional index (k) of a stop n on segment s of route r
q_k^{sr}	vector listing all stops on segment s of route r in order defined by positional index (k)
M^{sr}	set of intervals on segment s of route r, $m \in M^{sr}$
$\lvert M^{sr} \rvert$	total number of intervals on segment s of route r
p_{in}^{sr}	number of passengers assumed to be arriving in vehicle i at stop n on segment s of route r; this is derived from a previous OD matrix; the number of the first stop on route r is the initial number of passengers on the vehicle
e_{in}^{sr}	number of passengers entering vehicle i at stop n on segment s of route r
l_{in}^{sr}	number of passengers leaving vehicle i at stop n on segment s of route r
$d_i^{s_i r_i}$	length of segment s of route r on which vehicle i is moving
$d_{i,m}^{s_i r_i}$	distance of vehicle i on segment s of route r in interval m to the nearest transfer point
$d_s^{s_i r_i s_j r_j}$	length of shared road-segment s between segment s_i of route r_i and segment s_j of route r_j
t_{in}^{sr}	average running time of vehicle i on section between stop $n-1$ and stop n of segment s of route r, post-tactics
$\bar{t}_{in}^{sr}$	average running time of vehicle i on section between stop $n-1$ and stop n of segment s of route r, pre-tactics

Δt_{in}^{sr}	average running time difference of vehicle i on section between stop $n-1$ and stop n of segment s of route r between post-tactics and pre-tactics, i.e., $\Delta t_{in}^{sr} = t_{in}^{sr} - \overline{t}_{in}^{sr}$
$A(t)_{iq_k^{s_i r_i}}^{s_i r_i}$	arrival time of vehicle i of segment s_i of route r_i at transfer stop $q_k^{s_i r_i}$, post-tactics
$D(t)_{iq_k^{s_i r_i}}^{s_i r_i}$	departure time of vehicle i of segment s_i of route r_i at transfer stop $q_k^{s_i r_i}$, post-tactics
$\Delta D_{q_k^{s_i r_i}}^{m}$	dwell time of vehicle i of segment s_i of route r_i within the interval m, that is $\Delta D_{q_k^{s_i r_i}}^{m} = D(t)_{iq_k^{s_i r_i}}^{s_i r_i} - A(t)_{iq_k^{s_i r_i}}^{s_i r_i}$
$h_i^{q_{\lvert N^{s_i r_i}\rvert}^{s_i r_i}}$	holding time of vehicle i of segment s_i of route r_i at transfer stop $q_{\lvert N^{s_i r_i}\rvert}^{s_i r_i}$
$A(\overline{t})_{iq_k^{s_i r_i}}^{s_i r_i}$	arrival time of vehicle i of segment s_i of route r_i at transfer stop $q_k^{s_i r_i}$, pre-tactics
$D(\overline{t})_{iq_k^{s_i r_i}}^{s_i r_i}$	departure time vehicle i of segment s_i of route r_i at transfer stop $q_k^{s_i r_i}$, pre-tactics
v_{im}^{sr}	average speed of vehicle i in interval m on segment s of route r
$Vmin_{i,m}^{s_i r_i}$	minimum average speed of vehicle i in interval m on segment s of route r
$Vmax_{i,m}^{s_i r_i}$	maximal average speed of vehicle i in interval m on segment s of route r
$p_{iq_{\lvert N^{s_i r_i}\rvert}^{s_i r_i} jq_{\lvert N^{s_j r_j}\rvert}^{s_j r_j}}^{s_i r_i s_j r_j}$	number of passengers wishing to transfer from vehicle i of segment s_i of route r_i at transfer stop $q_{\lvert N^{s_i r_i}\rvert}^{s_i r_i}$ to vehicle j of segment s_j of route r_j at transfer stop $q_{\lvert N^{s_j r_j}\rvert}^{s_j r_j}$
$twt_{iq_{\lvert N^{s_i r_i}\rvert}^{s_i r_i} jq_{\lvert N^{s_j r_j}\rvert}^{s_j r_j}}^{s_i r_i s_j r_j}$	transfer walking time required from vehicle i of segment s_i of route r_i at transfer stop $q_{\lvert N^{s_i r_i}\rvert}^{s_i r_i}$ to vehicle j of segment s_j of route r_j at transfer stop $q_{\lvert N^{s_j r_j}\rvert}^{s_j r_j}$. It follows that if transfer stop $q_{\lvert N^{s_i r_i}\rvert}^{s_i r_i}$ and $q_{\lvert N^{s_j r_j}\rvert}^{s_j r_j}$ are a shared transfer stop, then $twt_{iq_{\lvert N^{s_i r_i}\rvert}^{s_i r_i} jq_{\lvert N^{s_j r_j}\rvert}^{s_j r_j}}^{s_i r_i s_j r_j} = 0$.

Distance-based dynamic speed-adjustment model

Consider a transfer point of two transit routes. The time-distance diagram of vehicles for fixed single-point encounter scenario and flexible road-segment encounter scenario are shown in Fig. 2.15(a) and 2.15(b), respectively. The time horizon of each route segment is divided into $|M^{sr}|$ intervals of equal size. The length of each interval is τ. Drivers will receive real-time advisory speed-adjustment information by a communication coordinator for each interval and holding time information for the last interval based on their distance to the transfer point.

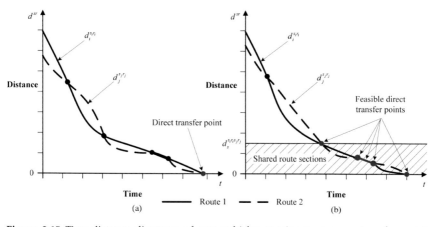

Figure 2.15 Time-distance diagrams of two vehicles moving to a same transfer point: (a) fixed single-point encounter transfer; (b) flexible road-segment encounter transfer (Source: Liu et al. 2014a).

Fixed single-point encounter scenario: The update rules for the average running speed adjustment, for the fixed single-point encounter scenario, are given as follows:

if $d_{i,m}^{s_i r_i} > d_{j,m}^{s_j r_j}$, $v_{i,m+1}^{s_i r_i} = v_{i,m}^{s_i r_i} + a$; $v_{j,m+1}^{s_j r_j} = v_{j,m}^{s_j r_j} - b$.

else if $d_{i,m}^{s_i r_i} < d_{j,m}^{s_j r_j}$, $v_{i,m+1}^{s_i r_i} = v_{i,m}^{s_i r_i} - b'$; $v_{j,m+1}^{s_j r_j} = v_{j,m}^{s_j r_j} + a'$.

else $d_{i,m}^{s_i r_i} = d_{j,m}^{s_j r_j}$

if $v_{i,m}^{s_i r_i} > v_{j,m}^{s_j r_j}$, $v_{i,m+1}^{s_i r_i} = \max\left(v_{i,m}^{s_i r_i} - \frac{1}{2}\left(v_{i,m}^{s_i r_i} - v_{j,m}^{s_j r_j} \right), \text{Vmin}_{i,m+1}^{s_i r_i} \right)$;

$v_{j,m+1}^{s_j r_j} = \min\left(v_{j,m}^{s_j r_j} + \frac{1}{2}\left(v_{i,m}^{s_i r_i} - v_{j,m}^{s_j r_j} \right), \text{Vmax}_{j,m+1}^{s_j r_j} \right)$.

else if $v_{i,m}^{s_i r_i} < v_{j,m}^{s_j r_j}$, $v_{i,m+1}^{s_i r_i} = \min\left(v_{i,m}^{s_i r_i} + \frac{1}{2}\left(v_{j,m}^{s_j r_j} - v_{i,m}^{s_i r_i} \right), \text{Vmax}_{i,m+1}^{s_i r_i} \right)$;

$$v_{j,m+1}^{s_jr_j} = \max\left(v_{j,m}^{s_jr_j} - \frac{1}{2}\left(v_{j,m}^{s_jr_j} - v_{i,m}^{s_ir_i}\right), \text{Vmin}_{j,m+1}^{s_jr_j}\right).$$

else $v_{i,m}^{s_ir_i} = v_{j,m}^{s_jr_j}, v_{i,m+1}^{s_ir_i} = v_{i,m}^{s_ir_i}; v_{j,m+1}^{s_jr_j} = v_{j,m}^{s_jr_j}$. end.

where a and a' are accelerations and b and b' are decelerations. The average running speed of vehicle i in interval m on segment s of route r, v_{im}^{sr} is defined as follows:

$$v_{im}^{sr} = \frac{d_{i,m+1}^{s_ir_i} - d_{i,m}^{s_ir_i}}{A(t)_i^{m+1} - A(t)_i^m - \sum \Delta D_{q_k^{s_ir_i}}^m} \tag{2.39}$$

For distance $d_{i,m}^{s_ir_i}$ larger than distance $d_{j,m}^{s_jr_j}$, vehicle i will be accelerated by a and vehicle j will be decelerated by b. If distance $d_{i,m}^{s_ir_i}$ is less than distance $d_{j,m}^{s_jr_j}$, then vehicle i will decelerate by b' and vehicle j will accelerate by a'. For distances $d_{i,m}^{s_ir_i}$ and $d_{j,m}^{s_jr_j}$ equal and speeds $v_{i,m}^{s_ir_i}$ and $v_{j,m}^{s_jr_j}$ not equal, the next speeds $v_{i,m+1}^{s_ir_i}$ and $v_{j,m+1}^{s_jr_j}$ are to be adjusted to attain equal speeds. If both the distance and speed are equal, the speeds remain unchanged. The values of a, b, a' and b' for each case are determined as follows:

Case I: $d_{i,m}^{s_ir_i} > d_{j,m}^{s_jr_j}$: Let $\lambda = \left|\dfrac{d_{i,m}^{s_ir_i} - d_{j,m}^{s_jr_j}}{\tau} - \left(v_{i,m}^{s_ir_i} - v_{j,m}^{s_jr_j}\right)\right|$, where τ is the length

of interval. The values of a and b are obtained by solving the following minimization problem.

$$\min |a + b - \lambda| \tag{2.40}$$

subject to:

$$v_{i,m}^{s_ir_i} + a \leq \text{Vmax}_{i,m+1}^{s_ir_i} \tag{2.41}$$

$$v_{j,m}^{s_jr_j} - b \geq \text{Vmin}_{j,m+1}^{s_jr_j} \tag{2.42}$$

$$a \geq 0 \tag{2.43}$$

$$b \geq 0 \tag{2.44}$$

The objective function of the model is to minimize the distance gap between $d_{i,m+1}^{s_ir_i}$ and $d_{j,m+1}^{s_jr_j}$ in interval $m + 1$, where $d_{i,m+1}^{s_ir_i} = d_{i,m}^{s_ir_i} - v_{i,m}^{s_ir_i}\tau$ and $d_{j,m+1}^{s_jr_j} = d_{j,m}^{s_jr_j} - v_{j,m}^{s_jr_j}\tau$, in order to achieve a direct transfer. The constraints are of minimum and maximum speed limits.

Let $\rho = |(\text{Vmax}_{i,m+1}^{s_ir_i} - \text{Vmax}_{j,m+1}^{s_jr_j}) - (v_{i,m}^{s_ir_i} - v_{j,m}^{s_jr_j})|$. Certainly the optimal solution for the above problem is $a + b = \lambda$. The rules for determining the values of a, and b, to minimize total passenger travel time, are as follows:

$$\text{if } \rho \le \lambda, a = \text{Vmax}_{i,m+1}^{s_i r_i} - v_{i,m}^{s_i r_i}; b = v_{j,m}^{s_j r_j} - \text{Vmin}_{j,m+1}^{s_j r_j}.$$

else $\rho > \lambda$

$$\text{if } \text{Vmax}_{i,m+1}^{s_i r_i} - v_{i,m}^{s_i r_i} \ge \lambda, a = \lambda; b = 0.$$

$$\text{else } \text{Vmax}_{i,m+1}^{s_i r_i} - v_{i,m}^{s_i r_i} < \lambda, a = \text{Vmax}_{i,m+1}^{s_i r_i} - v_{i,m}^{s_i r_i}; b = \lambda - \left(\text{Vmax}_{i,m+1}^{s_i r_i} - v_{i,m}^{s_i r_i} \right).$$

Case II: $d_{i,m}^{s_i r_i} < d_{j,m}^{s_j r_j}$: Similarly, let $\lambda' = \left| \left(v_{i,m}^{s_i r_i} - v_{j,m}^{s_j r_j} \right) - \dfrac{d_{i,m}^{s_i r_i} - d_{j,m}^{s_j r_j}}{\tau} \right|$, and a', b' are obtained by solving the following minimization programming problem.

$$\min |a' + b' - \lambda'| \tag{2.45}$$

subject to:

$$v_{j,m}^{s_j r_j} + a' \le \text{Vmax}_{j,m+1}^{s_j r_j} \tag{2.46}$$

$$v_{i,m}^{s_i r_i} - b' \ge \text{Vmin}_{i,m+1}^{s_i r_i} \tag{2.47}$$

$$a' \ge 0 \tag{2.48}$$

$$b' \ge 0 \tag{2.49}$$

The rules for determining the values of a', and b', to minimize total passenger travel time, are same as those in Case I.

Flexible road-segment encounter scenario: The update rules for average running speed adjustment, for the flexible road-segment encounter scenario, when the distances $d_{i,m}^{s_i r_i}$ and $d_{j,m}^{s_j r_j}$ are not equal, are same as for the fixed single-point encounter scenario. However, when the distances $d_{i,m}^{s_i r_i}$ and $d_{j,m}^{s_j r_j}$ are equal, the update rules for travel speed adjustment are a bit different. New rules for average running speed update are given as follows:

if $d_{i,m}^{s_i r_i} = d_{j,m}^{s_j r_j} > d_s$

$$\text{if } v_{i,m}^{s_i r_i} > v_{j,m}^{s_j r_j}, \ v_{i,m+1}^{s_i r_i} = \max \left(v_{i,m}^{s_i r_i} - \frac{1}{2} \left(v_{i,m}^{s_i r_i} - v_{j,m}^{s_j r_j} \right), \text{Vmin}_{i,m+1}^{s_i r_i} \right);$$

$$v_{j,m+1}^{s_j r_j} = \min \left(v_{j,m}^{s_j r_j} + \frac{1}{2} \left(v_{i,m}^{s_i r_i} - v_{j,m}^{s_j r_j} \right), \text{Vmax}_{j,m+1}^{s_j r_j} \right).$$

$$\text{else if } v_{i,m}^{s_i r_i} < v_{j,m}^{s_j r_j}, \ v_{i,m+1}^{s_i r_i} = \min \left(v_{i,m}^{s_i r_i} + \frac{1}{2} \left(v_{j,m}^{s_j r_j} - v_{i,m}^{s_i r_i} \right), \text{Vmax}_{i,m+1}^{s_i r_i} \right);$$

$$v_{j,m+1}^{s_j r_j} = \max \left(v_{j,m}^{s_j r_j} - \frac{1}{2} \left(v_{j,m}^{s_j r_j} - v_{i,m}^{s_i r_i} \right), \text{Vmin}_{j,m+1}^{s_j r_j} \right).$$

$\text{else } v_{i,m}^{s_i r_i} = v_{j,m}^{s_j r_j}, v_{i,m+1}^{s_i r_i} = v_{i,m}^{s_i r_i}; v_{j,m+1}^{s_j r_j} = v_{j,m}^{s_j r_j}. \text{ end,}$

$\text{else } d_{i,m}^{s_i r_i} = d_{j,m}^{s_j r_j} \leq d_s, \; v_{i,m+1}^{s_i r_i} = \text{Vmax}_{i,m+1}^{s_i r_i}; v_{j,m+1}^{s_j r_j} = \text{Vmax}_{j,m+1}^{s_j r_j}.$

The change of scenario occurs when the distances of the two vehicles to the end of their shared road-segment are equal and less than the length of the shared road-segment. In the next intervals, the vehicles will run at their maximum speeds in order to minimize the total passenger travel time.

Holding vehicles at transfer point: Holding vehicles at transfer points can improve the opportunity of direct transfers; but, it will increase the dwell time. It has both positive and negative impacts on the total passenger travel time. There is a trade-off between the number of transferring and on-board passengers.

Evaluation of performance: Evaluation of the performance of the proposed distance-based dynamic speed-adjustment model, IVC scheme based, takes place using two main measures: (i) total number of simultaneous arrivals and (ii) change of total passenger travel time. The total number of simultaneous arrivals of an observation time horizon can be obtained directly by comparing the arrival times of vehicles at a transfer point.

For the i-th vehicle, the change in total passenger travel time ($\Delta TPTT_i$) includes four elements: the change of in-vehicle passenger travel time (ΔPTT_i), the change of passenger waiting time (ΔPWT_i), the change of passenger transfer waiting time ($\Delta PTWT_i$) and the holding time at transfer points (H_i). The change of total passenger travel time of all vehicles is defined as:

$$\Delta TPTT = \sum_i \left(\Delta PTT_i + \Delta PWT_i + \Delta PTWT_i + H_i \right) \tag{2.50}$$

The change of in-vehicle passenger travel time (ΔPTT_i) is calculated by:

$$\Delta PTT_i = PTT_i^* - PTT_i = \sum_r \sum_s \sum_n \left(p_{in}^{sr} + e_{in}^{sr} - l_{in}^{sr} \right) \Delta t_{in}^{sr} \tag{2.51}$$

where PTT_i^* and PTT_i denote the total in-vehicle passenger travel time of post-tactics and pre-tactics, respectively. The change of downstream passenger waiting time ΔPWT_i is calculated by:

$$\Delta PWT_i = \sum_r \sum_s \sum_n \left(\sum_{k=PI_n^{sr}}^{|N^{sr}|} e_{ik}^{sr} \right) \Delta t_{in}^{sr} \tag{2.52}$$

The change of passenger transfer waiting time is defined by the following equation:

$$\Delta PTWT_i = PTWT_i^* - PTWT_i \tag{2.53}$$

where $PTWT_i^*$ and $PTWT_i$ denote the passenger transfer waiting time at the transfer point of post-tactics and pre-tactics, respectively. The $PTWT_i^*$ and $PTWT_i$ are obtained by the following Eqs. (54) and (55), respectively.

$$PTWT_i^* = \begin{cases} p_{iq_{|N^{s_ir_i}|}^{s_ir_i}jq_{|N^{s_jr_j}|}^{s_jr_j}}^{s_ir_is_jr_j}\left(D(t)_{jq_{|N^{s_jr_j}|}^{s_jr_j}}^{s_jr_j} - A(t)_{iq_{|N^{s_ir_i}|}^{s_ir_i}}^{s_ir_i} - twt_{iq_{|N^{s_ir_i}|}^{s_ir_i}jq_{|N^{s_jr_j}|}^{s_jr_j}}^{s_ir_is_jr_j} \right), \\[2em] p_{iq_{|N^{s_ir_i}|}^{s_ir_i}jq_{|N^{s_jr_j}|}^{s_jr_j}}^{s_ir_is_jr_j}\left(D(t)_{jq_{|N^{s_jr_j}|}^{s_jr_j}}^{s_jr_j} + \kappa h_{r_j} - A(t)_{iq_{|N^{s_ir_i}|}^{s_ir_i}}^{s_ir_i} - twt_{iq_{|N^{s_ir_i}|}^{s_ir_i}jq_{|N^{s_jr_j}|}^{s_jr_j}}^{s_ir_is_jr_j} \right), \end{cases} \tag{2.54}$$

$$\text{if } A(t)_{iq_{|N^{s_ir_i}|}^{s_ir_i}}^{s_ir_i} + twt_{iq_{|N^{s_ir_i}|}^{s_ir_i}jq_{|N^{s_jr_j}|}^{s_jr_j}}^{s_ir_is_jr_j} \leq D(t)_{jq_{|N^{s_jr_j}|}^{s_jr_j}}^{s_jr_j}$$

otherwise;

$$PTWT_i = \begin{cases} p_{iq_{|N^{s_ir_i}|}^{s_ir_i}jq_{|N^{s_jr_j}|}^{s_jr_j}}^{s_ir_is_jr_j}\left(\bar{D(t)}_{jq_{|N^{s_jr_j}|}^{s_jr_j}}^{s_jr_j} - \bar{A(t)}_{iq_{|N^{s_ir_i}|}^{s_ir_i}}^{s_ir_i} - twt_{iq_{|N^{s_ir_i}|}^{s_ir_i}jq_{|N^{s_jr_j}|}^{s_jr_j}}^{s_ir_is_jr_j} \right), \\[2em] p_{iq_{|N^{s_ir_i}|}^{s_ir_i}jq_{|N^{s_jr_j}|}^{s_jr_j}}^{s_ir_is_jr_j}\left(\bar{D(t)}_{jq_{|N^{s_jr_j}|}^{s_jr_j}}^{s_jr_j} + \kappa h_{r_j} - \bar{A(t)}_{iq_{|N^{s_ir_i}|}^{s_ir_i}}^{s_ir_i} - twt_{iq_{|N^{s_ir_i}|}^{s_ir_i}jq_{|N^{s_jr_j}|}^{s_jr_j}}^{s_ir_is_jr_j} \right), \end{cases} \tag{2.55}$$

$$\text{if } \bar{A(t)}_{iq_{|N^{s_ir_i}|}^{s_ir_i}}^{s_ir_i} + twt_{iq_{|N^{s_ir_i}|}^{s_ir_i}jq_{|N^{s_jr_j}|}^{s_jr_j}}^{s_ir_is_jr_j} \leq \bar{D(t)}_{jq_{|N^{s_jr_j}|}^{s_jr_j}}^{s_jr_j}$$

otherwise;

where h_{r_j} is the headway of route j, and κ is a integer variable that is used to minimize the passenger transfer waiting time. The holding time at transfer points (H_i) is defined by:

$$H_i = \left(p_{iq_{|N^{s_ir_i}|}^{s_ir_i}}^{s_ir_i} + e_{iq_{|N^{s_ir_i}|}^{s_ir_i}}^{s_ir_i} - l_{iq_{|N^{s_ir_i}|}^{s_ir_i}}^{s_ir_i} \right) h_i^{q_{|N^{s_ir_i}|}^{s_ir_i}} \tag{2.56}$$

By substituting Eqs. (2.54) and (2.55) with Eq. (2.53), and combining Eqs. (2.51), (2.52) and (2.56), the change of total passenger travel time is attained.

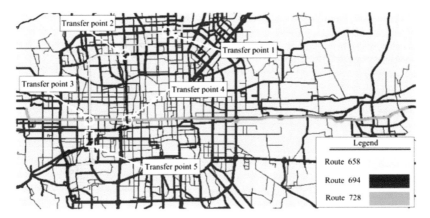

Figure 2.16 A bus network of Beijing with three bus routes and five transfer points (Note: Blue lines represent the road network) (Source: Liu et al. 2014a).

2.6.2 Case Study of the Optimization Model

The effectiveness of the optimization model and its potential for implementation is assessed by a case study of a real-life transit network at Beijing, China. The transit network is illustrated in Fig. 2.16; it comprises three bus routes with five transfer points. Route 658 and 694 run from north to south and route 728 runs from west to east. Transfer points 1, 2, 3, and 4 are fixed single-point encounter transfers and transfer point 5 is a road-segment encounter transfer.

Data Collection

Data of the three routes were collected from the Beijing Public Transport Holdings Ltd. The data include information of routes, vehicles and passengers.

Route information comprises route ID, route direction, number of stops, stop ID, stop latitude and longitude, distance between stops and pre-planned timetable.

Vehicle information All vehicles are equipped with GPS device and can share their real-time speed and location information with the control center. The information was updated using intervals of 30 seconds. The collected vehicle information comprises of vehicle ID, driver ID, departure and arrival time at each stop, planned headway, vehicle location and average running speed.

Passenger Information collected from IC (integrated circuit) cards, comprises the number of passengers boarding and alighting at each stop and the number of passengers transferring at each transfer stop.

Results

The optimization model described (Liu et al., 2014a) was applied to the case study with a comparison between the operation without using tactics and the operation with tactics. The optimization procedure for updating average running speeds for all the five transfer points was implemented, using MATLAB. The original time-distance diagram of the case study without using any operational tactics is depicted on the left hand side of Fig. 2.17 at all the five transfer points. The adjusted time-distance diagram, using the operational tactics (model outcome), are shown on the right hand side. The interval of speed adjustment, of the case study, was five minutes. The average transfer walking time was assumed to be one minute. The arrival time difference t is one minute when this difference is equal or less when one minute holding tactics are employed to increase the opportunity of direct transfers. The results of holding times of vehicles on each route at each transfer points are shown in Table 2.9 and range between zero (simultaneous arrivals of vehicles) and 47 seconds.

The results of the two main measures of comparison are shown in Table 2.10. These results indicate that without using any operational tactics, pre-planned direct transfers do not always occur. One direct transfer occurred at transfer point 4 for passengers transferring from route 694 to route 728. By implementing the operational tactics (model outcome) the number of successful direct transfers becomes 12, marking an increase of 1100 per cent. It may be noted that at transfer point 5 (see Fig. 2.17), which is a road-segment encounter transfer, simultaneous arrivals of vehicles occur at a shared road segment and contribute largely to this significantly improved result. The total passenger travel time with tactics is 547340 seconds and reduced by 83236 seconds, which is a reduction of 13.2 per cent compared with the figure 630576 seconds related to without using tactics. No doubt both results will make the service more attractive and reliable.

These encouraging results demonstrate that potential benefits existed for public transit systems with the implementation of IVC scheme and online operational tactics.

Table 2.9 Holding Times of Vehicles on Each Route at Each Transfer Point.

Transfer Point	Route 658	Route 694	Route 728
1	7 sec.	38 sec.	–
2	3 sec.	4 sec.	–
3	34 sec.	–	10 sec.
4	–	47 sec.	0 sec.
5	0 sec.	0 sec.	–

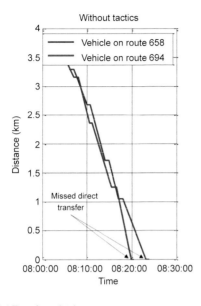

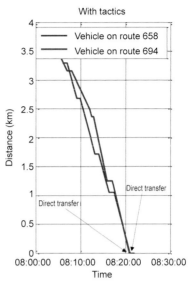

(a) Transfer-point 1

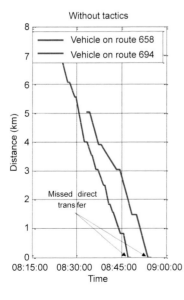

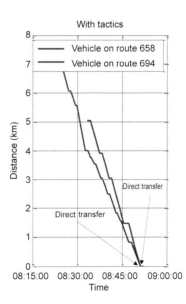

(b) Transfer-point 2

Figure 2.17 Cont....

Figure 2.17 Cont.

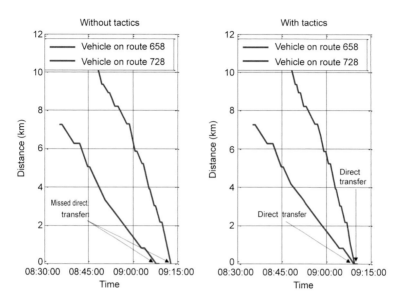

(c) Transfer-point 3

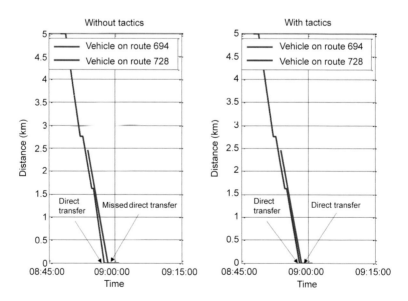

(d) Transfer-point 4

Figure 2.17 Cont....

Figure 2.17 Cont.

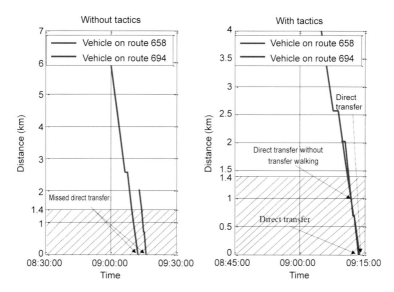

(e) Transfer-point 5

Figure 2.17 Time-distance diagrams of the five transfer points with and without tactics (Source: Liu et al., 2014a).

Table 2.10 Results of the Number of Direct Transfers and Change in Total Passenger Travel Time Without and With Using Tactics.

	Without Tactics	**With Tactics**
Number of direct transfers	1	12
Change of total passenger travel time	0	−83236 sec.

2.6.3 Predictive Control

The basis of the tactic-based predictive control methodology is an event-activity network extended from an original transit network. First, this section describes how to transform a transit network into a corresponding event-activity network in the first part; second, the predictive model is presented and third, the optimization procedure is provided for real-time tactic deployment.

Background of the event activity network

Given is a PT network represented by a directed graph $G = (N, A)$ with a set N of nodes and a set A of directed arcs connecting nodes. Then a corresponding event-activity network $\mathscr{G} = (\mathscr{E}, \mathscr{A})$ with a set $\mathscr{E}$ of events and a set $\mathscr{A}$ of activities can be constructed (Liu et al., 2015).

In an event-activity network, a set of nodes is used to denote a set $\mathscr{E}$ of events. There are two kinds of events: a departure event $\mathscr{E}_{dep}$ and an arrival event $\mathscr{E}_{arr}$.

$$\mathscr{E} = \mathscr{E}_{dep} \cup \mathscr{E}_{arr} \tag{2.57}$$

A departure event e_{dep}, $e_{dep} \in \mathscr{E}_{dep}$ is defined as the departure of a vehicle from a station/stop. An arrival event e_{arr}, $e_{arr} \in \mathscr{E}_{arr}$ is defined as the arrival of a vehicle at a station/stop.

A set $\mathscr{A}$ of arcs is used to describe activities. There are three kinds of activities: driving of vehicle $\mathscr{A}_{drive}$, waiting of vehicle $\mathscr{A}_{wait}$ and transferring of passengers $\mathscr{A}_{transfer}$.

$$\mathscr{A} = \mathscr{A}_{drive} \cup \mathscr{A}_{wait} \cup \mathscr{A}_{transfer} \tag{2.58}$$

A driving activity a_{drive}, $a_{drive} \in \mathscr{A}_{drive}$ is defined as the movement process of a vehicle from its departure time at one station/stop to its arrival time at the next station/stop. A waiting activity a_{wait}, $a_{wait} \in \mathscr{A}_{wait}$ is the wait process of a vehicle from its arrival time at a station/stop to its departure time to the next station/stop. A transferring activity $a_{transfer}$, $a_{transfer} \in \mathscr{A}_{transfer}$ is the transfer process of passengers at a transfer station/stop.

Predictive-control model

Without loss of generality, the following assumptions are made:

1. A pre-set maximum synchronized timetable has already been created for a given transit network with transfer points.

2. At each decision epoch the real-time estimation of passenger OD matrix and activity time matrix are available.

3. Transit operators can deliver suggested operational tactics information timely to drivers without a very long delay, using a communication system.

4. Drivers will comply with the suggested tactics provided by transit operators.

5. For long headway transit systems, passengers are aware of the scheduled bus arrival times at stations. Hence, they will time their arrival at the stations before the scheduled departure times. Delay caused by additional boarding occurs in short headway transit systems.

The tactic-based predictive control framework for controlling the movement of transit vehicles along their routes is graphically shown in Fig. 2.18.

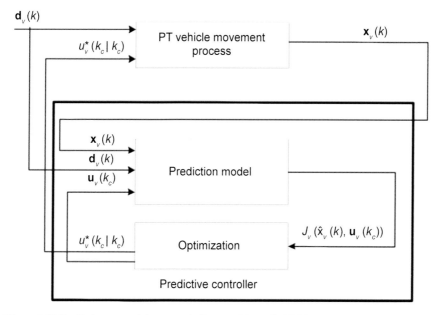

Figure 2.18 Predictive-control framework (Source: Liu et al., 2015).

The evolution of the transit vehicle movement process is described by a dynamic prediction model.

$$\mathbf{x}_v(k + 1) = f(\mathbf{x}_v(k), \mathbf{u}_v(k_c), \mathbf{d}_v(k)) \tag{2.59}$$

where $\mathbf{x}_v(k)$ is the state-variable vector, $\mathbf{u}_v(k_c)$ is the control vector and $\mathbf{d}_v(k)$ is the disturbance vector. The optimization problem at each predictive control epoch is given by:

$$\underset{\mathbf{u}_v(k_c)}{Min}\ J = \sum_{g \in G} \sum_{v \in V_g} J_v\left(\hat{\mathbf{x}}_v(k), \mathbf{u}_v(k_c)\right) \tag{2.60}$$

s.t.

$$\mathbf{x}_v(k + 1) = f(\mathbf{x}_v(k), \mathbf{u}_v(k_c), \mathbf{d}_v(k)) \tag{2.61}$$

$$\phi\left(\hat{\mathbf{x}}_v(k), \mathbf{u}_v(k_c)\right) = 0 \tag{2.62}$$

$$\varphi\left(\hat{\mathbf{x}}_v(k), \mathbf{u}_v(k_c)\right) \leq 0 \tag{2.63}$$

where $J_v(\hat{\mathbf{x}}_v(k),\ \mathbf{u}_v(k_c))$ is the objective function for vehicle v defined as the difference between the scheduled and the actual arrival times at a transfer point. Equation 2.43 is the prediction model; Eqs. 2.44 and 2.45 are the equality constraints and inequality constraints, respectively. The explicit expressions of the predictive state equations depend on the types of event, i.e., either arrival event or departure event and the specific tactics used.

To reduce the online computational complexity, control horizon N_c is defined, such that

$$u(k_c+j\,|\,k_c) = \varnothing, \text{ for } j = N_c, \text{ K, } N_p-1 \qquad (2.64)$$

where $\varnothing$ means without using any tactics at the decision epoch.

After solving the optimization problem, the optimal input tactics $\mathbf{u}_v^*(k_c)$ can be determined. The first element $u_v^*(k_c\,|\,k_c)$ is then applied to control the movement of the transit vehicle; the vehicle then moves to the next decision epoch $k_c + 1$. At this decision epoch, with new road traffic and passenger demand information, the optimization problem is solved again to generate new optimal input tactics $\mathbf{u}_v^*(k_c+1)$. Similarly, the next control element $u_v^*(k_c+1\,|\,k_c+1)$ is applied using the optimal tactics. The optimization process is repeated until the vehicle reaches the last decision epoch. This rolling horizon control scheme makes the predictive controller open to uncertainties and disturbances.

Procedure for real-time tactic deployment

The procedure for real-time tactics deployment includes seven components; it is explained step by step as follows:

Steps 1 (*Initialization*): Initialize parameters.
- Set $k = 1$. Set input tactic $u_v(k_c) = u_v(1) = u_v(\varnothing)$. That is, vehicle v moves without using any tactics. Decision epoch = N_c, prediction epoch = N_p and total decision epoch N.

Step 2 (*Estimation*): Estimate activity time and passenger demand.
- A activity time matrix $\hat{\mathbf{t}}_v(k)$ for vehicle v at event k, which includes the real-time link running times from the current location to the transfer point, is estimated based on the current traffic condition.
- A passenger OD matrix $\hat{\mathbf{w}}_v(k)$ at event k, which includes the number of boarding and alighting passengers at each stop of the vehicle, is estimated.
- These data can be estimated from both historical off-line and current on-line traffic and passenger information. The estimated waiting time depends on the estimated number of passengers boarding and alighting at the stop. It can be determined by a linear model as the one shown in Appendix A.

Step 3 (*Prediction*): Predict the arrival time of vehicle v at the transfer point.
- Based on the estimated link running time, passenger demand and input tactics $u_v^*(k_c\,|\,k_c)$, the arrival time of vehicle v at the transfer point is predicted.

Step 4 (*Comparison*): Compare the actual arrival time and the expected arrival time at the transfer point.

- The gap $\Delta AT(v)$ between the expected arrival time and the predicted arrival time is calculated. If $\Delta AT(v) \leq \varepsilon$, then go to *Step 7*; otherwise, go to *Step 5*.

Step 5 (*Optimization*): Optimize future input tactic set $\mathbf{u}_v(k_c)$.
- According to the objectives and constraints referring to event k, the future input tactic set $u_v^*(k_c \mid k_c)$ is determined.

Step 6 (*Update*): Update input tactic set $\mathbf{u}_v(k_c)$.
- The former input tactic set $\mathbf{u}_v(k_c)$ is updated by the optimized new future input tactic set $\mathbf{u}_v^*(k_c)$, i.e., $\mathbf{u}_v(k_c) = \mathbf{u}_v^*(k_c)$.

Step 7 (*Implementation*): Implement the updated input tactic.
- Based on the receding horizon principle, only the first tactic in the future input tactic set $\mathbf{u}_v(k_c)$ namely $u_v(k_c \mid k_c)$, is selected and implemented with reference to event k. After the implementation, the vehicle moves to the next event. Set $k = k + 1$. If $k = N$, end; otherwise, go to *Step 2*.

2.6.4 Case Study of Predictive-control Modelling

The MPC-based procedure for real-time tactics deployment was applied to examine a real-life bus route in Auckland, New Zealand. The route, known as Northern Express with a bus-only lane, links the North Shore and the Auckland CBD. The route starts from Albany Station and ends at Britomart Transport Centre. There are three main transfer stations. The studied route segment, stations and the travelling direction are shown in Fig. 2.19.

Smales Farm Station (see Fig. 2.19) is the transfer station examined. The scheduled travel times from Constellation Park & Ride to Sunnynook Station, and Sunnynook Station to Smales Farm Station are 2 minutes both ways. The estimated activity time matrix $\hat{\mathbf{t}}_v(k)$ and passenger OD matrix $\hat{\mathbf{w}}_v(k)$ are obtained from historical data. The actual passenger load profile and activity times were collected at each stop by checkers onboard. Two operational tactics, speed-up and skip-stop, are used.

The results obtained by the procedure are shown in Fig. 2.20. It can be seen that the arrival time gap can be reduced from 92 seconds to 27 seconds, and the change of total passenger travel time can be reduced from 964 passenger-seconds to 314 passenger-seconds. These promising results show that the MPC-based procedure can be applied to real-life transit systems to reduce both the total passenger travel time and the arrival time gap at transfer points in schedule-based transit networks.

It is a known phenomenon that synchronized transfers of transit networks do not always materialize because of uncertain and unexpected factors, such as traffic disturbances and disruptions, fluctuations in passenger demand and erroneous behavior of transit drivers. As a result,

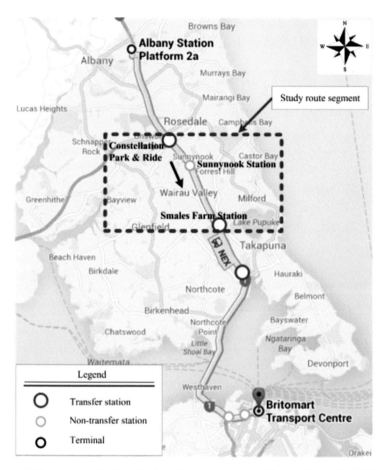

Figure 2.19 The bus route in Auckland used in the case study (Source: Liu et al., 2015).

missed direct transfers will not only frustrate the existing passengers, but also cause loss of potential new users.

This last section presents an MPC-based procedure to reduce the uncertainty of meetings between transit vehicles at transfer points by using some select online operational tactics. The MPC-based procedure, using real-time data, can analyze the combination effect of various online operational tactics on transit networks under dynamic and stochastic conditions.

The MPC scheme described can be extended in various ways. For instance, it can be applied to reduce bus bunching problems, and also to consider scenarios of small or nil number of passengers who want to make a transfer. Further applications of this approach, including multi-vehicle consideration, appear in Liu and Ceder (2016a, 2016b). Future studies

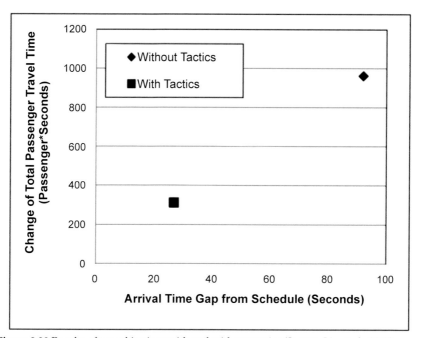

Figure 2.20 Results of two objectives with and without tactics (Source: Liu et al., 2015).

will handle large-scale real-life case studies using both optimization and simulation tools.

Acknowledgements

I would like to extend my gratitude to Mr. Mahmood Nesheli, and Mr. Tao Liu, both PhD students at the University of Auckland, for their valuable contribution to this chapter.

Keywords: multi-agent transit system; synchronized transfers; network simulation; reliability; direct transfers; operational tactics; optimization; Real-time control; inter-vehicle communication; V2V communication; model predictive control event-activity network.

References

AT. 2010. Auckland Regional Land Transport Strategy 2010–2040, Auckland Transport Report, Auckland, New Zealand.

Beijing Municipal Environmental Protection Bureau. 2014. Source apportionment of PM2.5 in Beijing. Retrieved from: http://www.bjepb.gov.cn/bjepb/372794/index.html (Accessed: September 22, 2015).

Bradshaw, J. M. 1997. Software Agents. MIT Press, Boston.

Caliper. 2013. TransModeler. Newton, Massachusetts, USA.

Ceder, A. 2007. Public Transport Planning and Operation: Theory, Modelling and Practice. Elsevier, Butterworth-Heinemann. Oxford, UK.

Ceder, A. 2016. Public Transit Planning and Operation: Modeling, Practice and Behaviour. Second Edition, CRC Press, Boca Raton, USA (appeared in July 2015).

Ceder, A. and N. H. Wilson. 1986. Bus network design. Transportation Research Part B 20: 331–344.

Ceder, A., B. Golany and O. Tal. 2001. Creating bus timetables with maximal synchronization. Transportation Research Part A 35: 913–928.

Ceder, A., Y. Hadas, M. McIvor and A. Ang. 2013. Transfer synchronization of public-transport networks. Transportation Research Record 2350: 9–16.

Davis, R. and R. G. Smith. 1983. Negotiation as a metaphor for distributed problem solving. Artificial Intelligence 20: 63–109.

Dessouky, M., R. Hall, A. Nowroozi and K. Mourikas. 1999. Bus dispatching at timed transfer transit stations using bus tracking technology. Transportation Research Part C 7: 187–208.

Dessouky, M., R. Hall, L. Zhang and A. Singh. 2003. Real-time control of buses for schedule coordination at a terminal. Transportation Research Part A 37: 145–164.

Directive. 2010. Directive 2010/40/EU of the European Parliament and of the Council of 7 July 2010 on the framework for the deployment of Intelligent Transport Systems in the field of road transport and for interfaces with other modes of transport, 7 July. Retrieved from: http://eur-lex.europa.eu/LexUriServ/LexUriServ. do?uri=OJ:L:2010:207:0001:0013:EN:PDF (Accessed: 16th June 2015).

Dueker, K. J., T. J. Kimpel, J. G. Strathman and S. Callas. 2004. Determinants of bus dwell time. Journal of Public Transportation 7: 21–40.

Eberlein, X. J., N. H. Wilson and D. Bernstein. 2001. The holding problem with real-time information available. Transportation Science 35: 1–18.

Hadas, Y. and A. Ceder. 2008a. Public transit simulation model for optimal synchronized transfers. Transportation Research Record 2063: 52–59.

Hadas, Y. and A. Ceder. 2008b. Multiagent approach for public transit system based on flexible routes. Transportation Research Record 2063: 89–96.

Hartenstein, H. and K. P. Laberteaux. 2008. A tutorial survey on vehicular ad hoc networks. IEEE Communications Magazine 46: 164–171.

Hickman, M. D. 2001. An analytic stochastic model for the transit vehicle holding problem. Transportation Science 35: 215–237.

Horn, M. E. T. 2004. Procedures for planning multi-leg journeys with fixed-route and demand-responsive passenger transport services. Transportation Research Part C 12: 33–55.

Ibarra-Rojas, O. J., F. Delgado, R. Giesen and J. C. Muñoz. 2015. Planning, operation, and control of bus transport systems: A literature review. Transportation Research Part B 77: 38–75.

IBM. 2012. ILOG CPLEX Optimization Studio. Somers. New York.

Levine, J., Q. Hong, G. E. Hug and D. Rodrigez. 2000. Impacts of an advanced public transportation system demonstration project. Transportation Research Record 1735: 169–177.

Liu, T., A. Ceder, J. Ma and W. Guan. 2014a. Synchronizing public-transport transfers using inter-vehicle communication scheme: Case study. Transportation Research Record 2417: 78–91.

Liu, T., A. Ceder, J. Ma and W. Guan. 2014b. CBVC-B: A system for synchronizing public-transport transfers using vehicle-to-vehicle communication. Procedia—Social and Behavioral Sciences 138: 241–250.

Liu, T., A. Ceder, J. Ma, M. M. Nesheli and W. Guan. 2015. Optimal synchronized transfers in schedule-based public-transport networks using online operational tactics. Transportation Research Record 2533: 78–90.

Liu, T. and A. Ceder. 2015a. Analysis of a new public-transport-service concept: Customized bus in China. Transport Policy 39: 63–76.

Liu, T. and A. Ceder. 2015b. Communication-based cooperative control strategy for public-transport transfer synchronization. Proceedings (CD-Rom) of the 13th Conference on Advanced Systems in Public Transport (CASPT), Rotterdam, The Netherlands, 19–23 July, 2015.

Liu, T. and A. Ceder. 2016a. Communication-based cooperative control strategy for public-transport transfer synchronization. To be presented at the 95th TRB annual meeting, January 10–14, 2016, Washington DC, and possibly published in Transportation Research Record of 2016.

Liu, T. and A. Ceder. 2016b. Synchronization of public-transport timetabling with multiple vehicle types. To be presented at the 95th TRB annual meeting, January 10–14, 2016, Washington DC.

Luo, J. and J. P. Hubaux. 2006. A survey of research in inter-vehicle communications. pp. 111–122. *In*: K. Lemke, C. Paar and M. Wolf (eds.). Embedded Security in Cars, Springer-Verlag Berlin.

Mees, P. (ed.). 2010. Transport for Suburbia: Beyond the Automobile Age. London, Sterling VA Earthscan.

Minsky, M. L. 1986. The Society of Mind. Simon and Schuster. New York.

Nesheli, M. M. and A. Ceder. 2014. Optimal combinations of selected tactics for public-transport transfer synchronization. Transportation Research Part C 48: 491–504.

Nesheli, M. M. and A. Ceder. 2015a. A robust, tactic-based, real-time framework for public-transport transfer synchronization. Transportation Research Part C 60: 105–123.

Nesheli, M. M. and A. Ceder. 2015b. Improved reliability of public transportation using real-time transfer synchronization. Transportation Research Part C (in press).

Nesheli, M. M., A. Ceder and V. Gonzalez. 2015. Real-time public-transport operational tactics using synchronized transfers to eliminate vehicle bunching. Proceedings (CD-Rom) of the 13th Conference on Advanced Systems in Public Transport (CASPT), Rotterdam, The Netherlands, 19–23 July, 2015.

Nesheli, M. M. and A. Ceder. 2016. Synchronized transfers in headway-based public transport service using real-time operational tactics. To be presented at the 95th TRB annual meeting, January 10–14, 2016, Washington DC, and possibly published in Transportation Research Record of 2016.

Pessaro, B. 2015. Evaluation of Automated Vehicle Technology for Transit. Final Report BDV26 977-07, National Center for Transit Research, University of South Florida. Florida, USA.

Raiffa, H. 1982. The Art and Science of Negotiation. Belknap Press of Harvard University Press. Cambridge, Massachusetts, USA.

Reichardt, D., M. Miglietta, L. Moretti, P. Morsink and W. Schulz. 2002. CarTALK 2000: Safe and comfortable driving based upon inter-vehicle-communication. pp. 545–550. *In*: IEEE Intelligent Vehicle Symposium, Vol. 2, IEEE, New York.

Schweiger, C. L. 2015. TRCP Synthesis 115 on open data: Challenges and opportunities for transit agencies. A synthesis of transit practice. Transportation Research Board, Washington, D.C. USA.

Sichitiu, M. L. and M. Kihl. 2008. Inter-vehicle communication systems: A Survey. IEEE Communications Surveys and Tutorials 10: 88–105.

Smith, R. G. 1980. The Contract net protocol: High-level communication and control in a distributed problem solver. IEEE Transactions on Computers 29: 1104–1113.

Sun, A. and M. Hickman. 2005. The real-time stop-skipping problem. Journal of Intelligent Transportation Systems 9: 91–109.

Timetables. 2015. Open Knowledge Foundation. Cambridge, U.K. Retrieved from: http://index.okfn.org/dataset/timetables/(Accessed: 16th June 2015).

Van Dyke Parunak, H. 1997. Go to the ant: Engineering principles from natural multi-agent systems. Annals of Operations Research 75: 69–101.

Zhao, J. M., S. Bukkapatnam and M. M. Dessouky. 2003. Distributed architecture for real-time coordination of bus holding in transit networks. IEEE Transactions on Intelligent Transportation Systems 4: 43–51.

3

A New Generation of Individual Real-time Transit Information Systems

A. Comi,[1,a,*] *A. Nuzzolo,*[1,b] *U. Crisalli*[1,c] and *L. Rosati*[2]

ABSTRACT

A new generation of transit trip planners, most of which are currently at the prototype phase, is being developed to overcome some of the current limits of such tools, aiming to meet the new requirements of multi-modal transit networks, to benefit from new opportunities offered by advances in telematics and to apply new methodologies in transit network modelling. This chapter points out the main characteristics and the limits of current trip planners and analyzes the current methods used for path generation, taking account of personal information and normative path suggestions on unreliable multi-modal transit networks. Finally, the effects on traveller's choices of predictive info on vehicle occupancy degree are explored. For each topic examined, the current state-of-the-art, open research questions and necessary developments are considered.

[1] Department of Enterprise Engineering, Tor Vergata University of Rome, via del Politecnico 1, 00133, Rome, Italy.
[a] E-mail: comi@ing.uniroma2.it
[b] E-mail: nuzzolo@ing.uniroma2.it
[c] E-mail: crisalli@ing.uniroma2.it
[2] Department of Civil Engineering and Computer Science Engineering, Tor Vergata University of Rome, via del Politecnico 1, 00133, Rome, Italy.
E-mail: rosati@ing.uniroma2.it
* Corresponding author

3.1 Introduction

Information based on the real-time network state may be an effective tool for improving the quality and effectiveness of transit services and thus diverting people to public transport modes (Ren and Lam, 2007; Wahba and Shalaby, 2009; Zhang et al., 2011; Cats et al., 2013). Systems that provide such information are referred to as *real-time traveller information systems* (RTIS). These systems, which can be grouped under the umbrella of intelligent transportation systems (ITS), are able to access, organise, summarise, process and display information to help travellers to plan their trips (Kenyon and Lyons, 2003; Rizos, 2010).

RTIS can provide travellers with information before trip departure (pre-trip) as well as during a trip (en-route). *Pre-trip information* systems are means of alleviating the uncertainty regarding transit schedules and routes, which are often cited by travellers as reasons for not using public transport. Providing accurate and timely information, such tools enable more informed decisions to be made about routes and departure times. *En-route information* systems offer a variety of information to travellers with a view to updating previous choices by taking into account current network states.

A possible classification of RTIS is reported below, based mainly on the functional aspects from the traveller's point of view. Initial classification criteria concern info content:

- timetable information, e.g., 'arrival time is scheduled at Y.YY a.m.';
- descriptive information of a quantitative type, e.g., 'the current bus schedule delay is X minutes';
- predictive information, giving travel attribute forecasts, e.g., 'the bus will arrive at the stop in X minutes'.

Further, RTIS can be designed to provide *shared* (or *collective*) information, i.e., information provided to groups of users (e.g., the arrival time of arriving runs at stops) or *individual* information, i.e., traveller-specific information (e.g., for a given origin-destination pair). Individual real-time systems may cover a single mode of transport (e.g., bus) or several transport modes for a multimodal journey (e.g., car, bus, metro, rail, including different transit services operated by different companies). Such systems are not only bound to supply information on the state of the network (e.g., waiting time at each stop, on-board time) but, given origin, destination and target time (i.e., desired departure or arrival time), they are also able to suggest the best path or a set of best paths. In some systems, the suggested paths are also ordered according to some criteria (e.g., travel time, number of transfers).

Individual information systems, which supply paths to destinations, can be classified into *trip planners, personal traveller advisors and route guidance tools*. Trip planners (TPs), also called route or journey planners, and personal traveller advisors (PTAs) are telematics applications to provide a set of path alternatives generated according to a set of criteria (e.g., minimum travel time, minimum walking distance). Personal traveller advisors are herein considered different from trip planners mainly because they allow interaction with the traveller, aiming to suggest *traveller-tailored* trip advice. PTAs use learning process mechanisms to give information according to personal travel habits. Finally, *route guidance* is an en-route tool designed to help travellers follow a given path with detailed instructions.

Trip planner path suggestions may be classified into:

- *multiple path advice*, i.e., some alternative paths are suggested, according to some path generation rules and the traveller chooses among them;
- *normative recommendations*, e.g., a path, which should be the absolute optimal option for traveller according to a normative approach, is recommended.

To date, several trip planners have been developed (e.g., GoogleTransit, Moovit, Citymapper). However, even if the performance required of such tools has been long established, for example by Chorus et al. (2006), current route planners, as argued in the course of this chapter, often have the following shortcomings:

- they fail to meet the new requirements arising from even more complex multi-modal transit networks;
- they do not apply the latest developments in communication technology and the potential of computer science (big data processing) and open source opportunity;
- they do not meet current needs in transit modelling research, for example relative to demand modelling, especially path-choice decision making, assignment modelling and reverse assignment.

Multi-modal transit networks, which are the target schemes for newly developing public transport worldwide, include not only several public transport services (bus, tram, metro, regional railway) but also some further modes used as alternatives, or to access and/or to egress from the traditional transit services, such as car, car-sharing, bike and bike-sharing and on-demand services. Therefore, planning a journey on such networks is becoming increasingly complex, especially given that booking and ticketing could also be involved.

Recent ITS developments and the implementation of city-wide transit ITS platforms allow several of the main traditional limitations of transit network modelling to be overcome. Such limitations concern data collecting which is usually costly, not always precise, increasingly complex (right of privacy) and often represents only certain mobility aspects of specific (limited) time periods. With transit network 'big data' collecting and processing, a large quantity of data can be obtained at low cost. These data can be used to improve implementation of transit assignment models which, for example, allow real-time forecasting of vehicle occupancy degrees (see Nuzzolo et al., 2015a). Further, better estimation is enabled either of origin-destination matrices or of model parameters; for example, through reverse assignment methods (Russo and Vitetta, 2011). Bi-directional communication between travellers and info centers generates data that can be used not only to estimate origin-destination matrices and model parameters in real time, but also to provide travellers with advice, taking into account personal preferences, as reported in Section 3.3.

Based on these new opportunities and requirements, a new generation of individual predictive trip planners is under development (most of them at the prototype or concept stage), as reported in the following. Among the new aspects of these TPs, new methods for generating recommended paths and additional provision of path attributes (e.g., degree of vehicle occupancy) will be analyzed.

As previously pointed out, existing TPs give path suggestions using rules for path generation, which do not consider that there may be a compensatory trade-off among several path attributes. In some cases, different weights are only applied for different travel-time components. Therefore, some new TPs generate paths through the estimation of the 'convenience' (utility) that the traveller assigns to each travel option, considering compensatory and non-compensatory relationships among characteristics or attributes of the suggested alternatives. In these cases, a path utility function is required, better if tailored to suit the traveller's personal preferences. Using new technologies, it is also possible to continuously upgrade this individual traveller utility function with respect to the revealed choices that the traveller makes during tool usage (traveller preference dynamic learning), as reported in Section 3.3.

Besides, in many cases, travellers move on a stochastic (unreliable) network, i.e., the network performances are random variables, forecasted with uncertainty. Therefore, path suggestions cannot be given only on the basis of path attribute forecasts, but also taking into account the probability distribution of these attributes, as indicated by decision theory. Furthermore, for origin-destination pairs that are connected through sub-graphs with diversion nodes (where path decisions are carried out according to occurrences of random events), suggestions according to an optimal travel strategy, rather than single paths, should be suggested, as argued in Section 3.4.

Finally, some surveys have shown extensive interest among transit travellers in receiving additional information, such as vehicle occupancy (Kenyon and Lyons, 2003; Bichard, 2015). This information may not require additional cost investment in technology for the transit agency, as shown in Section 3.5, but could affect the traveller's pre-trip and en-route choices, especially on congested networks.

The objective of this chapter is to review current developments on these topics and highlight the open research questions and necessary developments. Section 3.2 reports the characteristics and limits of current trip planners. Section 3.3 explores the new utility-based approach to giving path suggestions, also with the possibility of user-tailored advice. Section 3.4 introduces an approach proposing normative strategy-based path suggestions, while Section 3.5 focuses on the provision of vehicle occupancy degrees. Finally, some conclusions are drawn in Section 3.6.

3.2 Current Trip Planner Characteristics

The first trip planners were developed by transit agencies, owners of data on the scheduled transit service and of real-time data on the network state. As previously detailed, recent ITS developments and the implementation of city-wide transit ITS platforms allow transit network '*big data*' collecting and processing, and a large quantity of data can be obtained at low cost, often available as *open data*.

The availability of transit network data has allowed several non-institutional actors to develop applications for providing info, including trip planners. Among these tools, the most popular are, as reported above, Moovit (2009), Google Transit (2005) and Citymapper (2014). These route planners offer their services worldwide, entailing a necessary simplification in the modelling approach. Therefore, most of the current travel advisors use a *rule-based* method for generating recommended paths. This involves a selective approach in which a set of filters, able to reduce the choice set of all feasible paths, is applied to remove unrealistic paths (e.g., those exceeding maximum walking time or distance, number of transfers, maximum transfer time), as well as removing those paths which do not use preferred mode services (e.g., rail, metro, tram, bus).

A new family of trip planners using a different approach has recently emerged, which can be classified as a '*weighted time-based*' approach. After a set of paths is generated using rules similar to those presented above, the best paths are identified by minimising a function of weighted travel time components (such as access, waiting, transfer, on-board and so on). Rules and weights can be defined by the info provider and/or by the traveller. Two examples of these tools are Lazio-Mobility (2015) and Muovi-Roma (2015). The former, built in collaboration with the University of Rome 'Tor Vergata' and used in the region of Lazio, is based on a customised version of the OpenTripPlanner open-source platform. The latter, built by the Mobility Agency of the Municipality of Rome (Roma Servizi per la Mobilità), uses a proprietary system. Both of them allow paths to be generated through the estimation of the 'convenience' (utility) that the traveller assigns to each travel option. The different options are calculated with respect to some general rules used in the rule-based approach. Lazio-Mobility and Muovi-Roma now use journey utility which considers only travel time and its various components, such as waiting time at stops, on-board time, walking time and transfer time. Each component is appropriately weighted according to its importance in the user journey choice decisional process. For this reason, the approach used by these tools is termed 'weighted time-based'. Importantly, these two tools propose paths considering real-time information on waiting time at stops and on-board time on transit vehicles.

Given the opportunity offered by telematics development to collect individual data, the utility function could be improved by including other types of attributes and to estimate individual weights as argued in Section 3.3 below.

3.3 Utility-based Path Suggestions

A new generation of TPs uses a method to identify suggested paths, which can be classified as utility-based, since it refers to path 'cost' with a utility function of path attributes associated to each alternative. Different types of attributes, not only related to travel time, are considered, such as preferences for some modes (like metro), irregularity of services and schedule delay. The parameters of the utility function to be estimated may be average parameters, applied to homogeneous groups of travellers, or individual parameters tailored on the basis of traveller preferences, as explored in the next section.

The advantages (pros) and disadvantages (cons) of the three path generation approaches considered in this chapter are summarized in Table 3.1.

3.3.1 Individual Utility Function Modelling

In the case of individual parameters, learning process mechanisms have to be used to track profiled travellers in order to give them information according to their personal travel preferences. With this objective, smartphones equipped with GPS can provide the continuous exchange of data between the traveller and the info center. Bi-directional communication then allows real-time network information to be given, but also each traveller can be tracked, travellers' choices revealed and the values of attributes experienced by the traveller during his/her journey known.

Table 3.1 Pros and Cons of Path Generation Approaches in Trip Planners.

Type of Approach	Advantages (pros)	Disadvantages (cons)
Rule-based	• rules can be defined by the user • rapid implementation	• compensatory effects are not considered
Weighted time-based	• compensatory effects are considered (only among travel time components) • time component weights can be defined by the user	• few path attributes are considered • weight definition
Utility-based	• compensatory effects are considered • different types of path attributes are considered (not only travel time related) • modal preferences are considered • individual path attribute parameters can be used	• complexity in utility specification • complexity in individual parameter estimation

One of the main limits to obtaining an individual choice model can thus be solved, because many choices for each decision maker can be observed and each individual-level discrete choice model would successfully converge and yield reliable estimates, as proposed by Arentze (2013), and Nuzzolo and Comi (2014), and reported also in Chapter 1 (Nuzzolo, 2016).

Arentze (2013) proposes a Bayesian method to incorporate the learning of users' individual travel preferences in a multimodal routing system. Initial (and average) multinomial-logit utility parameters are obtained from choice observations of a sample of travellers (e.g., obtained from stated preference surveys) and a maximum likelihood (MaxLik) parameter estimation is performed. These estimated values provide the initial distribution of each individual parameter. Subsequently, when the choice of a specific traveller is observed, the system upgrades the individual parameter distribution.

The above approach as set out by Arentze requires preliminary estimation of the path utility function parameters. Besides, current Bayesian upgrading methods apply a sampling procedure from a multidimensional distribution, which can require long computation times. Therefore, as an approximation, Nuzzolo and Comi (2014) proposed to obtain the individual path utility function directly from a sample of choices of the same user. At the first step, the initial MaxLik estimation of individual parameters of a new user is obtained through a short SP (stated preference) survey. At the second step, the parameters are updated using a traveller preference learning procedure, based on the choice revealed when the traveller uses the TP. The parameter updating procedure applies a batch method which, after a given number of (user) choices is observed, provides a new estimate of model parameters by using all the available observations within a maximum likelihood estimation procedure.

The above critique shows that some studies have pointed out individual path utility, but further work has yet to be done. One issue concerns the type of model to be used as several model specifications can be calibrated, moving from the simplest multinomial logit to nested logit, in order to take into account correlations among overlapping path alternatives to a mixed logit (error component) model that allows to point out the heterogeneity (intra and inter users) or correlations due to use of the observations of the same traveller or due to path overlapping. This aspect is further explored in Section 3.3.2.

Another issue concerns the learning process. The relationship between the number of stated and revealed choices and path utility model performance has to be investigated, because one of the main challenges in the learning process is to minimize the time required to estimate initial and upgraded traveller-tailored model parameters in order to provide suitable path suggestions rapidly. This aspect of the learning process of the initial path utility function is examined in Section 3.3.2, whose results are based on some empirical evidences from the metropolitan and urban areas of Rome.

3.3.2 Individual Discrete Choice Modelling: Empirical Evidence

Several experiments were performed by Nuzzolo et al. (2015b) through individual SP surveys, to investigate the improvement resulting from the use of individual rather than average models. Further, the issues of the initial path utility model were considered, as it is important to investigate the minimum number of revealed choices, i.e., how many choice data are sufficient to obtain a statistically significant initial model, which is able to provide suggestions without missing users' actual experiences. Providing non-useful information could discourage travellers from continuing to use the tool.

A set of individual SP surveys, each with 150 choice scenarios, was used. Some university students were asked to choose the preferred path alternative belonging to each proposed scenario and characterised by different forecasted values of travel attributes. The analysis was performed for suburban and urban origins and destinations in order to detect differences in model specification and performance. For more details, refer to Nuzzolo et al. (2015b) and Nuzzolo and Comi (2016a).

The analyses were carried out using the maximum likelihood estimation procedure. The effectiveness of the models obtained was explicitly analyzed, since erroneous suggestions could have significant effects on TP trust and compliance (e.g., the traveller might be induced not to use it anymore). In relation to this aim, the *percentage-of-right* is used, that is, the percentage of observations in the sample for which the alternative actually chosen is that of maximum utility predicted by the model. Further, the *'percentage of the first and the second best'* was also considered, that is, the percentage of observations for which the alternative actually chosen is that of predicted maximum utility or the second best.

Different individual logit model forms were tested: multinomial, nested and mixed. Even if further analyses are in progress, the development of nested logit models did not provide satisfactory results, indicating that in the investigated case, the paths are perceived quite differently by travellers. Besides, mixed logit models showed that an intra-heterogeneity exists for users travelling in an extra-urban context (Nuzzolo and Comi, 2016a).

Individual versus average model parameters

Summarizing the results obtained, one of the main findings is that each user can have a different path utility specification and combine path attributes differently (e.g., the estimated parameters for the same attribute may statistically differ among users). For example, some users do not perceive on-board time differently with respect to transport modes, while others prefer surface modes (e.g., bus and train). By contrast, some users seem to have a large negative weight of waiting time.

Individual and average path utility models were estimated using the SP surveys of six different users in the suburban case and three for the urban case, and were compared through *percentage-of-right*. As reported in Tables 3.2 and 3.3, for some users the performances fell dramatically due to the sizeable behavioral differences among decision makers, which confirms the need to apply individual models in order to suggest paths in accordance with user preferences.

Table 3.2 Performance of Average and Individual Multinomial Logit Models: Suburban Case.

Applied to...	Individual Model		Average Model	
	%-of-right	*%-of-right incl. 1st + 2nd best*	*%-of-right*	*%-of-right incl. 1st + 2nd best*
User A	84%	97%	83%	98%
User B	81%	98%	51%	77%
User C	64%	100%	65%	89%
User D	95%	100%	91%	100%
User E	88%	100%	79%	95%
User F	90%	96%	85%	96%

Table 3.3 Performance of Average and Individual Multinomial Logit Models: Urban Case.

Applied to...	Individual Model		Average Model	
	%-of-right	*%-of-right incl. 1st + 2nd best*	*%-of-right*	*%-of-right incl. 1st + 2nd best*
User 1	91%	99%	72%	95%
User 2	78%	95%	66%	91%
User 3	80%	95%	79%	93%

Initial path utility model

After defining the best form of utility model to be used and the conclusion that better results can be obtained through an individual model, the further investigative step concerns the procedure for initializing individual utility parameters for new users. The main effort is to minimize the time needed to estimate an initial user-tailored utility model that allows suitable path suggestions to be rapidly provided. After a given number of user choices are observed, the estimation procedure provides a new model specification (i.e., number and type of attributes), which can change according to the number of total available observations. The investigation thus concerns the relationship between the number of revealed choices and performances of the path utility model.

The learning process was analyzed and the parameters of the initial utility function were estimated by varying the number of attributes and the number of observations, including 10 new observations at a time (*batch* updating).

The results for the suburban case, presented in Nuzzolo et al. (2015b), indicate that improving the learning process depends on the complexity of the transport network and hence on the complexity of the utility function. Indeed, with different transit service types (as in the investigated case study), the process is quite slow and several observations are necessary to obtain a statistically satisfactory model.

By contrast, in the urban case (Nuzzolo and Comi, 2016a), comparison of the path utility specifications shows that users travelling in urban contexts can have a higher heterogeneity in path utility specification (i.e., number and type of attributes) and in parameter estimates. On the other hand, in the urban test case, intra-user heterogeneity was not revealed. Therefore, for all these reasons, as also pointed out below, the learning process for urban travellers can be more easily implemented.

With regard to the procedure for identifying the initial individual path utility function through the same method as the suburban case, the results show that improvement in the upgrading process is not as slow as in the suburban case and few observations are sufficient to obtain a statistically good model.

Based on these results, a new traveller advisor system called the Tor Vergata personal traveller advisor (TVPTA) was developed. Its logical and functional architectures are briefly described in Section 3.3.3 below.

3.3.3 Example of an Individual Utility-based Traveller Advisor

TVPTA is an individual utility-based trip planner, under prototypical development at Tor Vergata University of Rome (Nuzzolo et al., 2014). It suggests the best path set according to each traveller's personal preferences.

Logical architecture of TVPTA

The overall logical architecture of TVPTA (Fig. 3.1) can be seen as four main components able to support the pre-trip and en-route information, as well as the initialization of personal traveller parameters and their updating according to their revealed preferences.

Each new user (traveller) is asked to fill in the registration form, which includes data on the usual trip: purpose, origin, destination, mode and desired arrival time at destination. The user is then asked to answer a

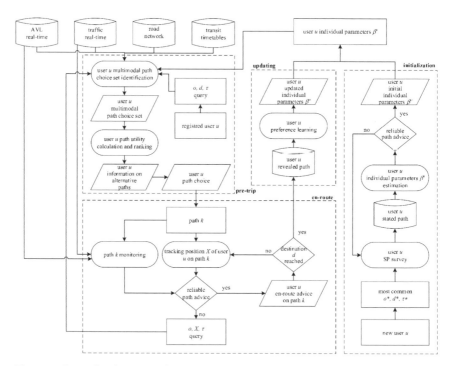

Figure 3.1 Logical architecture of TVPTA.

short stated preference (SP) survey, aiming to capture his/her preferences/ habits in mode and path choices, in order to initialize the main parameters required to apply the individual path choice modelling framework. The pre-trip module is enabled by a query of the registered traveller u, who is logged onto the system. At time τ in which traveller u asks for support to travel from origin o to destination d with a desired arrival time $\tau_{Ai'}$ TVPTA identifies and ranks the multimodal path choice set of user u based on his/her preferences and on traffic and transit real-time data.

Traveller u's path choice enables the en-route path information module, aiming to provide reliable advice on path k, which helps the traveller u to reach his/her destination. It is based on continuous tracking of the current position X of traveller u on path k and on continuous monitoring of path k, using available real-time information (including the occurrence of any unexpected event), considering the traveller u's behavior in following the instructions suggested by the system.

According to traveller usage, the system reveals the traveller's choices and updates the path choice parameters, including these new observations in the parameter estimation procedure.

Functional architecture of TVPTA

TVPTA provides personalised real-time multimodal information starting from open data sets and using only open source software. Figure 3.2 shows the functional architecture of TVPTA. The upper side of this figure reports the main input data sets, which are all open-data ones; the central part of the figure illustrates the processes which take place in the server's side of TVPTA, while the bottom part describes the main client TVPTA app services at the user's disposal.

Starting the description from the services offered on the client's side, the TVPTA user can be informed about traffic and transit condition near his/her position or near a desired point of interest (POI) through the *traffic condition informer* function (18) and that of *transit condition informer* (19). The information contents for functions (18) and (19) are provided by the server's side respectively with functions *real-time traffic condition information* (7) and *real-time transit condition information* (15). Function (7) uses the position of the requesting client, available through the *client position, speed, direction, transport mode* (20) and the road network data stored in the geodatabase (5) for information on average speed and traffic flow near the user or at a specific POI. The road network database (5), starting from a static OpenStreetMap road network, is continuously updated by the function *real-time car network updater* (6) using *real-time traffic information* (2) open data. Function (15) uses the position of the requesting client provided by function (20), and the data supplied by the function *real-time transit data integrator* (13), which uses *real-time transit*

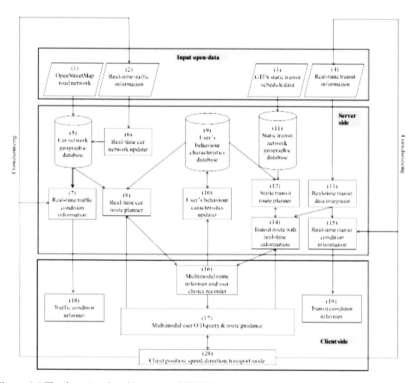

Figure 3.2 The functional architecture of TVPTA.

data (4) to process information on waiting time at stops, irregularity and disruption of transit services near the user or a specific POI. The client function called *Multimodal User O/D query & route guidance* (17) allows the user to request a multimodal origin/destination routing to the server side; the request is processed by the server functions *real-time car route planner* (8) and *transit route with real-time information* (14) using the individual user behavior characteristics supplied by the database (9). Only for transit the routing process is bi-level: routing is initially processed statically by the function *static transit route planner* (12) by using the static transit network representation contained in the *static transit network geographic database* (11) and supplied by *GTFS transit schedule data* (3). The static route is then revised by the functional *transit route with real-time information* (14), which uses real-time data supplied by function (13), allowing the static route to become a real-time dynamic transit route.

When the server's side has finished calculating a path choice set, made of different multimodal real-time paths which satisfy user requirements in terms of origin, destination and desired departure or arrival time, the level of service attributes (i.e., travel time, cost, arrival time at destination,

CO_2 produced, etc.) for each path are summarized in the client function *multimodal route informer* (16). Based on attributes of each proposed path, the user chooses his/her 'best' path using the function *user choice recorder* (16) and TVPTA will lead him/her through the selected path to the destination using the function *route guidance* (17). During the guidance of the user through the selected path, the server's side continuously monitors traffic conditions and transit operations; if problems on the chosen path arise, the server evaluates a new set of multimodal paths by functions (8) and (14) and proposes them to the user, through function (16), thus allowing the choice of an alternative path to reach the destination.

The proposed functional architecture is entirely based on open-source software integrated with proprietary scripts built in Perl language. The database server used is PostgreSQL 9.4 with PostGIS 2.0 extension for GIS functionality. The car route planner is made with the pgRouting 2.0 library for the PostgreSQL database while transit routes, static and real time, are built with a customised version of OpenTripPlanner version 0.9.

Currently, the prototype of TVPTA works in the municipal area of Rome, where different authorities have disclosed open-data sets related to mobility. The *real-time traffic information* (2) is open-data, provided by the Rome Provincial Authority every 10 minutes and contains data on mean speed and car flows for the main road network. *GTFS transit schedule data* (3) and *real-time transit information* (4) are supplied by the Rome Mobility Agency through some specific web services.

3.3.4 Concluding Remarks and Research Issues in Individual Utility-based Path Suggestion

One of the aims of the new generation of trip planners concerns generation of path alternatives through utility-based methods on the basis of individual traveller preferences, defined according to a learning process that uses the new information technology and telematics to track users and to register their path choices, through GPS devices (e.g., mobile smartphones).

Experimental evidence confirmed that use of individual, instead of average, utility path functions improves the path advice. The improved performance of individual models was pointed out, showing the increased *percentage-of-right* that can be obtained. Although intra-user heterogeneity was found in traveller behavior, the performance of the multinomial logit model remains quite statistically satisfactory. Considering its simplicity in terms of analytical tractability, it is preferable in a big data processing context.

Finally, the analyses show that improvement in the learning process depends on the complexity of the transport network. In more complex cases, such as the suburban O-D test recalled, where there are various transit services, the process is quite slow because several observations are required to obtain a statistically satisfactory path utility model. Therefore, in relation to such complexity, a different approach could be required; for example, starting from an average rather than an individual utility function. In other cases, as for the urban test case, few parameters may be required to reach satisfactory results. Applying the proposed method, few observations can thus be used to estimate the individual initial path utility function.

Further research developments mainly concern the investigation of effects due to correlation among the revealed choices of the same users and implementation of the whole estimation process through upgrading of the utility function, using en-route choices revealed during tool usage.

The traveller-tailored path suggestion entails that the individual path utility model be applied, but the review of individual path choice modelling shows that further substantial work has still to be done in this sector. One of the open issues concerns definition of the initial user-tailored utility function. In particular, research challenges entail design of the short individual SP survey (including definition of an efficient number of scenarios to propose to the initial user); identification of different model specifications according to different decision contexts (e.g., purpose of trip—work, pleasure; weather conditions); identification of a functional form to take into account the variations in user tastes and preferences over time; and finally, parameter estimation for taking into account correlations among choices revealed by the same user.

3.4 Normative Strategy-based Real-time Path Suggestion in Unreliable Networks

3.4.1 Introduction to Strategy-based Recommendation

As already discussed in Section 1, when the attributes (e.g., travel time, on-board crowding) characterising the route of some lines are random variables, the service network is classified as *unreliable or stochastic*. In an unreliable network, diversion nodes are present. In such nodes, path decisions have to be carried out according to occurrences of random events (e.g., order of arrival of lines at stops; transit vehicle crowding).

Recent research, using automated data collection, confirms that in unreliable networks with diversion nodes, travellers tend to apply a (subjective optimal) strategy, which allows maximisation of the (subjective) expected travel utility, as already indicated, for example, by Spiess and Florian (1989) (see also Chapter 5, Nuzzolo and Comi, 2016b). The number and kind of strategies which travellers may apply depend on the information available during the trip and on traveller socio-economic characteristics (Schmocker et al., 2013; Fonzone et al., 2013).

Given an O-D pair in a stochastic transit network with diversion nodes and the forecasted values of path attributes, at departure time the path of maximum predicted utility can be found by an individual information system and suggested to the traveller. However, the values of path attributes, forecast by the info system through statistical methods (see, for example, Moreira-Matias et al., 2015), are random variables (Lu et al., 2015) and thus, even with an information system, the uncertainty is generally not completely overcome. According to decision theory (Von Neumann and Morgenstern, 1947), the choice among available path alternatives on such networks is still a decision under risk or uncertainty. Therefore, on unreliable networks, in general, an info system should give suggestions following an *optimal travel strategy*, which allows travellers to reach the destination whilst maximising their *expected utility* (EU), rather than suggest a single path.

A travel strategy is defined by:

- a set of possible diversion nodes;
- for each diversion node, a set of attractive diversion links;
- the choice rule among these diversion links, according to random occurrences.

The topology of a travel strategy can be represented through hyperpaths (Nguyen and Pallottino, 1988; Fig. 3.3), with a set of diversion nodes, where a choice has to be made among diversion links, using a diversion rule and taking into account current and forecasted states of the service network. Once a diversion link has been suggested and chosen, at the following diversion node a new diversion choice is carried out (real-time and sequential *dynamic diversion link suggestions*).

As an example of a dynamic diversion link suggestion, consider Fig. 3.4, where the O-D pair 100–200 is connected by five paths composing a hyperpath with four diversion nodes (20, 40, 50). At origin node 100, the diversion link 100-10 has just one possible sub-hyperpath up to the destination, while diversion link 100-20 has four possible alternative paths consisting of 15 sub-hyperpaths. For a suggestion at time τ, the expected utility of each sub-hyperpath can be computed and, comparing these utilities, the optimal hyperpath can be found deterministically. If the optimal sub-hyperpath includes the diversion link 100-20, the traveller is guided to diversion node 20. When the traveller arrives at node 20, each time a run of line 2 or 3 arrives at the stop, a new search for the optimal total sub-hyperpath until destination is carried out. If the suggestion is to reach node 40, when the traveller arrives at node 40, a new optimal sub-hyperpath search is carried out until the destination.

Note that some approaches other than *EU* have been proposed, such as the prospect theory (Kahneman and Tversky, 1979; de Palma et al., 2008; Ramos et al., 2014, give a useful framework on this topic). In our opinion,

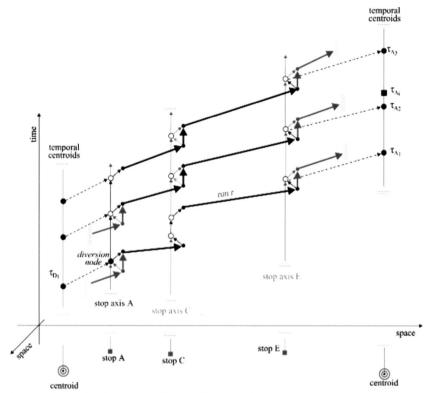

Figure 3.3 Example of a sequential binary choice mechanism at stop *A* for the arriving run *r* (Nuzzolo et al., 2016).

path recommendations should follow the *EU* approach, which leads to an objective optimal alternative. This should differ from the traveller's subjective optimal alternative, which is influenced by traveller cognitive and behavioral errors.

To the authors' knowledge, only the tool hyperpath (SISTeMA, 2015; Gentile, 2016a) uses a normative strategy-based approach for path suggestions and the optimal strategy is found according to an *adaptive-indifferent* rule, as proposed by Spiess and Florian (1989). The traveller boards the first arriving vehicle of a set of attractive lines, which maximize the expected utility. The tool hyperpath is a multimodal trip planner (TP). In the private transport trip planning, it includes real-time traffic information, such as traffic states and events on the network, while in the intermodal trip planning, it takes account of Park&Ride modality where the user starts on a private transport system (i.e., car, motorbike or bike), then he/she can park and finalize the trip on the transit network. For transit path suggestions, it handles both frequency-based and schedule-based transit services, modelling each service consistently with the

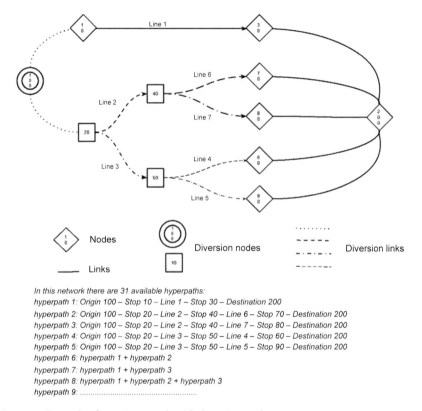

In this network there are 31 available hyperpaths:

hyperpath 1: Origin 100 – Stop 10 – Line 1 – Stop 30 – Destination 200

hyperpath 2: Origin 100 – Stop 20 – Line 2 – Stop 40 – Line 6 – Stop 70 – Destination 200

hyperpath 3: Origin 100 – Stop 20 – Line 2 – Stop 40 – Line 7 – Stop 80 – Destination 200

hyperpath 4: Origin 100 – Stop 20 – Line 3 – Stop 50 – Line 4 – Stop 60 – Destination 200

hyperpath 5: Origin 100 – Stop 20 – Line 3 – Stop 50 – Line 5 – Stop 90 – Destination 200

hyperpath 6: hyperpath 1 + hyperpath 2

hyperpath 7: hyperpath 1 + hyperpath 3

hyperpath 8: hyperpath 1 + hyperpath 2 + hyperpath 3

hyperpath 9: ....

Figure 3.4 Example of transit network with diversion nodes.

traveller's perception. In such a way, it manages to provide a trip strategy also when arrival and departure times are (supposed) known. This allows it to include in a travel strategy real-time transit or any service modification (disruptions, events). Currently, it is the only known tool with this advanced integration of functionalities. Hyperpaths adopt a utility-based approach which is highly customizable: any mode of travel is defined by a large set of preferences, which can be tuned to represent an average travel behavior. Moreover, the user can specify the whole set of parameters according to his/her own individual preferences in order to obtain user-tailored solutions. Further, hyperpath includes the functionality of the alternative routes: a given request is replied with a set of alternative paths obtained by combining preference randomization and graph coloring. In the first case, different paths are yielded by slightly randomizing the preferences of the selected mode, thus changing the overall costs of the paths, while in the second case, the network links building up a given solution are penalized in the successive searches by assigning an additional percentage cost to them. The two methods can be applied separately or combined together. A last feature, which is currently under development, is the implementation of a multi-label algorithm for path search to be used as an alternative to the standard algorithms: not only the best possible cost is associated to a given node but rather a distribution of points in the cost-time space. For more details, refer to Chapter 7 (Gentile, 2016b).

The diversion rule applied by hyperpath could be acceptable when no real-time individual predictive info is available, but at the current stage of evolution in information technology and telematics, the suggestion could be given according to an *adaptive intelligent* approach, which also uses the real-time predictive information available. The adaptive intelligent approach, still in the research development phase, is explored in Chapter 5 (Nuzzolo and Comi, 2016b). In the following section a heuristic search method for normative path suggestions is reported by applying an adaptive intelligent strategy-based approach.

3.4.2 A Heuristic Methodology for Normative Strategy-based Path Recommendation

In the methodology presented below, which can be applied when real-time predicted values of path attributes are available, two types of utilities of a strategy are considered:

- *expected utility EU*, deriving from path utilities which could be experienced following the considered strategy;
- *anticipated utility AU*, which combines expected utility and forecasted path attributes obtained with the methodology presented below.

At each diversion node m and at time τ_m of day t, the strategy anticipated utility AU is associated by the info system to each alternative sub-hyperpath of the choice set with a given diversion choice rule (χ), in order to support decision making. In this case, the *optimal diversion link search*, at diversion node m and time τ, all the sub-hyperpaths $K_{m,lj}$ for each diversion link l_j are considered and the best K_{ml}^* is found by comparing their *anticipated utilities AU*.

The AU of each hyperpath (K_{ml}) is computed by combining its expected utility ($EU_{K_{ml}}^{t,\tau_m}$) and the utility ($FU_{K_{ml}}^{t,\tau_m}$) forecasted with predictive path attributes:

$$AU_{K_{ml}}^{t,\tau_m} = \alpha \cdot FU_{K_{ml}}^{t,\tau_m} + \left(1 - \alpha\right) \cdot EU_{K_{ml}}^{t} \tag{3.1}$$

where $\alpha \in [0,1]$ is the weight given to forecasted utility.

The forecasted utility, $FU_{K_{ml}}^{t,\tau_m}$, of hyperpath K_{ml} can be assumed equal to the utility of the elementary path H_q (of the hyperpath) which has the maximum forecasted utility at time τ_m of day t:

$$FU_{K_{ml}}^{t,\tau_m} = max\left\{FU_{H_q}^{t,\tau_m};\ \forall H_q \in SQ\right\}$$ (3.2)

where SQ is the set of all paths within the hyperpath K_{ml}, $FU_{H_q}^{t,\tau_m}$ is the utility of path H_q forecasted at time τ_m of day t, which can be hypothesized as a linear function of predicted (forecasted) attributes $FX_{gH_q}^{t,\tau_m}$ (e.g., travel time, cost, schedule delay) computed at time τ_m of day t:

$$FU_{H_q}^{t,\tau_m} = \sum_g \beta_g \cdot FX_{gH_q}^{t,\tau_m}$$ (3.3)

with β_g the (average or individual) parameters (to be estimated). Therefore, the hyperpath forecasted utility $FU_{K_{ml}}^{t,\tau_m}$ corresponds to the utility that the traveller would experience following the path suggestions given by the info system, which only takes into account the predicted attributes computed at time τ_m of day t.

With the anticipated utility of the path suggestion, the experienced utility *EXU* (and its average or expected value *AEXU*), which would be experienced during the previous day z on sub-hyperpath K_{ml}, is a function $\psi(\alpha)$ of weight α (Fig. 3.5):

$$AU_{K_{ml}}^{z} \longrightarrow AEXU_{K_{ml}}^{z} = \psi\left(\alpha\right)$$ (3.4)

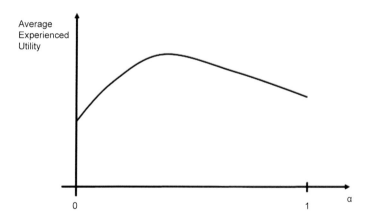

Figure 3.5 Average experienced utility as a function of α.

The optimal value of α to be used for a given strategy and related hyperpath is the α that maximizes the average experienced utility, depending on the error dispersion of forecasted attributes and the stochasticity of service networks. If the forecasts were perfect, past experiences could be omitted and suggestions provided only through forecasts (i.e., α equal to 1). If the forecasts were completely wrong, it would be better to use only expected values obtained from the past (i.e., α equal to 0). Further, for a given forecast accuracy, more service stochasticity could give service attributes very far from expected ones, making the forecasts more useful in the path choice. Therefore, the optimal weight α can be obtained by maximizing $\psi(\alpha)$, that is the average experienced utility, as plotted in Fig. 3.5. Besides, the value of optimal α could depend on diversion node m and on τ_m due to different characteristics of the service network and forecast provision.

In order to obtain the optimal strategy, a methodology can then be used which includes two steps: 1) search for the optimal strategies conditional to α varying between 0 and 1; 2) select the best strategy as the one of maximum *average experienced utility*.

Example of optimal α search

The above-presented method of optimal α search was applied to the service transit network described in Fig. 3.6, with seven lines which connect origin 100 with destination 200. The hyperpath comprises all five different elementary paths. One path (line 1) directly connects the origin to the destination, while the remaining four paths, starting from stop 20,

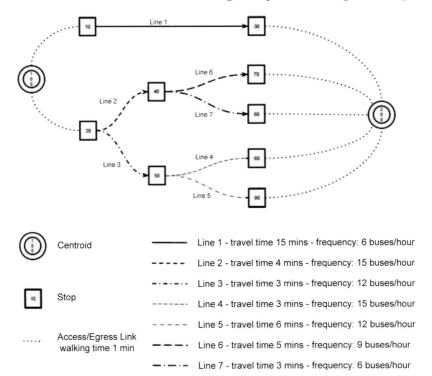

Figure 3.6 Simulated transit service network.

contain stop 40 (line 2) or stop 50 (line 3) as diversion nodes. The possible spatio-temporal dependency of α is not considered in the application and a unique α is sought.

For a given α, the average experienced utility was obtained by considering different levels of service stochasticity through a simulation-based assignment model. Starting from line topology and their scheduled services, for the supply model representation a diachronic graph with about 1,100 links and 600 nodes was implemented, considering 151 runs which describe the transit service in the network simulation period (from 7.00 a.m. to 10.00 a.m.). During the within-day evolution of the transit system, at each minute a service irregularity is simulated, perturbing the vehicle link running times through a Monte Carlo realization from a truncated Gaussian distribution with an average equal to the scheduled running time and a coefficient of variation cv. The new link running times are used to calculate the simulated run travel times and arrival times at stops.

In parallel, in order to provide the forecasted utilities $FU_{H_q}^{t,\tau_m}$ for each path H_q at time τ_m of day t, prediction of service attributes is carried out. In this application, the path attributes considered in forecasted utility evaluation, described in Eq. 3.3, are stop waiting time, on-board time, transfer time and access/egress walking time, duly weighted. The forecast of link vehicle running time is based on a moving average of the last three observed running times of vehicles of whichever line on the links.

The dynamic context of the transit demand was simulated through time-dependent O-D matrices, each related to a demand with a specific desired arrival time to destination (DTD). In all, 60 time-dependent O-D matrices were considered, each describing the demand level from origin 100 to destination 200 with a DTD corresponding to a specific minute between 8.00 a.m. and 9.00 a.m.

When the user reaches stop m at time τ_m, according to travellers' request, i.e., origin (100), destination (200) and DTD (between 8.00 and 9.00 a.m.), the system answers with the first diversion node to reach (in our case stop 10 or stop 20, considering the anticipated utility of stop 10 $AU_{K_{100-10}}^{t,\tau_{100}}$ and stop 20 $AU_{K_{100-20}}^{t,\tau_{100}}$ calculated through Eq. 3.1. For example, the system may suggest going to stop 20 ($AU_{K_{100-20}}^{t,\tau_{100}} > AU_{K_{100-10}}^{t,\tau_{100}}$) and the traveller arrives there. Every time a run of line 2 or 3 arrives at the stop, the system suggests whether or not to board, comparing the anticipated utility of the arriving run (e.g., $AU_{K20-40}^{t,\tau20}$ if a run of line 2 is arriving) and the anticipated utility related to the next runs (e.g., $AU_{K20-50}^{t,\tau20}$ for line 3). This mechanism is repeated for each user at each diversion node (stop) until the user reaches his/her destination. When the within-day simulation of day t ends, the experienced utility of each sub-hyperpath potentially usable by users for satisfying their trip in day t is used to update the expected utility $EU_{K_{ml}}^{t}$ and calculate the $EU_{K_{ml}}^{t+1}$ which will be used in the simulation framework of day $t + 1$.

Several simulations were carried out for different levels of service stochasticity obtained with different values of simulated running time dispersion and different values of parameter α. The average experienced utility $AEXU$ is computed for 60 simulation days. In Table 3.4, the results are summarized, reporting the different values of the running time variation coefficient and of α tested. The best values of average experienced utility ($AEXU$) for each pair of cv and α are in bold. As expected with increasing transit service unreliability, and hence with increasing forecasting failures, the best overall performances are obtained with a normative strategy-based real-time path suggestion which uses a low α parameter, in order to consider much more hyperpath expected utility than forecasted attribute values.

Table 3.4 Simulation Results.

Running Time cv	α	Average Value of EXU	Running Time cv	α	Average Value of EXU
	0.00	−517.64		0.00	−503.46
	0.25	−509.41		**0.25**	**−490.46**
0.20	0.50	−509.90	0.75	0.50	−502.03
	0.75	**−507.51**		0.75	−505.80
	1.00	−509.92		1.00	−501.05
	0.00	−513.58		0.00	−505.20
	0.25	−501.15		**0.25**	**−487.60**
0.50	**0.50**	**−500.10**	1.00	0.50	−511.84
	0.75	−505.51		0.75	−505.02
	1.00	−505.50		1.00	−511.62

3.5 Vehicle Occupancy Degree

As reported in Chapter 1 (Nuzzolo, 2016), among the further path attributes that can be provided, an important real-time predictive information is the occupancy degree of vehicles, relevant both from the travellers' and operations control point of view. Indeed, travellers at stops can choose to skip overloaded runs and wait for less crowded ones, trading a longer waiting-time off against higher on-board comfort. On the other hand, in control strategy applications relative to at-stop and on-board situations, such information can be used by operations controllers both to improve the regularity of transit vehicle trips and mitigate the crowding phenomenon.

Providing such information requires real-time estimation and the short-term prediction of the number of on-board passengers and at-stop travellers waiting for each vehicle. Such variables can be obtained by using two possible prediction approaches: the first entails aggregate-statistical data-driven methods as reported in Chapter 1 (Section 1.1; Nuzzolo, 2016); the second, used in this section, pertains to transit network modelling, reported in several chapters herein.

The effects of introducing info on occupancy degree, in addition to that on travel time components, to help travellers make choices were analyzed by Nuzzolo et al. (2016) on a test network based on the structure of a real transit network of a district in Naples (Italy), which was appropriately modified to capture the effects of different scenarios. Given a traveller information system with individual predictive information about travel time components, the simulated scenarios refer to the presence (WITH scenario) or otherwise (WITHOUT scenario) of real-time information about on-board crowding. Two levels of demand were simulated, as represented by the LOW-crowding scenario and the HIGH-crowding one. The LOW-crowding scenario considers the presence of crowding that is not so high as to result in fail-to-board events; this is the case in which crowding entails only a reduction in traveller comfort. The HIGH-crowding scenario takes into account a transport demand for which fail-to-board events are not unusual. The results obtained (Table 3.5) show the positive effects of this type of info, because:

- on-board crowding info influences departure time: due to travellers departing earlier with respect to their desired target time to anticipate crowding, an increase in early schedule delay (14–15 per cent) with a significant reduction (i.e., 26.1 per cent and 24 per cent respectively in LOW and HIGH-scenario) in the late schedule delay was obtained;
- in HIGH-crowding, on-board crowding info leads to a fail-to-board reduction of about 7 per cent, which is also a side effect of anticipating additional crowding;
- info on crowding allows increasing gains as crowding increases; for LOW-crowding 8.5 per cent in waiting time and 14 per cent in transfer time, which become 10 per cent and 15 per cent, respectively, in the case of HIGH-crowding;
- the benefits in terms of total travel time are quite limited (1.7 per cent in the best case scenario of HIGH-crowding), but the presence of information about on-board crowding reduces the average disutility of 4 per cent for LOW-crowding to 7 per cent for HIGH-crowding.

The research developments required by the provision of on-board occupancy degree in order to improve forecasting results mainly concern improvements in transit assignment methods to be implemented (see Chapter 6, Comi and Nuzzolo, 2016) and in reverse assignment models for joint updating of demand and path choice model parameters (see Chapter 8, Russo and Vitetta, 2016).

Table 3.5 Effects of On-board Crowding Degree: Simulation Results.

Crowding Scenario	Information about On-boad Loads	Total Waiting Time (hours)	Total On-board Time (hours)	Total Transfer Time (hours)	Total Travel Time (hours)	Total Early Schedule Delay (hours)	Total Late Schedule Delay (hours)	Avg Traveller (dis)untility	Avg On-board Load at Stop (pass)
LOW	WITH	123.1	870.2	5.5	993.3	132.31	53.3	0.0865	24.9
	WITHOUT	134.2	871.9	6.1	1006.1	116.1	72.0	0.0899	25.4
	DIFF (%)	-8.3%	-0.2%	-9.8%	-1.3%	13.7%	-26.1%	-3.8%	-1.8%
HIGH	WITH	331.3	1774.8	15.6	2 106.1	256.6	208.2	0.1038	41.1
	WITHOUT	366.8	1 774.9	18.3	2 141.7	224.3	274.0	0.1112	41.3
	DIFE (%)	-9.7%	0.0%	-14.8%	-1.7%	14.4%	-24.0%	-6.6%	-0.5%

3.6 Concluding Remarks and Future Work

The development of new tools for traveller info seems very promising as bi-directional communication and big data collecting and processing can lend a substantial contribution. Besides, as shown throughout the paper, for the provision of info to be effective, several factors with respect to the transit network characteristics, the traveller's perception of the transit service and the structure of available paths are taken into consideration.

In particular, the development of real-time and short-term forecasting methods, and of traveller information tools shows that major changes are required in transit modelling methods, such as the use of more advanced transit assignment and reverse assignment methods. Besides, according to the provision of path suggestions, in order to set up a user-tailored utility function, further efforts should be addressed to investigate the issues related to its definition. For example, an effective number of scenarios to propose to the initial user needs to be defined; different specifications according to different decision contexts have to be identified, as does the functional form to allow for variations in user tastes and preferences over time; and parameter estimation has to be carried out to take into account correlations among choices revealed by the same user.

Finally, since the use of strategies gives a wide range of possibilities to support different kinds of passenger behavior in detail, research (see also Chapter 5) should focus on dynamic real-time generation of the available strategy set and dynamic real-time expected utility computation for each strategy (pointing collective or individual preferences).

Keywords: transit real-time info; individual info provision; transit trip planner; unreliable transit networks; travel strategy.

References

Arentze, T. A. 2013. Adaptive, personalized travel information systems: A Bayesian method to learn users' personal preferences in multi-modal transport networks. *In*: Proceedings of Transportation Research Board Annual Meeting, Washington, USA.

Bichard, N. 2015. ITS on the passenger side. The point of view of agencies and operators: Advanced real time information for tube passengers. *In*: Public Transport Passenger Flows in the Era of ITS. Final Conference COST Action 1004, Paris.

Cats, O., H. N. Koutsopoulos, W. Burghout and T. Toledo. 2013. Effect of real-time transit information on dynamic passenger path choice. CTS Working Paper 2013: 28.

Chorus, C. G., E. J. E. Molin and B. Van Wee. 2006. Use and effects of advanced traveller information services (ATIS): A review of the literature. pp. 127–149. *In*: Transport Reviews: A Transnational Transdisciplinary Journal 26(2), Taylor & Francis.

Citymapper. http://citymapper.com, Retrieved February 8, 2016.

Comi, A. and A. Nuzzolo. 2016. A dynamic strategy-based path choice modelling in real-time transit simulation. *In*: A. Nuzzolo and W. H. K. Lam (eds.). Modelling Intelligent Multi-modal Transit Systems, CRC Press.

de Palma, A., M. Ben-Akiva, D. Brownstone, C. Holt, T. Magnac, D. McFadden, P. Moffatt, N. Picard, K. Train, P. Wakker and J. Walker. 2008. Risk, uncertainty and discrete choice models. *In*: Market Letters 19: 269–285.

Fonzone, A., J. D. Schmöcker, F. Kurauchi, S., Hassan, M. 2013. Strategy choice in transit networks. *In*: Proceedings of the Eastern Asia Society for Transportation Studies, Vol. 9.

Gentile, G. 2016a. Formulating and solving transit assignment. *In*: G. Gentile and K. Nokel (eds.). Modelling Public Transport Passenger Flows in the Era of Intelligent Transport Systems, Springer International Publishing.

Gentile, G. 2016b. Dynamic routing on transit networks. *In*: A. Nuzzolo and W. H. K. Lam (eds.). Modelling Intelligent Multi-modal Transit Systems, CRC Press.

Google Transit. http://maps.google.it. Retrieved February 8, 2016.

Kahneman, D. and A. Tversky. 1979. Prospect theory: An analysis of decision under risk. *In*: Econometrica 47(2): 263–291.

Kenyon, S. and G. Lyons. 2003. The value of integrated multimodal traveller information and its potential contribution to modal change. pp. 1–21. *In*: Transportation Research Part F 6, Elsevier.

Lazio-Mobility. 2015. http://www.laziomobility.it. Retrieved February 8, 2016.

Lu, Y., F. C. Pereira, R. Seshadri, A. O'Sullivan, C. Antoniou and M. Be-Akiva. 2015. DynaMIT2.0: Architecture design and preliminary results on real-time data fusion for traffic prediction and crisis management. pp. 2250–2255. *In*: IEEE 18th International Conference on Intelligent Transportation Systems.

Moovit. http:// www.moovitapp.com, Retrieved February 8, 2016.

Moreira-Matias, L., J. Mendes-Moreira, J. F. de Sousa and J. Gama. 2015. Improving mass transit operations by using AVL-based systems: A survey. *In*: IEEE Transactions on Intelligent Transportation System, DOI 10.1109/TITS.2014.2376772.

Muovi-Roma. 2015. http://www.muovi.roma.it. Retrieved February 8, 2016.

Nguyen, S. and S. Pallottino. 1988. Equilibrium traffic assignment for large-scale transit networks. *In*: European J. Oper. Res. 37: 176–186.

Nuzzolo, A. 2016. Introduction to modelling of multimodal transit systems in ITS context. *In*: A. Nuzzolo and W. H. K. Lam (eds.). Modelling Intelligent Multi-modal Transit Systems, CRC Press.

Nuzzolo, A. and A. Comi. 2014. Advanced public transport systems and ITS: New tools for operations control and traveler advising. pp. 2549–2555. *In*: IEEE Proceedings of the 17th International IEEE Conference on Intelligent Transportation Systems, DOI: 10.1109/ITSC.2014.6958098, IEEE.

Nuzzolo, A. and A. Comi. 2016a. Individual utility-based path suggestions in transit trip planners. *In*: IET Intelligent Transport System, DOI: 10.1049/iet-its.2015.0138, The Institution of Engineering and Technology.

Nuzzolo, A. and A. Comi. 2016b. Real-time modelling of normative travel strategies on unreliable dynamic transit networks: a framework analysis. *In*: A. Nuzzolo and W. H. K. Lam (eds.). Modelling Intelligent Multi-modal Transit Systems, CRC Press.

Nuzzolo, A., A. Comi, U. Crisalli and L. Rosati. 2014. A new advanced traveler advisory tool based on personal user preferences. pp. 1561–1566. *In*: 2014 IEEE 17th International Conference on Intelligent Transportation Systems (ITSC 2014), DOI: 10.1109/ITSC.2014.6957915, IEEE.

Nuzzolo, A., U. Crisalli, A. Comi and L. Rosati. 2016. A mesoscopic transit assignment model including real-time predictive information on crowding. *In*: Journal of Intelligent Transportation Systems: Technology, Planning, and Operations, DOI: 10.1080/15472450.2016.1164047; Taylor & Francis.

Nuzzolo, A., U. Crisalli, L. Rosati and A. Comi. 2015a. DYBUS2: a real-time mesoscopic transit modeling framework. pp. 303–308. *In*: 2015 IEEE 18th International Conference on Intelligent Transportation Systems (ITSC 2015), DOI: 10.1109/ITSC.2015.59, IEEE.

Nuzzolo, A., U. Crisalli, A. Comi and L. Rosati. 2015b. Individual behavioural models for personal transit pre-trip planners. pp. 30–43. *In*: Transportation Research Procedia 5, DOI: 10.1016/j.trpro.2015.01.015, Elsevier Ltd.

Ramos, G. M., W. Daamen and S. Hoogendoorn. 2014. A state-of-the-art review: Developments in utility theory, prospect theory and regret theory to investigate travellers' behaviour in situations involving travel time uncertainty. *In*: Transport Reviews: A Transnational Transdisciplinary Journal 34(1): 46–67. DOI: 10.1080/01441647.2013.856356.

Ren, H. and W. H. K. Lam. 2007. Modeling transit passenger travel behaviors in congested network with en-route transit information systems. *In*: Journal of the Eastern Asia Society for Transportation Studies 7: 670–685.

Rizos, A. C. 2010. Implementation of Advanced Transit Traveler Information Systems in the United States and Canada: Practice and Prospects. Thesis for the Degree of Bachelor of Science in Planning and Master in City Planning, MIT, Boston, USA.

Russo, F. and A. Vitetta. 2011. Reverse assignment: calibrating link cost functions and updating demand from traffic counts and time measurements. *In*: Inverse Problems in Science and Engineering 19(7): 921–950.

Russo, F. and A. Vitetta. 2016. Real-time reverse dynamic assignment for multiservice transit systems. *In*: A. Nuzzolo and W. H. K. Lam (eds.). Modelling Intelligent Multi-modal Transit Systems, CRC Press.

Schmocker, J. D., H. Shimamoto and F. Kurauchi. 2013. Generation and calibration of transit hyper-paths. *In*: Transportation Research C 36: 406–418.

SISTeMA. 2015. Hyperpath, www.hyperpath.it

Spiess, H. and M. Florian. 1989. Optimal strategies. A new assignment model for transit networks. *In*: Transportation Research. Part B: Methodological 23B(2): 83–102.

Von Neumann, J. and O. Morgenstern. 1947. Theory of Games and Economic Behavior. Princeton: Princeton University Press.

Wahba, M. and A. Shalaby. 2009. MILATRAS. A new modelling framework for the transit assignment problem. pp. 171–194. *In*: N. H. M. Wilson and A. Nuzzolo (eds.). Schedule-Based Modeling of Transportation Networks: Theory and Applications, Kluwer Academic Publisher.

Zhang, L., J. Li, K. Zhou, S. D. Gupta, M. Li, W. B. Zhang, M. A. Miller and J. A. Misener. 2011. Traveler information tool with integrated real-time transit information and multi-modal trip planning. *In*: Transportation Research Record: Journal of the Transportation Research Board 221, Washington DC, USA.

4

Real-time Operations Management Decision Support Systems
A Conceptual Framework

Oded Cats

ABSTRACT

Public transport operations are subject to inherent uncertainties. In the era of Advanced Public Transport Systems (APTS), operators can collect, process and analyze real-time system conditions as well as deploy real-time operations management strategies. Proactive service management requires operators to continuously monitor service performance and assess the implications of alternative interventions by forecasting and simulating how the system will evolve under various scenarios. Operations management decisions need to consider the prevailing and evolving supply and demand uncertainties as well as the capability of both service provider and service users to adapt to changing conditions. Despite the importance of considering these interactions, there is a pronounced division in the public transport modelling sphere between public transport operations and public transport

Delft University of Technology and Royal Institute of Technology (KTH), Department of Transport and Planning, Delft University of Technology, P.O. Box 5048, 2600 GA Delft, The Netherlands.
E-mail: o.cats@tudelft.nl

assignment models. A conceptual framework for a real-time operations management decision support system (RT-OMDDS) is presented in this chapter along with its prospective components. Modelling requirements, system architecture, challenges involved in the real-time deployment of such a system and potential applications are discussed. The envisaged RT-OMDDS consists of the following modules: network initializer, traffic flow, passenger flow, real-time strategies, scenario design and scenario evaluation. These modules reconstruct the current system state, predict the dynamics and distribution of vehicles and passengers and a toolbox for testing a series of real-time strategies. Advances in modelling public transport dynamics and online public transport assignment, such as agent-based simulation models, facilitate the development of a RT-OMDDS. This chapter concludes with an outlook on the prospects of RT-OMDDS and related research questions.

4.1 Towards Decision Support Tools in Real-time Operations

4.1.1 Real-time Operations Management

Public transport planning is conventionally divided into strategic, tactical and operational decision making. Each planning level involves the consideration of a different decision horizon, geographical scope and performance indicators. Furthermore, the different planning levels entail different planning and modelling issues. For example, strategic network design requires estimating future demand levels and forecasting the distribution of passenger flows over the network, whereas tactical planning involves solving large-scale crew and fleet scheduling requirements. While effective strategic and tactical planning are critical for service performance, the uncertainty associated with the public transport operation environment leads to a discrepancy between planning and operations.

The real-time operations management of public transport systems is concerned with collection, processing and analysis of real-time conditions and the assessment and deployment of alternative control, fleet management and information provision strategies (see Chapter 2 of this book). The operation of public transport systems is overseen by service providers in real-time in order to monitor and control its performance. Service providers may intervene in the operations if a deviation from plan or a deterioration in the provisioned level-of-service is observed. Interventions include changes in service provision (e.g., transfer coordination, vehicle rescheduling) and information provision (e.g., prescriptive journey planner). The deployment of such measures has economic implications for both operators and passengers, as well as society as a whole.

The increasing importance of service reliability and robustness calls for the development of more effective and efficient tools for real-time operations management. The abundance of instantaneous data concerning vehicle positions and passenger flows paves the way to such developments that will address the growing need for adopting proactive and adaptive strategies for public transport management. These strategies may involve real-time control and fleet management strategies, and possibly, even alteration in the scheduling and routing decisions made at the planning stages. In the absence of tools to assess the consequences of alternative measures, operators in the control room rely solely on expert local knowledge and experience.

The amount of data streaming into the control center and the complex relations between a large number of factors that influence system performance hinders the capability of the human mind to process all the information, analyze the current conditions and future scenarios, consider alternative possible actions and evaluate their consequences within a short time interval under stressful working conditions. As a result, control center dynamics are dominated by a narrow definition of schedule constraints rather than service provision and are largely myopic (Carrel et al., 2010).

While operations management has been traditionally steered by infrastructure and rolling stock considerations, there is an increasing shift towards demand-driven operations. This shift is reflected in changes in the performance indicators and incentive schemes (Cats, 2014b). For example, the Dutch parliament passed in 2015 a bill that implies that service reliability should be measured and evaluated based on passenger experience rather than vehicle-based indicators. This trend is expected to result in an increasing importance of real-time operations management geared towards passenger reliability, such as ensuring transfer coordination, service regularity and congestion mitigation measures. Hence, decision support systems need to account for the impacts of the prevailing traffic conditions as well as real-time strategies on passenger flows and travel experience.

4.1.2 Decision Support Systems for Real-time Operations Management

Decision support systems were developed in a wide field of applications to provide decision makers with tools to monitor, analyze, interact and deploy measures to improve system performance in complex environments. Decision support systems were deployed in the transportation sector, including in the public transport context (Adamski and Tumau, 1998; Törnquist, 2005). These systems enable monitoring service performance and rolling stock management. In the context of the urban traffic networks, Gentile and Mescchini (2011) developed a tool that calibrates in real-time a dynamic assignment model and allows reconstruction of current traffic conditions and prediction of future conditions to support real-time decision making. Burgholzer et al. (2013) proposed a decision support model for analyzing the impacts of disruptions in multi-modal networks. However, there is currently no decision support system which encompasses predictions concerning traffic and passenger flows of public transport systems that allow service providers to evaluate the implications of alternative real-time strategies.

The aim of this chapter is to outline the architecture, potential features and applications and consequently the modelling requirements that are associated with real-time operations management decision support systems (RT-OMDDS). The envisioned system will support the control center of the responsible public transport agency or operator in daily operations.

By integrating big data analytics, prediction schemes, transport simulation and assignment models, a RT-OMDDS will facilitate the evaluation and selection of intervention measures. Figure 4.1 presents a conceptual framework for developing such a system. The real-time operations management process is concerned with the continuous interaction between the transport system (left) and the operations management system (right). Real-time data concerning the prevailing transport conditions (top left) such as public transport vehicle positions, traffic and passenger flow data (counts, fare collection, plate recognition,

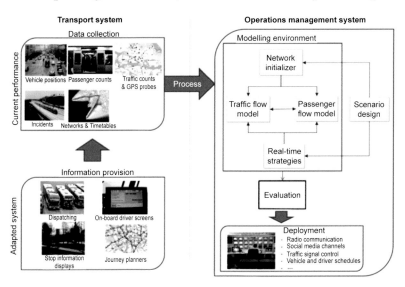

Figure 4.1 A conceptual framework of a real-time operations management decision support tool.

social media) and infrastructure sensors are processed and compared with the planned service and historical databases.

The operations management system (right) includes a modelling environment which analyzes current system performance and generates predictions concerning future system states, such as vehicle arrival times and passenger flows. This modelling environment could be embedded in a decision support system to enable the specification, analysis and evaluation of the consequences of deploying alternative measures as opposed to a do-nothing scenario. Control center staff will then implement the selected strategies (e.g., rescheduling services, allocating reserve resources) using information and communication technologies with vehicles and infrastructure. The implementation of such measures will result in an adapted transport system (bottom left) for which new data is collected in a continuous monitoring and control loop.

The remainder of this chapter describes and discusses how the above-mentioned conceptual framework could be realized. First, developments in modelling the dynamics of public transport systems that underpin any implementation of RT-OMDDS are described in Section 4.2. Section 4.3 discusses the modules involved in the modelling architecture and related modelling requirements and challenges in the context of a real-time deployment of such a system. Section 4.4 presents how the modelling environment could be embedded in a decision support system to allow scenario design and evaluation. Finally, Section 4.5 discusses potential applications of the RT-OMDDS and provides an outlook on system development and deployment.

4.2 Dynamic Modelling of Public Transport System Evolution

4.2.1 Public Transport as a Dynamic System

Public transport systems encompass various components that interact through several processes. The development of a model for public transport as is true for any involves the fundamental decision on which relevant factors need to be included in the model and at what level of detail, depending on model objectives. Operation management is primarily concerned with service reliability and measures to mitigate sources of uncertainty, such as delays caused by disturbances and disruptions (e.g., unplanned technical failure, a planned temporal service cancellation, route change). Mitigation measures include public transport signal priority, holding, expressing and short-turning, or transfer coordination (see a review of real-time strategies by Muñoz et al., 2013). In order to model whether such interventions are needed and their prospective effects, it is essential to represent the sources of uncertainty that are inherent to the public transport operations environment. Operations management decisions need to consider the prevailing and evolving supply and demand uncertainties as well as the capability of both service provider and service users to adapt to changing conditions. Despite the importance of considering these interactions, there is a pronounced division in the public transport modelling sphere between public transport operations and public transport assignment models.

Public transport systems evolve through dynamic interactions between traffic, passenger demand and operations management. Traffic conditions are determined by the infrastructure (rail- or road-bound, design speed, capacity) and the interaction with other traffic (mixed-traffic, separate lane, blocking times). Travel time variability could be the result of either exogenous or endogenous processes. Exogenous sources include traffic congestion, traffic incidents, weather conditions, a vehicle that breaks down, defected infrastructure or events that attract large crowds. Processes that are endogenous to the public transport system are terminal operations and dispatching, the bunching phenomenon and the implementation of control strategies. Since travel time variability is an important determinant of service reliability and hence level-of-service, it is essential to reproduce and predict travel time variability. The explicit modelling of these processes facilitates a more realistic reproduction of supply uncertainty rather than generating it based on statistical distributions and assembling independent models of separate system components (Toledo et al., 2010).

In addition to vehicular traffic flows, the performance of public transport systems is determined by the evolution of passenger flows. The impacts of alternative operation management strategies depends on passengers' en-route responses to changing service conditions. Static assignment models are therefore, not suitable for this analysis as they cannot capture the time-dependent variation in public transport supply and demand and passengers' capability to adapt their plans accordingly. In contrast, a dynamic assignment model could potentially replicate the inter-related dynamic processes that drive service unreliability, crowding conditions, the impact of real-time information and both operators' and passengers' reaction to system conditions. Furthermore, due to the size and complexity of public transport systems—in particular in the era of ITS—it is unrealistic to apply global analytical models. Simulation-based assignment models offer an alternative approach for modelling public transport performance.

4.2.2 The Agent-based Approach to Public Transport Assignment

An agent-based simulation approach to public transport assignment models allows modelling individual passengers that make a sequence of travel decisions based on their preferences and expectations, where the latter depend on the information available to them when making a decision. Each agent in the system carries out decisions, interacts with other agents and so affects the way the system evolves over time. For example, service uncertainty will affect passenger waiting time and crowding levels at stations and on-board. While the interactions between agents are constrained in time and space, the impacts of these interactions may propagate over public transport lines and service times and result in spill-over effects. Examples of network-wide applications of such a model are described in Chandakas et al. (2016).

The agent-based modelling approach helps to explicitly account for variation and heterogeneity in travel demand and preferences. Fonzone et al. (2010) provide empirical evidence based on an international survey that suggests that the vast majority of public transport users exercise discrepancies in their typical trips in at least one of the following aspects: departure stop, deviation from preferred line, transfer stop and changing line once on board. A probabilistic choice model with taste variations could potentially represent travel demand more realistically. In addition, the choice model can be enhanced by considering cognitive limitations and biases, network-related knowledge and limited adaptation. Furthermore, agent-based modelling supports the representation of inter-dependencies in passenger decisions due to discomfort and capacity considerations.

Simulation-based assignment models often consist of within-day and day-to-day assignment loops. The former is concerned with the dynamic path choice process rather than equilibrium conditions, whereas the latter results in an iterative dynamic loading that could be regarded as a learning process. Nagel and Marchal (2003) argue that while this iterative process does not satisfy the mathematical definition of equilibrium conditions, the learning process may under certain conditions correspond to user equilibrium or stochastic user equilibrium conditions. This argument also underlines the learning function and steady-state conditions examined in the context of a microscopic dynamic traffic simulation by Liu et al. (2006), in a schedule-based public transport assignment model by Nuzzolo et al. (2011) and in a learning-based public transport assignment model by Wahba and Shalaby (2014) and Cats and Gkioulou (2015).

4.2.3 Modelling Public Transport Reliability and Information Provision

The interaction between service uncertainty and passenger flows is most pronounced in the context of operation management that is concerned with the impacts of information provision, passenger congestion and service disruptions. These three application domains will be hereby discussed. Modelling these phenomena requires representing passengers' en-route decisions as system performance evolves through interaction between service reliability, rerouting decisions, and discomfort and capacity constraints. Nevertheless, most of the methods for rescheduling public transport services in case of disturbances assume that traveller behavior is deterministic and independent of the operational plan. Similarly, existing solutions to the real-time transfer coordination problem consider passenger flows as an exogenous variable that is estimated on the basis of average historical values (Dessouky et al., 2003; Guevara and Donoso, 2014). A noticeable exception is a recent study by Schmidt and Schöbel (2015) which investigates the influence of integrating the passenger routing decision in the transport service scheduling process with the objective to minimize the overall passenger travel time in the network.

Passenger congestion effects are intimately linked to service uncertainty due to the positive feedback loop between supply and demand variability. A dynamic public transport model, and in particular an agent-based assignment model, facilitates a more realistic analysis of passenger on-board discomfort, passengers left behind due to capacity constraints and prolonged waiting times due to delays and irregularity (Cats et al., 2016). In contrast, a static assignment model will result in an underestimation of the congestion effects and may result in unrealistic waiting time and passenger loads on both over- and under-utilized public transport lines (Schmöcker et al., 2008).

The analysis of information provision strategies requires a dynamic representation of public transport systems. Static public transport assignment models that investigate the impact of real-time information on passengers' decisions assume that the information is perfectly correct and that passengers perceive the information to be perfectly credible (e.g., Nuzzolo et al., 2001). A new generation of individualized and self-learning information services (see Chapter 3) further stress the need to account for their effects on passengers' choices. Modelling the impact of real-time information using static assignments is, therefore, equivalent to the case that passengers have perfect knowledge concerning the provisioned travel attributes. Conventional approaches are, therefore, inadequate for analyzing the impacts of information. Furthermore, while considering the impacts of real-time information on en-route travel decisions, MILATRAS, a public transport simulation model, assumes that the information is universally available and is perceived as credible (Wahba and Shalaby, 2014). However, a dynamic simulation of public transport operations allows modelling of algorithms that are used for generating real-time information in an uncertain operations environment and their effects on en-route passengers' decisions. Moreover, the availability and impact of real-time information provision can vary across the network and passengers' population. By integrating the generation and dissemination of real-time information into an agent-based public transport assignment, the relation between service reliability and information reliability and their effects could be investigated

(Cats and Gkioulou, 2015). The potential time savings from providing real-time information were investigated using BusMezzo, a public transport operations and assignment simulation model (Cats et al., 2011).

Understanding how public transport systems evolve as a result of deviations from plans is especially crucial in the case of service disruptions. While network reliability is concerned with small and inherent disturbances in service provision, network vulnerability refers to the capability of the service to absorb severe disruptions (typically caused by factors exogenous to the public transport system) and recover to normal operation. A disruption will not only affect vehicle traffic on the disrupted link as delays may propagate across the network, especially on links directly upstream of the disruption. Furthermore, in addition to the passengers restrained on-board a vehicle positioned on a blocked link passengers waiting further downstream may reconsider their travel decisions and secondary effects could be caused by either supply knock-down effects (e.g., vehicle scheduling) or passengers' rerouting decisions (e.g., delays, denied boarding). All of these effects were modelled using BusMezzo by applying a non-equilibrium assignment model, for identifying critical links, the impact of disruptions and the value of reserve capacity (Cats and Jenelius, 2014, 2015).

4.3 Modelling Architecture

The growing availability of instantaneous traffic and travel data accompanied by the rapid and consistent increase in computational power facilitates the development and deployment of dynamic public transport models that could support the real-time operations and control of multimodal public transport networks. While emulating the dynamics of public transport systems is essential to enable the evaluation of real-time strategies, embedding such models in a RT-OMDDS requires the development of several modules. Moreover, assignment models have traditionally been used solely for offline evaluation of alternative network designs and forecasting the respective distribution of passenger flows to support project appraisal. In the following sub-sections, the design and specification of several modules envisioned in the system architecture of a RT-OMDDS are described.

4.3.1 Modelling Environment Components

Real-time operations applications require modelling how the public transport system and its performance will evolve in a relatively short-decision horizon; for example, generating every 5 minutes predictions concerning the next 30 minutes. The modules described in this section correspond to the general operation management process which consists of reconstructing the current system state, forecasting future states, considering a limited set of actions, forecasting the performance attained under each of these scenarios, action selection and specification, and then monitoring its execution and consequences. The following discusses how a decision support system based on simulation, prediction and optimization techniques can be embedded into this general operations management process.

The deployment of dynamic public transport models in real-time operations impose several requirements on their modelling capabilities. Most importantly, the temporal features of public transport services must be considered along with their implications on service reliability and crowding. Moreover, the impacts of these service aspects as well as real-time information provision on passengers' route choice should be accounted for when modelling system dynamics.

The development of a decision support tool in real-time operations requires short-term prediction of public transport system performance. Predicting future states of the system under alternative scenarios will facilitate a proactive public transport operation. Short-term predictions of travel times and passenger flows could be obtained through analytical models, machine learning techniques (e.g., Ma et al., 2014) or using assignment models. Only the latter is capable of explicitly accounting for the inter-dependencies between passenger flows on different lines and trips by representing passenger route-choice behavior.

The RT-OMDDS comprises of the modelling environment, the scenario design and evaluation modules as shown in Figure 4.1. This section details the four modules included in the modelling environment, namely *network initializer, traffic flow, passenger flow* and *real-time strategies*. The data collected from transport operation is transmitted and provided as input to the initialization phase of the modelling tool. Network initializer produces outputs that are then fed to both vehicular and passenger flow modules which represent how the network evolves under alternative scenarios. Finally, the real-time strategies module simulates the control center operations and the specification of these strategies. While the strategy is typically concerned with either supply-side or demand-side interventions, it carries implications for both vehicle and passenger progress due to their inter-dependencies. These three core modules interact either iteratively or within an integrated modelling environment.

4.3.2 Network Initializer

The *Network initializer* module is responsible for reconstructing current system conditions based on real-time data. The RT-OMDSS needs to efficiently process and fuse instantaneous data in order to reconstruct the current traffic and passenger flow conditions. The specification of the initialization conditions seeks to replicate as realistically as possible the current system conditions based on the information available in real-time. The module receives instantaneous feeds from the respective application area, such as traffic counts, vehicle positions and passenger counts, depending on data availability. These data sources need to be processed, fused and integrated with static information sources, such as geographical road and rail layers, public transport network and service and planned timetables. The representation of current network conditions serves as an initialization phase for the dynamic public transport model.

The development of this module involves the deployment of techniques to process instantaneous data as well as methods to estimate information which may not be available in real-time by deploying state estimation techniques. The latter is particularly relevant for generating initial demand data since information concerning demand is often limited and is only seldom available in real-time. Passenger distribution over stops and vehicles needs therefore to be estimated. Estimates for the number of waiting passengers could be derived from a time-dependent OD matrix whereas estimates for on-board passenger loads could be obtained either from historical passenger counts or schedule-based or agent-based assignment results. In both cases, estimates could be adjusted on the basis of observed headways/delays in case the observed traffic conditions divert significantly from the planned ones. This module results in an estimated initial network state which is then simulated to forecast future system states.

4.3.3 Traffic Flow

Any decision concerning public transport operational management needs to take into consideration the prevailing traffic conditions and how they are expected to evolve due to their implications on public transport performance. In particular, the progress and future positions of public transport vehicles need to be predicted. Bus arrival predictions can be generated on the basis of statistical models, aggregate flow-density relations or machine learning techniques (Shalaby and Farhan, 2004). The prediction schemes might be either based on a combination of historical and instantaneous traffic data (Fadaei Oshyani and Cats, 2014) or based on the simulation of traffic dynamics, if time-dependent private car demand data is available. Predictions for urban rail-bound services can be made based on link-specific speed profiles (Cats, 2014a).

Given modelling requirements, traffic simulation models are well-suited to represent complex system dynamics and the implications of ITS applications. Simulation environments can exploit available data concerning the current state of the system in order to evaluate, in real-time, the impacts of alternative scenarios. Several simulation models, such as MITSIMLab and DynaMIT (Ben-Akiva et al., 2002), were developed to facilitate the analysis of ITS operating strategies. Traffic simulations are conventionally classified into three classes, according to their level of detail and aggregation: macroscopic, microscopic and mesoscopic. Macroscopic models represent traffic as a continuous flow based on flow-density functions without the explicit modelling of lanes or vehicles. At the other end, microscopic models represent traffic at the most detailed level: individual vehicles are represented and their behavior depends on their interactions with other vehicles, geometry, lane assignments, etc. As a result of computational constraints, there is an inverse proportionality between the level of details and network size on one hand and complexity and the feasibility of their analysis on the other hand. A third group of models exists on this spectrum—mesoscopic models—which represent individual vehicles but avoid detailed modelling of their second-by-second movement. Since real-time strategies are concerned with individual vehicles, the traffic simulation model should be either microscopic or mesoscopic.

The traffic flow model needs to include multi-modal flows that follow distinctive traffic regimes in order to represent the traffic conditions that prevail for public transport in mixed-traffic, dedicated lanes, segregated ways and road- vs. rail-bound services. Different public transport modes are also characterized by distinctive flow-dependent dwell-time functions. These sets of operational attributes yield different levels of reliability and capacity depending on service design and right-of-way, which could be accounted for in the public transport simulation model (Toledo et al., 2010). The coupling of road and rail traffic simulation models along with their distinguished traffic management features would allow analyzing metropolitan or national multi-modal systems. This is especially important in cases where a disruption on one of these networks might lead to modal shift or could be relieved by offering a replacement service using another network.

4.3.4 Passenger Flow

While the *traffic flow* module represents the vehicular movement, the *passenger flow* module results in the assignment of passengers to individual vehicles. In order to predict the distribution of passengers over the public transport network, passengers' choices—in particular, route choice—and responses to changing service conditions have to be modelled. Moreover, the passenger flow model should account for passengers' adaptation to real-time strategies.

A substantial research effort was devoted in the last three decades to the development of dynamic traffic assignment (DTA) models. In their comprehensive review of DTA, Peeta and Ziliaskopoulos (2001) highlighted the limitations involved in analytical approaches for developing a DTA model for general networks and the unrealistic representation of traffic dynamics that they entail. In contrast, the simulation-based approach has substantial advantages in the development of DTA models that are practical for realistic networks. Moreover, simulation models help to incorporate multi-user classes and their respective interactions in the transport network, information provision and decision processes. They concluded that simulation models are more suitable for studying system robustness and for incorporating sources of randomness that yield the stochastic DTA problem. The main drawback of simulation models is their inability to form mathematical functions that describe the system properties in order to get some insights. De Palma and Marchal (2002) discussed modelling issues related to DTA simulation models. They concluded that the combination of event-based mesoscopic modelling of the supply side along with a disaggregate demand modelling of individual decision makers yields the best conditions for analyzing large-scale systems, in particular with advanced traffic management systems and advanced traveller information systems applications. An important advantage of this approach is that it enables the behavioral modelling of decision makers based on time-dependent origin-destination matrices.

Developments in the field of traffic assignment models point to the potential role that simulation models can play in the context of public transport assignment models. Simulation models provide an appropriate platform to enhance the realization of public transport system modelling. These include the capabilities to reproduce time-dependent trip generation; the dynamic evolution of network conditions; the interaction between supply and demand; the variation among travellers and their adaptive behavior; representing operational management strategies, and; emulating the generation and dissemination of passenger information services. Liu et al. (2010) concluded in their review that the developments in public transport route choice and assignment lag behind the counterpart developments in car traffic networks. While the two problems have important similarities and modelling issues, there are also important differences which limit the transferability of developments in the car traffic network sphere to the public transport network domain. Most important among these differences is the additional service layer which implies limited temporal and spatial availability of the public transport network and results in a more complex definition of path alternatives and consequently, the sequence of travel decisions.

Public transport assignment models are conventionally classified into frequency-based and schedule-based models—differing in their network supply representation and its implications on passenger loading procedure. The frequency- and schedule-based classes are denominated in Chapter 6 of this book as line-oriented and trip- or vehicle-oriented public transport assignment models, respectively. Passengers are assigned to common line corridors in frequency-based models while schedule-based models assign passengers to specific vehicle trips. Since service reliability issues and the respective operations management strategies often pertain to individual vehicles, the dynamic public transport model should represent the movement and assign passengers to individual vehicle runs. While schedule-based models represent individual vehicle runs, the evolution of service reliability cannot be fully captured and en-route choices of individual passengers are prohibitive due to demand representation. In their review, Liu et al. (2010) asserted, based on the developments in traffic assignment models, that they expect multi-agent non-equilibrium models to emerge in the public transport domain as well. The main modelling issues are supply uncertainties and adaptive user decisions. They identified dynamic loading process and multi-agent-based simulation as two potential approaches for modelling complex public transport systems.

Following developments in the sphere of traffic assignment models, there are indeed few recent corresponding efforts in the public transport domain. Although still in its early stages, agent-based simulation models emerged recently as an alternative approach to public transport assignment models. A review of the simulation-based approach to public transport assignment models and description of the features of the main models developed in this domain in recent years is available in Gentile et al. (2016). The so-called agent-based approach used in a range of sciences is aimed at modelling complex systems by representing the strategies of individual agents and the dynamics between agents and the environment as well as interactions between agents. Agent-based models represent complex systems using a bottom-up modelling approach where each individual entity is represented as an agent. Simulation models can facilitate the dynamic loading of passengers over a dynamic representation of the public transport system. Wahba and Shalaby (2006) discuss the potential advantages of a multi-agent simulation framework for modelling the public transport assignment problem, in particular in the context of ITS.

Frequency-based assignment models offer a robust modelling framework for long-term strategic planning. However, they are not capable of modelling system dynamics and support real-time applications. The schedule-based approach facilitates the modelling of congestion effects at the individual vehicle trip rather than on a common corridor. However, similarly to frequency-based models, it has a limited capability to capture the dynamics of service reliability and its evolution along the line. Furthermore, the static and aggregate representation of passenger demand prevents the consideration of en-route travel decisions at the individual level. In contrast, agent-based simulation models facilitate the dynamic representation of individual passengers and the emergence of dynamic network effects based on numerous inter-dependent local decisions, as explained in Section 4.2.

Conventional offline public transport assignment models obtain passenger flows by solving equilibrium conditions for a public transport network graph. However, in the context of real-time applications, there is no reason to assume that equilibrium conditions will prevail due to the short time-frame. In reality, service perturbations are expected to result in non-equilibrium assignment conditions. Hence, the passenger flow module is solely concerned with within-day dynamics rather than day-to-day network evolutions. Notwithstanding, in the absence of complete real-time information concerning passengers' locations and travel plans, a day-to-day dynamic network loading is an essential component of network initialization in order to estimate passenger departure times as well as waiting and on-board flows (e.g., impact of expected reliability and congestion levels on route choice for which their assessment requires an iterative day-to-day assignment).

In some cases where the RT-OMDDS is concerned with the prediction and evaluation of the performance of real-time strategies which are applied to an isolated line or sub-network, the modelling environment does not necessarily have to perform an assignment if there are no viable route-choice alternatives. If applicable, this could also be effective in reducing model running time. The short-term prediction model of on-board passenger loads based on machine-learning techniques proposed by Ma et al. (2014) and Morriea-Matias and Cats (2016) might be used in such cases. The advances in assignment methods presented in Chapters 7 and 8 of this book can also facilitate the real-time applicability of short-term predictive public transport assignment models.

With existing computational capabilities, model running times still pose a significant challenge to the development of RT-OMDDS. In order to become operational in real-time applications, dynamic public transport assignment models need to run within a matter of seconds. This is especially prohibitive in case the real-time strategy may alter passengers' choice-set, as choice-set generation algorithms are exceptionally computationally expensive. This issue is further amplified by the need to perform a number of simulation replications in order to obtain statistically robust results (i.e., provide predictions with confidence intervals) given the stochastic properties of traffic and assignment simulation models. A possible mitigation measure is assigning a representative sample of the entire passengers' population and then adjusting the results proportionally as was suggested by Nagel and Marchal (2003). In addition, the number of choice model instances which can become excessive when modelling journeys as a sequence of travel decisions for a large number of passengers, could be reduced by splitting flows based on probabilities or limiting the number of travel dimensions (e.g., choosing an alighting stop upon boarding).

4.3.5 Real-time Strategies

The combination of the *network initializer*, *traffic flow* and *passenger flow* modules enables generation of predictions concerning future system states. The role of the *real-time strategies* module is to execute and test alternative futures by simulating potential interventions in public transport operations and management strategies and thus allow a 'what if' analysis. The integration of optimization techniques into this module will enable service providers to optimize in real-time the design of control and management strategies. The service provider may influence operations by disseminating travel information, applying fleet management strategies and disrupting mitigation measures. This module corresponds to the control center units and how it interacts with infrastructure, vehicles/drivers and passengers using information and communication technologies.

The real-time strategies module determines how information is provisioned to service users. Real-time information could be disseminated through sign displays, on-board screens or smartphone apps. The RT-OMDDS allows testing the implication of disseminating various information, provided that the passenger flow model captures the behavioral effects of such strategies. For example, the impact of disseminating information concerning delays or crowding conditions could be incorporated into the route choice utility model to predict its impact on passengers' rerouting decisions. Furthermore, some control strategies require informing passengers due to their implications on passengers' travel plans. For example, when implementing real-time stop skipping, expressing or short-turning strategies, passengers on-board the vehicle as well as passengers waiting at relevant downstream stops must be informed about it, so that they may choose to reroute as a result.

Control strategies that require communication with vehicles or infrastructure might range from local heuristics (e.g., schedule-based or headway-based holding control) to system-wide optimization (e.g., transfer coordination, signal priority). The real-time strategies module computes the control decision and monitors its progress. In very involving control contexts, it might be relevant to simulate control center dynamics to endogenously determine decision rules concerning the application of real-time strategies.

4.4 Embedding the Dynamic Public Transport Model in a Decision Support System

The aforementioned modelling environment needs to be embedded in the RT-OMDDS by complementing it with *scenario design* and *scenario evaluation* modules (see Fig. 4.1). These two modules are responsible for the human-machine interaction and allow system operators to interact with the RT-OMDDS and make it useful for decision makers. These two modules are described in the following sub-sections:

4.4.1 Scenario Design

This module allows the RT-OMDDS operator to specify and control the input and parameters related to the modelling environment. The *scenario design* module could be realized in the form of a user-interactive toolbox. The real-time strategy menu can include the default values and allow the user to specify alternative values or tick/untick options that differ from default ones.

The implementation of each of the modelling environment modules involves the specification of relevant parameters. Network initializer involves the initial positioning of all public transport vehicles and the properties of vehicular and passenger flows (e.g., speed per link, number of passengers waiting at each stop). System operator may specify changes that are endogenous (e.g., unplanned station blockage) or exogenous (e.g., weather conditions, large-scale event) to the system prior to running the RT-OMDDS in order to adjust the expected car traffic or passenger demand levels, travel conditions and the functionality of network elements.

Traffic and passenger flow models involve the specification of a large number of parameters. However, these are determined by model estimation and calibration rather than subject to scenario design experiments. In exceptional cases, the system operator may want to specify a value other than the default values. For example, in case a replacement service is needed, system operator may remove the capacity constraint element from this particular service to assess the overall demand for it and deduce from the unconstrained result what is the capacity that needs to be allocated to accommodate passenger demand for this replacement service.

The most important feature of scenario design is the specification of the real-time strategies module. The set of real-time strategies that is of interest is likely to change for different application contexts. For example, bus operators in many cities do not consider limited boarding— determining the desired dwell-time at a certain stop and enforcing it by not allowing the remaining passengers to board the bus—due to its low acceptability and the possible consequences on service image and travel satisfaction. Similarly, applying expressing, stop-skipping or short-turning as real-time strategies (as opposed to their inclusion in the planning and scheduling phase) might not be acceptable in certain contexts since it implies that some passengers may not be able to alight at their intended destination stop because it will not be served. Moreover, even for a given application context, the range of real-time strategies that the system operator is willing to consider may be subject to contextual factors (e.g., disable transfer coordination at a certain station due to information on station capacity limitations). System operator should be allowed to select strategies that are of interest from the strategies toolkit.

4.4.2 Scenario Evaluation

The modelling environment will result in a set of outputs for each scenario. In order to assist decision makers, in this case primarily the control center staff, in their daily work and enable them to make more informed decisions, these outputs need to be conveyed in a clear, systematic and straightforward manner that allows them to assess the impacts of a number of alternatives within a very short time. To this end, model results need to be processed, visualized and aggregated into a manageable set of measures of performance.

The most important measures of performance need to be decided on the basis of the overall policy objectives. In general, both measures of service effectiveness and service efficiency should be considered to reflect both service users' and service providers' perspectives. Among the most important scenario evaluation indicators is total passenger travel time (both in nominal and generalized terms). This indicator encompasses the expected delays, waiting times, on-board crowding, reliability and transfers. However, a non-compensatory multi-criteria evaluation approach might be adopted to avoid selecting a scenario which leads to very poor results with respect to one of these travel experience dimensions. From service provider point of view, impacts on operational costs due to vehicle and crew scheduling consequences because of the allocation of additional resources or contractual constraints (e.g., penalties for deviating from the planned timetable or production-based incentives) constitute important performance measures.

A multi-criteria evaluation approach is also instrumental in providing the decision maker with room for judgment based on experience, expertise and contextual factors which may give greater weight to certain performance aspects than others, under certain circumstances. Furthermore, the robustness of system performance to the real-time strategy considered should be assessed in terms of sensitivity to changes (e.g., by using a confidence interval) and the reversibility of the control action.

An interactive graphical interface of the RT-OMDDS will allow control-center staff to quickly assess the expected spatial and temporal consequences of alternative strategies. This could be done, for example, by displaying the predicted volume over capacity ratios across the network and allowing dispatchers to simulate its evolution over the analysis period. In order to reduce the information load for control-center staff, the indicators could be communicated by using a color spectrum illustrating their relative performance when compared with other alternatives (including the do-nothing option).

In order to introduce the RT-OMDDS into control center daily routine, it is necessary to establish new working procedures. For example, a general pool of resources, such as reserve buses and drivers, need to be managed collectively in order to avoid myopic local decisions. This is also true under current working routines, but needs to be reconsidered when introducing the new system. In addition, control-center staff should be trained to interpret and use the predictions generated by the RT-OMDDS. Deficiencies need to be address by implementing self-learning mechanisms, for example, with respect to the travel and demand prediction schemes whenever the predictions could be validated against actual measurements. Such a closed-loop self-learning system will in itself evolve by generating predictions, evaluating scenarios, implementing actions and monitoring their effects and updating the modelling and prediction schemes.

4.5 The Road Ahead: Future Prospects

This chapter presented a framework for a decision support system of public transport operations management that will facilitate the implementation of real-time strategies. The development of models that are able to represent the dynamics of public transport systems, the evolution of both supply and demand, and respond to changes in system conditions is essential for embedding the modelling environment described above in a decision support system. Embedding dynamic public transport models in a real-time decision support system will potentially lead to a shift towards more adaptive operations by considering the implications of alternative measures on the overall system performance and passengers' experience. The analysis and evaluation of passengers' experiences when evaluating system performance will assist service providers in moving beyond single-operator fleet management considerations to passenger door-to-door travel perspective.

As the discussion in previous sections made evident, there are substantial modelling and technical challenges involved in the development, design, estimation and implementation of the envisioned system. In particular, the behavioral modelling components and traffic prediction schemes need to be carefully estimated based on local conditions. The rapid increase in the availability of traffic and passenger data facilitates the estimation and validation of multimodal traffic and public transport passenger route choice models. The RT-OMDDS should therefore be developed by a diverse multi-disciplinary group of experts, including transport modellers, software engineers, information and communication technology developers, machine learning and data scientists, logistics and human-machine interaction professionals. The development of such a system should be made jointly with potential users in order to ensure that it is relevant and useful in their daily operational needs. Model validity and fidelity are important concerns that still need to be addressed in the context of real-time applications.

For a RT-OMDDS to become operational and useful, it is not sufficient to develop the necessary modelling capabilities. It needs to be introduced into the working routine of control-center staff as well as drivers, dispatchers, conductors and other support staff. The system operator should be able to determine the activation mode of the decision support system. Three modes could be defined: time-based—the system runs on a regular basis based on the selected time interval; event-based—the system will be triggered every time that a certain event from a pre-defined set of monitoring events occurs (e.g., observed delay or missed connection) or; self-actuated—the system is activated upon direct request by system operator. The latter two options risk triggering the system when it is already too late to respond or completely miss an event that carries importance to system performance whereas the objective of the RT-OMDDS is to facilitate proactive rather than responsive operations management. Notwithstanding, a too frequent activation of the RT-OMDDS risks becoming an annoyance to control-center staff, especially if it results in conflicting successive advice. A good compromise might be to run the network initializer, traffic flow and passenger flow regularly in a time-based fashion but define a set of 'expected disturbance alarms' that will result in an event-based interaction with the system operator, who will only then need to interact without compromising the proactive approach to operation management. In addition to these alarms, the system operator may activate the system on his/her own initiative whenever the needs arise.

The development of decision support systems for real-time operations management of public transport systems is still in its early stages. The incorporation of public transport assignment models into real-time applications has great potential to transform public transport management and calls for further research to elevate the knowledge and gain experience on how to develop and operationalize such tools.

Keywords: decision support system; control center; real-time operations; agent-based models; reliability; control; fleet management; adaptive service planning; traffic simulation models; traffic management.

References

Adamski, A. and A. Tumau. 1998. Simulation support tool for real-time dispatching control in public transport. Transportation Research Part A 32(2): 73–87.

Ben-Akiva, M., M. Bierlaire, H. K. Koutsopoulos and R. Mishalani. 2002. Real-time simulation of traffic demand-supply interactions with DynaMIT. pp. 19–36. *In*: M. Gendreau and P. Marcotte (eds.). Transportation and Network Analysis: Current Trends.

Burgholzer, W., G. Bauer, M. Posset and W. Jammernegg. 2013. Analyzing the impact of disruptions in intermodal transport networks: A micro simulation-based model. Decision Support Systems 54(4): 1580–1586.

Carrel, A., R. G. Mishalani, N. H. M. Wilson, J. P. Attanucci and A. B. Rahbee. 2010. Decision factors in service control on a high-frequency metro line and their importance in service delivery. Transportation Research Record 2146: 52–59.

Cats, O. 2014a. Real-time predictions for light rail systems. IEEE conference on Intelligent Transportation Systems (ITSC), China, 1535–1540.

Cats, O. 2014b. Regularity-driven bus operations: Principles, implementation and business models. Transport Policy 36: 223–230.

Cats, O. and Z. Gkioulou. 2015. Modelling the impacts of public transport reliability and travel information on passengers' waiting time uncertainty. EURO Journal of Transportation and Logistics. In press, DOI 10.1007/s13676-014-0070-4.

Cats, O., H. N. Koutsopoulos, W. Burghout and T. Toledo. 2011. Effect of real-time transit information on dynamic path choice of passengers. Transportation Research Record 2217(2): 46–54.

Cats, O. and E. Jenelius. 2014. Dynamic vulnerability analysis of public transport networks: Mitigation effects of real-time information. Networks and Spatial Economics 14: 435–463.

Cats, O. and E. Jenelius. 2015. Planning for the unexpected: The value of reserve capacity for public transport network robustness. Transportation Research Part A 81: 47–61.

Cats, O., J. West and J. Eliasson. 2016. A Dynamic stochastic model for evaluating congestion and crowding effects in transit systems. Transportation Research Part B 89: 43–57.

Chandakas, E., F. Leurent and O. Cats. 2016. Applications and future developments: modeling software and advanced applications. pp. 521–560. *In*: G. Gentile and K. Nökel (eds.). Modeling Public Transport Passenger Flows in the Era of Intelligent Transport Systems. Springer International Publishing. ISBN 978-3-319-25082-3.

De Palma, A. and F. Marchal. 2002. Real cases applications of the fully dynamic METROPOLIS tool-box: An advocacy for large-scale mesoscopic transportation systems. Networks and Spatial Economics 2: 347–369.

Dessouky, M., R. Hall, L. Zhang and A. Singh. 2003. Real-time control of buses for schedule coordination at a terminal. Transportation Research Part A 37: 145–164.

Fadaei Oshyani, M. and O. Cats. 2014. Real-time bus departure time predictions: Vehicle trajectory and countdown display analysis. IEEE conference on Intelligent Transportation Systems (ITSC), China, 2556–2561.

Fonzone, A., J-D. Schmöcker, M. Bell, G. Gentile, F. Kurauchi, K. Nökel and N. H. M. Eilson. 2010. Do 'hyper-travelers' exist? – Initial results of an international survey on public transport user behaviour. Proceeding of 12th World Congress on Transport Research, July 2010, Lisbon, Portugal.

Gentile, G., M. Florian, Y. Hamdouch, O. Cats and A. Nuzzolo. 2016. The theory of transit assignment: Basic modelling frameworks. pp. 287–386. *In*: G. Gentile and K. Nökel (eds.). Modeling Public Transport Passenger Flows in the Era of Intelligent Transport Systems. Springer International Publishing. ISBN 978-3-319-25082-3.

Gentile, G. and L. Meschini. 2011. Using dynamic assignment models for real-time traffic forecast on large urban networks. MT-ITS Leuven, June 22–24.

Guevara, C. and G. Donoso. 2014. Tactical design of high-demand bus transfers. Transport Policy 32: 16–24.

Liu, Y., J. Bunker and L. Ferreira. 2010. Transit users' route choice modeling in transit assignment: A review. Transport Reviews 30(6): 753–769.

Liu, R., D. Van Vliet and D. Watling. 2006. Microsimulation models incorporating both demand and supply dynamics. Transportation Research Part A 40: 125–150.

Morriea-Matias, L. and O. Cats. 2016. Towards an AVL-based Demand Estimation Model. Transportation Research Record, 2544, in press.

Ma, Z., J. Xing, M. Mesbah and L. Ferreira. 2014. Predicting short-term bus passenger demand using a pattern hybrid approach. Transportation Research Part C 39: 148–163.

Muñoz, J. C., C. Cortés, R. Giesen, D. Sáez, F. Delgado, F. Valencia and A. Cipriano. 2013. Comparison of dynamic control strategies for transit operations. Transportation Research Part C 28: 101–113.

Nagel, K. and F. Marchal. 2003. Computational methods for multi-agent simulations of travel behavior. Presented at the 10th International Conference on Travel Behaviour Research (IATBR), Luzern.

Nuzzolo, A., U. Crisalli and L. Rosati. 2011. A schedule-based assignment model with explicit capacity constraints for congested transit networks. Transportation Research Part C 20(1): 16–33.

Nuzzolo, A., F. Russo and U. Crisalli. 2001. A doubly dynamic schedule-based assignment model for transit networks. Transportation Science 35(3): 268–285.

Peeta, S. and A. K. Ziliaskopoulos. 2001. Foundations of dynamic traffic assignment: The past, the present and the future. Networks and Spatial Economics 1: 233–265.

Schmidt, M. and A. Schöbel. 2015. The complexity of integrating passenger routing decisions in public transportation models. Networks, DOI 10.1002/net.21600.

Schmöcker, J. D., M. G. H. Bell and F. Kurauchi. 2008. A quasi-dynamic capacity constrained frequency-based transit assignment. Transportation Research Part B 42(10): 925–945.

Shalaby, A. and A. Farhan. 2004. Prediction model of bus arrival and departure times using AVL and APC data. Journal of Public Transportation 7(1): 41–61.

Toledo, T., O. Cats, W. Burghout and H. N. Koutsopoulos. 2010. Mesoscopic simulation for transit operations. Transportation Research Part C 18(6): 896–908.

Törnquist, J. 2005. Computer-based decision support for railway traffic scheduling and dispatching: A review of models and algorithms. 5th Workshop on Algorithmic Methods and Models for Optimization of Railways.

Wahba, M. and A. Shalaby. 2006. MILATRAS: A microsimulation platform for testing transit-ITS policies and technologies. IEEE Conference on Intelligent Transportation Systems (ITSC), Canada, 1495–1500.

Wahba, M. and A. Shalaby. 2014. Learning-based framework for transit assignment modeling under information provision. Transportation 41(2): 397–417.

5

Real-time Modelling of Normative Travel Strategies on Unreliable Dynamic Transit Networks

A Framework Analysis

A. Nuzzolo[1,a,]* and *A. Comi*[1,b]

ABSTRACT

After an introduction on travel strategies and a relatively brief state of the art, the chapter starts by recalling the main factors influencing path choice decision making and focuses on unreliable dynamic service networks, on which a strategy-based path choice should be used. Travel strategies, with their related hyperpaths and diversion rules, together with the different types of optimal strategies, are then defined and analyzed. The search methods of the normative optimal strategies are hence presented, taking due consideration of their applications in a real-time predictive info context. Finally, some conclusions are drawn and further necessary research developments are indicated.

[1] Department of Enterprise Engineering, Tor Vergata University of Rome, via del Politecnico 1, 00133, Rome, Italy.
[a] E-mail: nuzzolo@ing.uniroma2.it
[b] E-mail: comi@ing.uniroma2.it
* Corresponding author

5.1 Introduction

Transit path choice models have been quite extensively covered in the literature, mainly in the context of transit assignment modelling (Gentile and Nokel, 2016). A new impulse was recently given by the need to take into account the presence of traveller information systems and the availability of a large quantity of data derived from automated data collecting and bidirectional communication between travellers and information centers.

Path choice decision-making is influenced by several factors, as explained in Section 2 below. In this context, one of the more complex situations is when travellers move on an unreliable network where, at certain nodes (*diversion nodes*), some travel decisions have to be made according to the occurrence of random events (e.g., bus arrivals at stops, on-board crowding). Even if there were a system available to predict information on network states and travellers could be provided with such info, due to the uncertainty of such forecasts, the choice at departure time of a specific path up to destination might not be the best decision. Rather, a travel strategy should be used like a sequence of choices to make in a set of diversion nodes, according to random occurrences and in order to minimize expected travel costs.

In general, two types of strategy can be used in path choice modelling: *normative strategy* and *descriptive strategy.* Their definitions derive from philosophy: *normative* statements make claims about how things should or *ought* to be, while *descriptive* statements attempt to describe how they are in *reality.*

A *normative strategy or objective optimal strategy* is one that travellers should apply to objectively maximize their travel utility objectively. Due to cognitive limitations and due to psychological constraints, travellers use an optimal strategy, hereafter called *descriptive or subjective optimal strategy,* which can differ from the *normative* one and which should be modelled to reproduce traveller behavior. Recent research using data collected through new ticketing technologies indicates that on unreliable transit networks with diversion nodes, travellers tend to apply strategies, even if the number and kind of strategies depend on the information that is available and on user socio-economic characteristics (Kurauchi et al., 2012; Schmocker et al., 2013; Fonzone et al., 2013).

Descriptive optimal strategy-based path choice models have been quite extensively developed in the literature (Bovy, 2009; Prato, 2009; Liu et al., 2009). Descriptive strategy-based models for transit networks with individual predictive information were recently presented by Cats et al. (2011) and Nuzzolo et al. (2016), and in an extensive interpretation, also by Wahba and Shalaby (2009). Further, a descriptive strategy-based path choice model is reported in Chapter 6 (Comi and Nuzzolo, 2016).

A normative strategy-based approach is used when paths are suggested in some innovative transit trip planners, as reported in Chapter 3 (Comi et al., 2016). For example, the transit route planner Hyperpath (SISTeMA, 2015) applies a strategy-based approach in path suggestion. Further, the normative strategy-based approach can be used to simplify descriptive path choice modelling in transit assignment, such as in some frequency-based assignment models (see, for example, Gentile et al., 2016) because faster search algorithms can be applied, especially for large networks, as reported in Chapter 7 (Gentile, 2016a).

To the authors' knowledge, a framework paper on normative travel strategy on transit networks is currently lacking. Therefore, this chapter presents an analysis of the strategy approach from a theoretical point of view, defining the different types of optimal strategies and classifying the related search methods, also in relation to availability or not of real-time forecasts of path attributes. In particular, the normative strategy is examined in-depth, so that the travel strategy should guarantee for travellers absolute maximisation of their expected travel utility.

In synthesis, this chapter is organized as follows. Section 5.2 recalls the main factors influencing transit path choice decision making and focuses on unreliable networks. In Section 5.3, the definitions of travel strategy on dynamic unreliable networks are recalled and analyzed. Since different optimal strategies can be defined according to the diversion rules used, in Section 5.4 the search methods of the optimal strategies conditional on given diversion rules are presented, while Section 5.5 is devoted to a normative strategy search, given the availability or not of real-time predictive info on path attributes. Finally in Section 5.6, some conclusions are drawn and some major research issues connected to optimal strategy are pointed out.

5.2 Factors Influencing Travel Decision Making

In order to improve the effectiveness of path choice models and information provided to travellers, several factors influencing path choice decision making are pointed out:

- trip frequency on the origin-destination relation;
- traveller's perception of the type of service (schedule-based or frequency-based);
- service regularity;
- presence of diversion nodes with random occurrences (vehicle arrival and on-board crowding);
- time-dependency of network performance (dynamic networks);
- type of available information system.

Given an origin-destination relation, a trip can be classified as *frequent or recurrent* if a traveller often undertakes this trip and therefore is quite familiar with the characteristics of the services supplied. Otherwise, the traveller is *non-usual* (occasional, non-habitual) and may have no experience of the transit system in question.

A dichotomy for investigating factors that can influence travel decisions arises if the traveller perceives the services as *schedule* or *headway/frequency*-based. If the operator publishes the full timetable and the scheduled arrival and departure times of all runs at all stops are regular, a complete trip plan can be defined before departure, taking into account the available timetables (schedule-based services). On the other hand, if services are quite irregular and/or frequent (e.g., a metro vehicle passes on an average every 2 minutes), the traveller may be induced not to consider timetables (frequency-based services).

If service departure times are irregular and attributes characterising the routes (e.g., travel time, on-board crowding) are random variables, the network is classified as *stochastic* or *unreliable*. Given an unreliable service network and an O-D pair *od*, during the trip there could be *diversion nodes* where travellers may have to make choices (*diversion choices*) according to the occurrences of random events (e.g., different order of arrival of lines at stops; different transit vehicle crowding). To travel through such an unreliable network, if travellers wish to maximize their expected travel utility, they may no longer rely on a single path, selected at travel departure, but could use an *optimal strategy*, with a sequence of decisions at some nodes of the network.

Another factor influencing path choice decision making is the *dynamicity* of the network, with within-day time-dependency of its performances, such as variations in service frequency, on-board crowding, route travel time irregularity and so on.

The choice results also depend on the type of available forecasts of travel attributes which, in relation to the contents of this chapter, can be classified into:

- *collective/shared* (given to a set of travellers, such as bus arrival times at stops) or *individual* (for specific origin-destination and trip departure times);
- *timetable information*, e.g., 'arrival time is scheduled at Y.YY a.m.' or *predictive information*, giving real-time travel attribute forecasts, e.g., 'the bus will arrive at the stop in X minutes';
- *alternative path advice*, e.g., some alternative paths are given according to some rules, or *normative path recommendations or suggestions*, e.g., a path or hyperpath is recommended according to a normative approach.

In the following, if not otherwise reported, travellers are assumed to be *frequent* and transit systems are assumed *frequency-based, dynamic* and *unreliable*, with *diversion nodes* and *individual predictive information*.

5.3 Travel Strategies

5.3.1 Uncertainty and Optimal Choice in Decision Theory

In this section, some basic concepts of decision theory are recalled, which may be useful to introduce and review the definition of *optimal strategy.* In decision theory, decision makers choose among a finite number of alternatives (for example, for transit networks, travellers choose a path among n available paths from origin o to destination d), and are assumed to have a rational behavior, i.e., they compare the possible consequences of the choice of each alternative and choose the best in terms of such consequences. Based on the axioms of rational behavior, von Neumann and Morgenster (1947) propose the existence of a utility function U that describes the preference scheme of a decision maker. This function can be identified and is unique (less than linear transformations).

When for each alternative only one possible state can occur, and therefore, only one consequence, if a_i is the i-th element of the set of all possible actions (alternatives) and c_i is the consequence (result) of each action a_i, the alternative with the maximum utility U_c has to be chosen:

$$a_i \geq a_k \Leftrightarrow U_{c_i} \geq U_{c_k} \tag{5.1}$$

On the other hand, if the consequence of each action a_i is not unique but m states are possible, let

- a_i be the i-th element of the set of all possible actions (alternatives);
- Θ_{ij}, the j-th possible status for action i;
- $P(\Theta_{ij})$, the (objective or subjective) probability that the status Θ_{ij} happens;
- c_{ij}, the consequence (result) of each status j for action i.

If the *objective* or the *subjective* probabilities of the random events are known, $P(\Theta_{ij})$, the decision is called under *risk* or *uncertainty* and the alternative with the *maximum expected utility* has to be chosen:

$$a_i \geq a_k \Leftrightarrow \sum_{j=1}^{m} P(\theta_{ij}) \cdot U_{c_{ij}} \geq \sum_{j=1}^{m} P(\theta_{kj}) \cdot U_{c_{kj}} \tag{5.2}$$

In other words, in risk or uncertainty context, it is not possible to maximize the result of each decision, but it is possible to maximize the expected utility over a long sequence of decision making.

Non-expected utility approach

There is much empirical evidence against EU (Kahneman and Tversky, 1979): e.g., Allais and Ellsberg paradoxes. Therefore, based on these experimental results, *non-expected* oriented utility theories have been developed (Starmer, 2000; de Palma et al., 2008; Ramos et al., 2014), of which the most widespread are prospect theory (PT) (Kahneman and Tversky, 1979) and regret theory (Bell, 1982). However, in the authors' opinion, when objective optimal strategies are considered, as in the following part of this chapter, the expected utility approach should be applied, while in the cases of descriptive strategies, the non-expected utility approach can be applied. Therefore, the non-expected utility approach will be dealt with in Chapter 6 (Comi and Nuzzolo, 2016), relative to the descriptive travel strategies.

5.3.2 Path Choice and Travel Strategies on Unreliable Networks

When services are regular and hence their costs are constant and an individual info system with predictive path attributes is available, the path allowing users to reach their destination with maximum utility can be determined exactly. In the case of a stochastic transit network with diversion nodes, it should be considered that the values of path attributes, forecasted (e.g., through statistical methods, Moreira et al., 2015) by an info system, are random variables. Thus, even with an information system, the uncertainty is not completely overcome. Therefore, on stochastic unreliable networks in order to minimize travel cost or to maximise utility, even if individual predictive information is available, a complete path cannot be chosen pre-trip, but a strategy has to be followed (see, for example Spiess and Florian, 1989). A strategy is a plan that allows travellers to reach their destination. With non-deterministic events outside the control of the traveller, the possible executions of the plan form a tree. At every node (*diversion node)* the traveller has to determine the appropriate action (*link diversion choice*). Therefore, a strategy includes:

- a set of diversion nodes;
- for each diversion node, a set of useful diversion links;
- a diversion rule, to choice amongst these diversion links, according to random outcomes.

For example, two different diversion rules could be used at diversion nodes, namely:

- *indifferent adaptive behavior*: the traveller boards the first run of the 'attractive' line set arriving at the stop;

- *comparative (or intelligent) adaptive behavior*: when a run of an "attractive" line arrives at the stop, the traveller compares the travel utility of boarding that run versus waiting for the next runs, and chooses what to do.

In the network represented in Fig. 5.1, a strategy could be 'From the Origin go to node A and use the first line arriving of either lines 2 or 4; if line 2 is chosen, go to node G; then use line 3 up to node D; if line 4 is chosen, go to node H, then use line 10 up to node D; from node D go to the destination'.

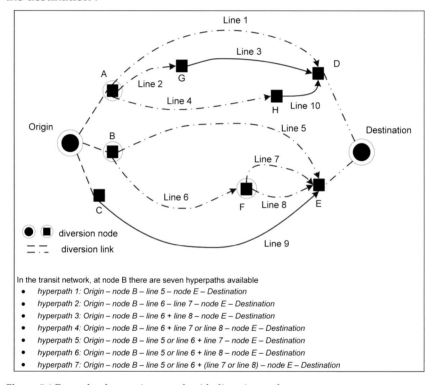

In the transit network, at node B there are seven hyperpaths available
- hyperpath 1: Origin – node B – line 5 – node E – Destination
- hyperpath 2: Origin – node B – line 6 – line 7 – node E – Destination
- hyperpath 3: Origin – node B – line 6 + line 8 – node E – Destination
- hyperpath 4: Origin – node B – line 6 + line 7 or line 8 – node E – Destination
- hyperpath 5: Origin – node B – line 5 or line 6 + line 7 – node E – Destination
- hyperpath 6: Origin – node B – line 5 or line 6 + line 8 – node E – Destination
- hyperpath 7: Origin – node B – line 5 or line 6 + (line 7 or line 8) – node E – Destination

Figure 5.1 Example of a transit network with diversion nodes.

The topology of a travel strategy can be graphically represented through a hyperpath (Gallo et al., 1993; Gentile, 2016b), as explored in Chapter 7 (Gentile, 2016a), to which readers can refer for more details.

It should be noted that different strategies can be used on the same network for the same O-D pair *od*, with different diversion nodes and/or diversion links (hence different hyperpaths), and/or diversion rules. For example, for the network represented in Fig. 5.1, where the origin and nodes A, B and F are diversion nodes, at the origin, 127 different hyperpaths are available, while at node B, seven different sub-hyperpaths are available.

Different strategies could have different travel utilities and, therefore, the strategy which maximizes the expected travel utility (*optimal strategy*) should be used.

Now, we have to define what means 'strategy which maximizes the expected travel utility' and introduce the concept of *expected experienced travel utility* of a strategy.

5.3.3 Expected Experienced Utility of a Strategy

In this section, the definition of *expected experienced utility of a strategy* for dynamic unreliable networks is introduced. Let $S_{ml}^{\tau_m}$ be the generic travel strategy, with a defined diversion rule to reach destination d from diversion node m at time τ_m, including the diversion link l, topologically represented by hyperpath $K_{ml}^{\tau_m}$.

We define *experienced utility* $U_{S_{ml}}^{z,\tau_m}$ of strategy S_{ml}^{z,τ_m} and *experienced utility* $U_{K_{ml}}^{z,\tau_m}$ of its hyperpath K_{ml}^{z,τ_m} at time τ_m of day z as the utility $U_{H_q}^{z,\tau_m}$ of the path H_q that the traveller would experience if strategy S_{ml} is followed. For example, referring to the network in Fig. 5.1, let us suppose that to reach the destination from node A, the following decision rule is used—'board the first arriving vehicle'. At time τ_A of day z, the traveller boards line 4, if it is the first line arriving at node (stop) A. Subsequently, he/she reaches node H and boards line 10; then the destination is reached. In this case, the experienced path H_q is relative to the path 'Origin – node A – line 4 + line 10 – node D – Destination'. According to the above definition, the experienced utility of the strategy ($U_{S_{Al}}^{z,\tau_A}$) for time τ_A of day z is given by the utility ($U_{H_{4+10}}^{z,\tau_A}$) that traveller would have experienced using this path H_{4+10}.

The *expected experienced utility* or *expected travel utility EU* of strategy $S_{ml}^{\tau_m}$ and of its hyperpath $K_{ml}^{\tau_m}$ is thus the expected value of utilities experienced in using the paths H_q of the hyperpath K_{ml}:

$$EU\left[U_{S_{ml}}^{\tau_m}\right] = EU\left[U_{K_{ml}}^{\tau_m}\right] = \sum_{q\in Q} p\left[H_q,\tau_m\right]\cdot EU_{H_q}^{\tau_m} \tag{5.3}$$

where:

- $EU_{H_q}^{\tau_m}$ is the expected value of the experienced utility of path H_q belonging to hyperpath K_{ml} at time τ_m;
- $P[H_q, \tau_m]$ is the frequency or the probability of experiencing path H_q following strategy S_{ml} at time τ_m;
- Q is the set of paths available following strategy S_{ml} at time τ_m.

Although the travel utility function can have different forms, for analytical and statistical convenience, it is usually assumed (Ettema and Timmermans, 2006), as in the following, that the utility U_{H_q} of path alternative H_q is a linear function of a set of characteristics (attributes) X_{gH_q} of alternative H_q:

$$U_{H_q}^{z,\tau_m} = \sum_g \beta_g \cdot X_{gH_q}^{z,\tau_m}$$
(5.4)

where parameter β_g can be considered the weight given by decision makers (travellers) to attribute $X_{gH_q}^{z,\tau_m}$ (e.g., travel time and/or cost, scheduled delay).

If the parameters of the utility function (5.4) for determining utility $U_{H_q}^{z,\tau_m}$ are estimated using an individual modelling approach, the parameters β_g become individual parameters (β_g^u) and reflect the weight given by the individual decision maker u to path attributes g (see Section 3.3 of Chapter 3; Comi et al., 2016).

5.3.4 Optimal Strategies

Applying the above results of the decision theory for the risk or uncertainty context on an unreliable network, the strategy to be chosen (*optimal strategy*) should be the one of maximum expected experienced (or travel) utility.

Objective, subjective and normative optimal strategies

Different types of optimal strategy can be considered, as anticipated in Section 5.1. One is the *objective optimal strategy*, i.e., a strategy that objectively maximizes expected travel utility. Due to limits of the cognitive process and to psychological aspects, travellers use a *subjective (or descriptive) optimal strategy*, which can differ from the objective one.

The descriptive optimal strategies (i.e., what people actually do or have done) should be applied when we try to simulate the actual behavior of travellers, as in the assignment models, while the objective approach for example should be used in trip planner path recommendations or to simplify optimal strategy search algorithms in assignment models. In the following part of this chapter, only the *objective optimal strategies* are considered, while Chapter 6 (Comi and Nuzzolo, 2016) is devoted to the *descriptive optimal strategy*.

Conditional and absolute optimal strategies

Given a diversion rule, different hyperpaths could be used and, given a hyperpath, different rules could be applied with different expected travel utilities *EU*, because for each hyperpath/rule pair, *EU* and probability *p* can change.

Let us define two types of optimal strategies, in relation to the relative hyperpaths and diversion rules:

- *optimal strategy conditional on a given rule*, which is the strategy of maximum expected travel utility using that rule; subsequently, the

optimal hyperpath is the hyperpath which maximizes the expected travel utility using that rule;
- *absolute optimal strategy*, with the hyperpath and the rule which jointly give the absolute maximum expected travel utility.

It has to be noted that, while the number of alternative hyperpaths of a network is finite, and therefore, given a rule it is possible to apply a search method of the *optimal strategy conditional on that rule* (see next Section 5.4), in the case of *absolute optimal strategy*, in theory all the possible rules should be considered, but in practice the search refers to a set of a few considered alternative rules, as reported in the following.

5.4 Search Methods of an Objective Optimal Strategy Conditional on a Given Rule

5.4.1 Search Method Classification

Several methods can be used to obtain the objective optimal strategy (and related hyperpath) conditional on a given rule. They can be classified into two main classes:

- *with hyperpath explicit enumeration*: for each feasible hyperpath, the expected utility is computed or estimated and the strategy with maximum expected utility hyperpath is considered the conditional optimal strategy;
- *without hyperpath explicit enumeration*, where a suitable search algorithm is used to obtain directly the hyperpath with the maximum expected utility and hence, the conditional optimal strategy.

5.4.2 Methods with Hyperpath Explicit Enumeration

At diversion node m and at time τ_m of day z and for each diversion link l, each feasible hyperpath K_{ml}^{z,τ_m} and the relative $EU[K_{ml}^{z,\tau_m}]$ are considered. The strategy $S_{ml}^{z,\tau_m}*$ with the maximum expected utility hyperpath is then considered as the optimal strategy conditional on the adopted rule.

This method requires knowledge of the expected utility of each hyperpath, which can be obtained, for example, by analytical methods or data processing and simulation-based methods reported below.

a) Analytical methods for computation of hyperpath expected utility

The $EU[U_K]$ has to be estimated through (5.3) with respect to a given day z, determining

- the expected value $EU_{H_q}^{z,\tau_m}$ of the experienced utility of path H_q belonging to hyperpath K_{ml} at time τ_m of day z;

- the probability $P[H_q, \tau_m]$ of experiencing path H_q following strategy S_{ml} at time τ_m of day z.

In general, the analytical formulations of EU and p could be extremely difficult, as the occurrence of a path in a hyperpath, and hence its experienced utility can depend on several factors: time of day, number and characteristics of the paths belonging to the hyperpath (besides, some of these characteristics can be random variables), decision rule and so on, but, if suitable hypotheses on service functioning and on diversion choice rules are assumed to analytically compute EU and p, $EU[U_k]$ is doable.

For example, Spiess and Florian (1989), assumed a completely random pattern (i.e., negative exponential) of arrival at-stops for vehicles and a uniform pattern for travellers in a non-congested transit system, with known average link travel time and without real-time traveller info. Suppose travellers board the first arriving line of the hyperpath. In this case, for example, the probability $p(l)$ that line l is the first arriving at the stop and hence the probability to use line l can be easily computed as a function of line frequencies (assumed constant during the time):

$$p(l) = \varphi_l \bigg/ \sum_{j' \in I_m} \varphi_{j'} \tag{5.5}$$

where

- φ_l is the frequency (number of arrivals per time unit) of line l;
- $\varphi_{l'}$ is the frequency (number of arrivals per time unit) of generic line l' belonging to the set of *attractive* lines at stop (diversion node) m, I_m.

Consider, for example, the simple strategy represented by hyperpath $K4$ as reported in Fig. 5.2 'Origin – stop B – Line 2 or 3 – stop D – Destination' and with decision rule 'use of the first arriving vehicle of line 2 or 3 (no fail-to-board exists)'. If the above hypotheses hold, the probabilities of the two paths including line 2 and including line 3 can be computed through their frequencies.

More complex formulations are necessary if some of the above simplifying hypotheses are removed, such as an extension to congested networks and at-stop waiting time info, as reported by Gentile (2016a). Importantly, the analytical method can also be applied in the search methodology without hyperpath enumeration.

When, with different assumptions, the analytical determination of the expected utility of a hyperpath is a very difficult task, one of the methods reported below can be applied:

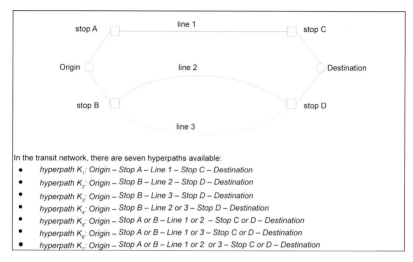

In the transit network, there are seven hyperpaths available:
- *hyperpath K_1: Origin – Stop A – Line 1 – Stop C – Destination*
- *hyperpath K_2: Origin – Stop B – Line 2 – Stop D – Destination*
- *hyperpath K_3: Origin – Stop B – Line 3 – Stop D – Destination*
- *hyperpath K_4: Origin – Stop B – Line 2 or 3 – Stop D – Destination*
- *hyperpath K_5: Origin – Stop A or B – Line 1 or 2 – Stop C or D – Destination*
- *hyperpath K_6: Origin – Stop A or B – Line 1 or 3 – Stop C or D – Destination*
- *hyperpath K_7: Origin – Stop A or B – Line 1 or 2 or 3 – Stop C or D – Destination*

Figure 5.2 Example of hyperpath choice set.

b) Data processing and simulation-based methods for computing hyperpath expected utility

If the network has been monitored, for example by an automatic vehicle location (AVL) system, and the performance data have been collected and made available, such that for a large number of days T the realizations of the random service network are available, the path H_q (potentially) experienced each day z at time τ can be obtained through a simulation-based approach. For example, a transit simulation system, such as that found in BUSMEZZO (Cats, 2011) and in DYBUSRT (Nuzzolo et al., 2016), imposing a strategy-based path choice model with the diversion rule of the strategy to be optimized, can be used.

The *expected utility* of strategy S_{ml}^{t,τ_m} at time τ_m of day t, $EU[U_{S_{ml}}^{t,\tau_m}]$ is computed directly as the expected utility $EU[U_{K_{ml}}^{t,\tau_m}]$ of hyperpath $K_{ml'}$ according to the utilities that would have been experienced in using hyperpath K_{ml} in previous days z at time τ_m:

$$EU\left[U_{K_{ml_j}}^{t,\tau_m}\right] = 1/T \cdot \sum_{z \in ST} U_{H_q}^{z,\tau_m} \tag{5.6}$$

where:

- T is the (large enough) number of days with available data (measured or simulated);
- ST is the set of days in which data are available;
- $U_{H_q}^{z,\tau_m}$ is the utility on path H_q that would have been experienced at time τ_m of day z following the strategy S_{ml}^{z,τ_m} ($z \in \{1, \dots, t-1\}$).

The number of sub-hyperpaths and paths to evaluate grows with the increase in complexity of the transit network and it could become impractical to provide real-time results, if required, through methods which entail path/hyperpath enumeration. Methods without explicit path enumeration in these cases have then to be used.

5.4.3 Methods Without Hyperpath Enumeration for Direct Conditional Optimal Strategy Search

In order to avoid hyperpath enumeration and directly obtain the optimal strategy conditional on a given rule, the following methods can, for example, be applied:

- dynamic routing methods; for this approach, readers can refer to Chapter 7 (Gentile, 2016b);
- artificial intelligence methods, reported in the appendix of this chapter.

Note that, at the current stage of research and application, these methods require specific assumptions on service functioning and/or on diversion rules, which could be far removed from the actual service functioning and/or on diversion rules for real-time application on transit systems with real-time individual predictive info, and hence these methods could lead to approximate optimal solutions.

5.5 Normative Travel Strategy

5.5.1 Normative Strategy Search Methods

A methodology of normative strategy search should find the pair of hyperpath and diversion rule that give the absolute objective maximum expected utility (e.g., what people should do, in theory). The optimal pair may depend on the presence or not of real-time predictive info. Without traveller predictive info, the optimal diversion rule should be based only on the past experienced utility. When a real-time predictive info system is present, the issue arises as how to utilize the available info. Therefore, the normative strategy search will be explored, considering the two types of transit systems: without and with predictive info systems.

a) Transit systems without real-time predicted path attributes

As no real-time predicted information is available on path attributes, among strategies with different hyperpaths and diversion rules, the absolute optimal strategy can be found by comparing the alternative hyperpaths in terms of maximum expected travel utility. In this case, the search includes the following steps (Nuzzolo and Comi, 2015):

- at diversion node m and time τ_m of day t, all the sub-hyperpaths K_{m,l_j}^{t,τ_m} for each diversion link l_j are considered, and the best $K_{m,l_j}^{t,\tau_m}*$ is found by comparing their expected utilities EU, obtained with the methods presented in Section 5.4;
- the optimal $K_{m,l}^{t,\tau_m}*$ of the different diversion links are compared and the maximum expected utility sub-hyperpath $K_m^{t,\tau_m}*$ is the optimal hyperpath.

For example, considering Fig. 5.1, at node B two diversion links and three elementary paths are available (5, 6 + 7 and 6 + 8). Then, at time τ_B of day t, if a vehicle of line 5 is arriving, the first choice entails boarding a run of line 5 or waiting for a run of line 6. The comparison should be performed between the expected experienced utilities up to destination of the elementary path of line 5 and of the composite hyperpath germinating from boarding line 6 and due to the presence of diversion node F.

Suppose that, from the collected data, it is possible to compute the average utility EU of the path on line 5 (i.e., $EU[U_{H_5}^{t,\tau_B}]$), on lines 6 + 7 (i.e., $EU[U_{H_{6+7}}^{t,\tau_B}]$) and on lines 6 + 8 (i.e., $EU[U_{H_{6+8}}^{t,\tau_B}]$) conditional on time τ_B and their frequency of use conditional on time τ_B. These values can be used as estimates of corresponding expected utilities and probabilities of use.

Therefore, for example, the expected utility of waiting for a run of line 6 at node B can be expressed, applying (5.3), as:

$$EU\left[U_{K_{B,line6}}^{t,\tau_B}\right] = p\left[6+7\right]\cdot EU\left[U_{6+7}^{t,\tau_B}\right] + p\left[6+8\right]\cdot EU\left[U_{6+8}^{t,\tau_B}\right] \tag{5.7}$$

where $p[.]$ are the probabilities of using paths on lines 6 + 7 or lines 6 + 8. Comparing the values of expected utilities of the previous three identified hyperpaths, the best can be found.

b) Transit systems with real-time predicted path attributes

Even if real-time predicted values of path attributes are available, as anticipated, there is the problem of the best diversion rule to be used, which takes into account these predicted values. In the case of the path choice of the traveller decision maker in the presence of information, it is highly likely that travellers' knowledge/perception of network performance will vary depending, amongst other things, on their past experience and accessibility to ATIS (Advanced Traveller Information Services). The role of information integration and learning in the decision making process has been extensively studied in recent years and different approaches have been developed (Jha et al., 1998; Bogers, 2009): moving average (Horowitz, 1984; Srinivasan and Guo, 2004; Tian et al., 2010); exponential filter (Cantarella and Cascetta, 1995; Cantarella, 2013; Wu et al., 2013); Markov process representation of learning (Ben-Akiva et al., 1991; De Palma and Marchal, 2002); Bayesian updating (Jha et al., 1998). However, little has been done on the normative approach.

Among the methods of normative strategy search, in that reported below, which is an advancement of that presented in Nuzzolo and Comi (2015), the *anticipated utility AU* is considered. It is the utility associated with each alternative sub-hyperpath of the choice set, in order to support decision making, obtained by combining, with a given rule *(utility anticipation rule)*, expected experienced utility *EU* and forecasted utility *FU*, obtained applying forecasted attributes. The *forecasted utility*, $FU[U_{K_{ml}}^{t,\tau_m}]$, of hyperpath K_{ml} can be assumed equal to the utility of the elementary path H_q (of the hyperpath) which has the maximum forecasted utility at time τ_m of day t:

$$FU\left[U_{K_{ml}}^{t,\tau_m}\right] = max\left\{FU\left[U_{H_q}^{t,\tau_m}\right]; \quad \forall H_q \in SQ\right\} \tag{5.8}$$

where $FU[U_{H_q}^{t,\tau_m}]$ is the utility of path H_q (belonging to path set SQ) forecasted at time τ_m of day t which, as done above, can be hypothesized as a linear function of predicted (forecasted) attributes $FX_{gH_q}^{t,\tau_m}]$ (e.g., travel time, cost, schedule delay) computed at time τ_m of day t:

$$FU\left[U_{H_q}^{t,\tau_m}\right] = \sum_g \cdot FX_{gH_q}^{t,\tau_m} \tag{5.9}$$

with β_g the (average or individual) parameters (to be estimated).

Therefore, the strategy forecasted utility $FU[U_{K_{ml}}^{t,\tau_m}]$ corresponds to the utility that the traveller would experience following the path which has the maximum utility, taking into account only the attributes predicted at time τ_m of day t.

In this case, at diversion node m and time τ, all the sub-hyperpaths $K_{m,lj}$ for each diversion link l_j are considered and the best K_{ml}^* is found, comparing their real-time *anticipated utilities AU*.

Different anticipation rules can be considered, and which is the best is still a matter of debate. Below, a possible approach that seems more suitable for transit users is reported.

A heuristic decision rule

In the heuristic decision rule proposed by Nuzzolo and Comi (2015), the anticipated utility of each optimal hyperpath (K_{ml}) is computed by combining its expected utility $(EU[U_{K_{ml}}^{t,\tau_m}])$ and the utility $(FU[U_{K_{ml}}^{t,\tau_m}])$ forecasted with predictive path attributes:

$$AU\left[U_{K_{ml}}^{t,\tau_m}\right] = \alpha \cdot FU\left[U_{K_{ml}}^{t,\tau_m}\right] + \left(1-\alpha\right) \cdot EU\left[U_{K_{ml}}^{t,\tau_m}\right] \tag{5.10}$$

where $\alpha \in [0,1]$ is the weight given to the forecasted utility. The value of α depends on the stochasticity of service networks and on forecasted attribute error dispersion. If the forecasts were perfectly respected, the utilities of the past experiences could be omitted and optimal strategies obtained through forecasts alone, with α equal to *1*. If completely wrong, α should be equal to *0*. With a certain degree of prediction exactness, the contribution of forecasting should increase as network stochasticity increases and so does α.

The expected experienced utility is a function of α, $\psi(\alpha)$ (Fig. 5.3), with a maximum value, which gives the best α. In order to obtain the optimal strategy, a methodology can then be used which includes two steps: (1) search for the optimal strategies conditional to α varying between *0* and *1*; (2) select the best strategy as the one of maximum *average experienced utility*. The optimal value of α could depend on where the info is provided (i.e., diversion node m) and on $\tau_{m'}$.

As anticipated, some applications of this methodology are reported in Chapter 3 (Comi et al., 2016).

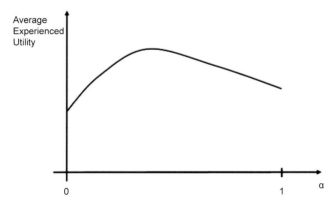

Figure 5.3 Example of expected experienced utility variation with alpha.

5.5.2 Dynamic Search for a Normative Strategy

Until now, the normative strategy at a certain node m has been considered, according to updated info available and random occurrences. As reported above, the topology of a travel strategy has a set of successive diversion nodes, where a choice has to be made among diversion links, using a diversion rule. Traveller compares the optimal strategies connecting the node to the destination, taking into account current and real-time forecasted states of the service network, and chooses the link belonging to the hyperpath of the absolute optimal strategy. Once a diversion link has been chosen and used, at the following diversion node a new diversion choice is carried out (real-time and sequential *dynamic diversion link choice*).

An example of optimal dynamic diversion link suggestion derives from a real-time *optimal diversion link search* and includes the following steps:

- at diversion node m and time τ, all the sub-hyperpaths $K_{m,lj}$ for each diversion link l_j are considered and the best K_{ml}^* is found by comparing their expected utilities *EU*;
- the optimal K_{ml}^* of the different diversion links are compared and the maximum expected utility sub-hyperpath K_m^* is found;
- the diversion link and the next diversion node derived of the optimal K_m^* are suggested at node m.

At the first boarding stop and at each transfer stop (if any), the info system should use a *sequence of binary choices*, that is: *to board the arriving run* or *to wait*, in order to suggest the optimal diversion link (see Fig. 5.4). That is, when a run r arrives at stop m, the info system has to find the optimal sub-hyperpath $K_{m,lr}^*$ up to destination, including the boarding link of the run (l_r) and the optimal hyperpath K_w^* up to destination, with respect to waiting for the arrival of another run r', including the waiting link. It thus suggests boarding r if the anticipated utility $AU_{K_{m,lr}^*}$ of sub-

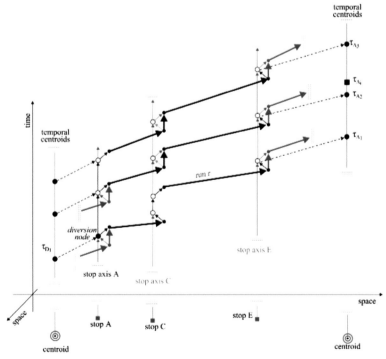

Figure 5.4 Example of a sequential binary choice mechanism at stop *s* for the arriving run *r* (Nuzzolo et al., 2016).

hyperpath $K_{m,lr}*$ is greater than the anticipated utilities $AU_{K_w^*}$ of optimal hyperpath K_w^*. If run r is not boarded, the choice is reconsidered when the next run of the choice set arrives and so on. The diversion link choice set at stop s depends on the user arrival time τ at stop m and on the current transit operations and those predicted at time τ_m of day t.

In general, at each stop, the info system could suggest whether in-vehicle travellers should stay on board or alight, according to real-time predictive info. Besides, upon arriving at each pedestrian diversion node, the info system could make a new decision among the attractive links. Neither case is considered in this chapter.

As an example of dynamic diversion link choice, consider Fig. 5.1, where the O-D pair *od* is connected by seven paths composing, among others, a hyperpath with four diversion nodes (O, A, B, F). At the origin node O, the diversion link O-C has just one possible sub-hyperpath up to the destination, while diversion link O-B has seven possible alternative sub-hyperpaths. For a decision at time τ_o, the expected utility of each sub-hyperpath can be computed and, for a given diversion rule, comparing these utilities, the optimal hyperpath can be found deterministically. If the optimal sub-hyperpath includes the diversion link O-B, the traveller has to travel up to diversion node B. When the traveller arrives at node B, each time a run of line 5 or 6 arrives at the stop, a new search is carried out for the optimal total sub-hyperpath until destination. If the best is to reach node F, when the traveller arrives at node F, a new optimal sub-hyperpath search until destination is carried out.

5.5.3 Real-time Search for a Normative Strategy

The real-time search for a normative strategy for normative path suggestions or real-time transit assignment, as real-time procedures adopt one-second simulation steps (see Chapter 4—Cats, 2016—on transit simulation systems) means that the search has to be performed in that time window. Therefore, the fast 'short hyperpath' search algorithms have to be applied, as considered in Chapter 7 (Gentile, 2016a) on dynamic routing on transit networks.

5.6 Conclusions and the Road Ahead

In the above framework analysis of real-time modelling of a normative travel strategy on unreliable dynamic transit networks, the definition of normative strategy was explored in depth and its search methods were classified and analyzed, when real-time predictive info on path attributes is or not available.

Some research issues arise, in particular relative to the diversion rule to be used in the absolute optimal strategy, such as the suitable real-time anticipated utility to be used when predictive on-path attribute info is available. For this aim, applications for example based on Bayesian decision theory and fast 'short hyperpath' search algorithms in real-time context should be useful tools.

APPENDIX

Artificial Intelligence Methods for Optimal Strategy Search

As a further way to search for the optimal strategy, artificial intelligence methods, especially as the *automated planning* problem (Ghallab et al., 2004), can be applied. Given an initial state, a desired goal and a set of possible actions, the planning problem is to find a plan that generates a sequence of actions, typically for execution by intelligent agents (travellers in our case) that lead to the target state. With non-deterministic events outside the control of the agent, the possible executions form a tree and plans have to determine the appropriate actions for every node of the tree. Probabilistic automated planning can be solved with iterative methods, such as *value iteration* and *policy iteration*, when the state space is sufficiently small. Discrete-time *Markov decision processes (MDPs)* are planning problems with non-deterministic events with known probabilities, maximization of a reward function and a single agent. In other words, an MDP is a *discrete time stochastic control* process, where at each time step, the process is in some state p, and the decision maker may choose any action a that is available in state p. The process responds at the next time step by randomly moving into a new state, p' and giving the decision maker a corresponding reward, $RV_a(p, p')$. The core problem of MDPs is to find a *'policy'* for the decision maker, that is a rule that the agent follows in selecting actions, given the state it is in: a function π that specifies the action $\pi(p)$ which the decision maker will choose when in state p. A policy therefore is a way of defining the agent's action selection with respect to changes in the environment. The goal is to choose a policy π that will maximize some cumulative function of the random rewards, typically the expected discounted sum over a potentially infinite horizon. If the probabilities of rewards are unknown, the problem is one of *reinforcement learning* (Sutton and Barto, 1998). For this purpose, it is useful to define a further function, which corresponds to taking action p and then continuing optimally (or according to whatever policy one currently has):

$$Q(p,a) = \sum_{p'} P_a(p,p') \cdot \left[RV_a(p,p') + \gamma \cdot V(p') \right] \qquad \text{(A.1)}$$

where

- $P_a(p, p')$ is the probability that action a in state p at time t will lead to state p' at time $t+1$;
- $V(p')$ is the value function of the policy, defined as the expectation of the return given that the agent acts according to that policy;
- $\gamma \in [0,1)$ is the discount rate used to calculate the long-term return.

While this function is also unknown, experience during learning is based on (p, a) pairs (together with the outcome p'); that is, 'I was in state p and I tried doing a and p' happened'. Thus one has an array Q and uses experience to update it directly. This is known as *Q learning*, a model-free *reinforcement learning* technique. Specifically, Q-learning can be used to find an optimal action-selection policy for any given (finite) MDP. It works by learning an *action-value function* that ultimately gives the expected utility of taking a given action in a given state and following the optimal policy thereafter. When such an action-value function is learned, the optimal policy can be constructed by simply selecting the action with the highest value in each state.

A learning process can be assumed (Wahba, 2008), with an exploration phase and an exploitation phase. During the exploration phase, the traveller randomly chooses an alternative hyperpath K_o and exploits it for a certain number of days (e.g., seven in Wahba, 2008). During the exploitation phase, the traveller moves on the chosen hyperpath, updating the subjective expected utility of the hyperpath, which, as reported above, is a function of all the path utilities experienced in the previous days. At the end of the process, if the stochastic process of service network states is stationary, the traveller chooses the hyperpath K^* of maximum subjective expected (or non-expected) utility.

This artificial intelligence approach is still at an initial state of research. A version has been implemented in the assignment model of the transit simulation system MILATRAS (Wahba, 2008) to reproduce the traveller's Markovian decision process in path choice. In this case a descriptive rather than normative application of automated planning is carried out.

Keywords: strategy-based transit path choice models; unreliable transit network; individual predictive traveller information; normative transit path choice models.

References

Bell, D. E. 1982. Regret in decision making under uncertainty. *In*: Operations Research 30(5): 961–981.

Ben-Akiva, M., A. De Palma and K. Isam. 1991. Dynamic network models and driver information systems. *In*: Transportation Research Part A 25(5): 251–266.

Bogers, E. A. I. 2009. Traffic Information and Learning in Day-to-Day Route Choice. Ph.D. Thesis, TRAIL Thesis Series no. T2009/5, the Netherlands TRAIL Research School.

Bovy, P. H. L. 2009. On modelling route choice sets in transportation networks: A synthesis. pp. 43–68. *In*: Transport Reviews: A Transnational Transdisciplinary Journal 29(1), Taylor & Francis.

Cantarella, G. E. 2013. Day-to-day dynamic models for intelligent transportation systems design and appraisal. *In*: Transportation Research Part C 29: 117–130.

Cantarella, G. E. and E. Cascetta. 1995. Dynamic process and equilibrium in transportation networks: Towards a unifying theory. *In*: Transportation Science 29: 305–329.

Cats, O. 2016. Real-time operations management decision support systems. *In*: A. Nuzzolo and W. H. K. Lam (eds.). Modelling Intelligent Multi-modal Transit Systems, CRC Press.

Cats, O. 2011. Dynamic Modelling of Transit Operations and Passenger Decisions. Ph.D. Thesis, KTH, School of Architecture and the Built Environment (ABE), Transport Science, Traffic and Logistics, Lindstedtsvägen 26, KTH, Stockholm.

Cats, O., H. N. Koutsopoulos, W. Burghout and T. Toledo. 2011. Effect of real-time transit information on dynamic path choice of passengers. *In*: Transportation Research Record: Journal of the Transportation Research Board, No. 2217: 46–54.

Comi, A. and A. Nuzzolo. 2016. A dynamic strategy-based path choice modelling in real-time transit simulation. *In*: A. Nuzzolo and W. H. K. Lam (eds.). Modelling Intelligent Multi-modal Transit Systems, CRC Press.

Comi, A., A. Nuzzolo, U. Crisalli and L. Rosati. 2016. A new generation of individual real-time transit information systems. *In*: A. Nuzzolo and W. H. K. Lam (eds.). Modelling Intelligent Multi-modal Transit Systems, CRC Press.

de Palma, A. and F. Marchal. 2002. Real cases applications of the fully dynamic METROPOLIS tool-box: An advocacy for global large-scale mesoscopic transportation systems. *In*: Networks and Spatial Economics 2(4): 347–369.

de Palma, A., M. Ben-Akiva, D. Brownstone, C. Holt, T. Magnac, D. McFadden, P. Moffatt, N. Picard, K. Train, P. Wakker and J. Walker. 2008. Risk, uncertainty and discrete choice models. *In*: Market Letters 19: 269–285.

Ettema, D. and H. Timmermans. 2006. Costs of travel time uncertainty and benefits of travel time information: Conceptual model and numerical examples. *In*: Transportation Research Part C 14: 335–350.

Fonzone, A., J. D. Schmöcker, F. Kurauchi, S., Hassan, M. 2013. Strategy choice in transit networks. *In*: Proceedings of the Eastern Asia Society for Transportation Studies, Vol. 9.

Gallo, G., G. Longo, S. Nguyen and S. Pallottino. 1993. Directed hypergraphs and applications. Discrete Applied Mathematics 42: 177–201.

Gentile, G. 2016a. Dynamic routing on transit networks. *In*: A. Nuzzolo and W. H. K. Lam (eds.). Modelling Intelligent Multi-modal Transit Systems, CRC Press.

Gentile, G. 2016b. Formulating and solving transit assignment. *In*: G. Gentile and K. Noekel (eds.). Modelling Public Transport Passenger Flows in the Era of Intelligent Transport Systems, Springer International Publishing.

Gentile, G., M. Florian, Y. Hamdouch, O. Cats and A. Nuzzolo. 2016. The theory of transit assignment: Basic modelling frameworks. pp. 287–386. *In*: G. Gentile and K. Noekel (eds.). Modelling Public Transport Passenger Flows in the Era of Intelligent Transport Systems: COST Action TU1004 (TransITS), DOI: 10.1007/978-3-319-25082-3_6, Springer Tracts on Transportation and Traffic 10, Springer, Switzerland.

Gentile, G. and K. Nokel (eds.). 2016. Modelling Public Transport Passenger Flows in the Era of Intelligent Transport Systems. Springer International Publishing.

Ghallab, M., D. S. Nau and P. Traverso. 2004. Automated Planning: Theory and Practice. Morgan Kaufmann.

Horowitz, J. 1984. The stability of stochastic equilibrium in a two-link transportation network. *In*: Transportation Research B 18(1): 13–28.

Jha, M., S. Madanat and S. Peeta. 1998. Perception updating and day-to-day travel choice dynamics in traffic networks with information provision. *In*: Transportation Research Part C 6: 189–212.

Kahneman, D. and A. Tversky. 1979. Prospect theory: An analysis of decision under risk. *In*: Econometrica 47(2): 263–291.

Kurauchi, F., J. D. Schmöcker, A. Fonzone, S. M. H. Hemdan, H. Shimamoto and M. G. H. Bell. 2012. Estimation of weights of times and transfers for hyperpath travellers. *In*: Transportation Research Records 2284: 89–99.

Liu, Y. and J. M. Bunker and L. Ferreira. 2009. Modelling urban public transit users' route choice behaviour: Review and outlook. *In*: Rethinking Sustainable Development: Planning, Infrastructure Engineering, Design and Managing Urban Infrastructure, Queensland University of Technology, Brisbane, Queensland.

Moreira-Matias, L., J. Mendes-Moreira, J. F. de Sousa and J. Gama. 2015. Improving mass transit operations by using AVL-based systems: A survey. *In*: IEEE Transactions on Intelligent Transportation System, DOI 10.1109/TITS.2014.2376772.

Nuzzolo, A. and A. Comi. 2015. Transit trip planners: Real-time strategy-based path recommendation. pp. 196–201. *In*: 18th IEEE International Conference on Intelligent Transportation Systems (ITSC 2015), DOI: 10.1109/ITSC.2015.41.

Nuzzolo, A., U. Crisalli, A. Comi and L. Rosati. 2016. A mesoscopic transit assignment model including real-time predictive information on crowding. *In*: Journal of Intelligent Transportation Systems: Technology, Planning, and Operations, DOI: 10.1080/15472450.2016.1164047, Taylor & Francis.

Prato, C. G. 2009. Route choice modeling: Past, present and future research directions. pp. 65–100. *In*: Journal of Choice Modelling 2(1), Elsevier.

Ramos, G. M., W. Daamen and S. Hoogendoorn. 2014. A state-of-the-art review: Developments in utility theory, prospect theory and regret theory to investigate travellers' behaviour in situations involving travel time uncertainty. *In*: Transport Reviews: A Transnational Transdisciplinary Journal 34(1): 46–67, DOI: 10.1080/01441647.2013.856356.

Schmocker, J. D., H. Shimamoto and F. Kurauchi. 2013. Generation and calibration of transit hyper-paths. *In*: Transportation Research C 36: 406–418.

SISTeMA. 2015. Hyperpath, www.hyperpath.it, Retrieved February 8, 2016.

Spiess, H. and M. Florian. 1989. Optimal strategies. A new assignment model for transit networks. *In*: Transportation Research. Part B: Methodological 23B(2): 83–102.

Srinivasan, K. K. and Z. Guo. 2004. Day-to-day evolution of network flows under route choice dynamics in commuter decisions. *In*: Transportation Research Record 1894.

Starmer, C. 2000. Developments in non-expected utility theory: The hunt for a descriptive theory of choice under risk. *In*: Journal of Economic Literature 38(2): 332–382.

Sutton, R. S. and A. G. Barto. 1998. Reinforcement Learning: An Introduction (Adaptive Computation and Machine Learning). The MIT Press.

Tian, L., H. Huang and T. Liu. 2010. Day-to-day route choice decision simulation based on dynamic feedback information. *In*: Journal of Transportation Systems Engineering and Information Technology 10(4): 79–85.

Von Neumann, J. and O. Morgenstern. 1947. Theory of Games and Economic Behavior. Princeton: Princeton University Press.

Wahba, M. 2008. MILATRAS - MIcrosimulation Learning-based Approach to Transit Assignment. Ph.D. Thesis. Department of Civil Engineering University of Toronto, Canada.

Wahba, M. and A. Shalaby. 2009. MILATRAS. A new modelling framework for the transit assignment problem. pp. 171–194. *In*: N. H. M. Wilson and A. Nuzzolo (eds.). Schedule-Based Modeling of Transportation Networks: Theory and Applications, Kluwer Academic Publisher.

Wu, J., H. Sun, D. Z. W. Wang, M. Zhong, L. Han and Z. Gao. 2013. Bounded-rationality based day-to-day evolution model for travel behavior analysis of urban railway network. *In*: Transportation Research Part C 31: 73–82.

6

A Dynamic Strategy-based Path Choice Modelling for Real-time Transit Simulation

A. Comi[1,a,]* and *A. Nuzzolo*[1,b]

ABSTRACT

The main objective of this chapter is to present a behavioral assumption and model formulation framework for transit path choice modelling with a descriptive travel strategy approach. Such modelling is suitable for real-time run-oriented simulation-based mesoscopic assignment models, which can support real-time predictive info and real-time operations control when considering unreliable transit systems, with predictive info on the characteristics of the services. The introductory section is devoted to classifying transit assignment models in relation to their use in real-time simulation and is followed by a section with the concepts of strategy, of the dynamic link diversion choice rule and of anticipated utility, as a combination of experienced and forecasted travel attributes. The second part deals with the path choice model formulation and the relative hyperpath choice set

[1] Department of Enterprise Engineering, Tor Vergata University of Rome, Via del Politecnico 1, 00133, Rome, Italy.
[a] E-mail: comi@ing.uniroma2.it
[b] E-mail: nuzzolo@ing.uniroma2.it
* corresponding author

generation issue. The path choice models of three mesoscopic transit simulation tools presented in the literature, namely MILATRANS, BUSMEZZO and DYBUSRT are also recalled and analyzed within this presented framework. Finally, some concluding remarks and research prospects for this topic are reported.

6.1 Introduction

In recent years transit agencies have been increasingly supported by information technology and telematics to achieve more advanced public transportation systems (APTS), improving real-time transit operations control and supplying real-time traveller information. Operations control variables concern travel speeds and delays, route and link schedule adherence, as well as numbers of passengers on board and waiting at stops. Traveller information variables in the case of individual information (e.g., about specific origin-destination and departure time) mainly concern path travel time components (access, waiting, on-board, transfer, egress) and costs, as well as on-board crowding at the single vehicle level. Therefore, real-time operations control and traveller information require *real-time* estimation and *short-term* prediction of transit vehicle travel times, on-board loads and other state variables representing system functioning. These variables are obtained through transit simulation systems, where a major role is played by dynamic path choice models, which are able to reproduce the traveller's choice behavior so as to replicate, in *dynamic transit assignment* models, the way in which travellers propagate on the network, according to transit service configurations.

As reported in detail in Chapter 1 (Nuzzolo, 2016), transit assignment models can be classified into *frequency-based* or *line-oriented* models and *schedule-based* or *run-oriented* models. *Frequency-based* or *line-oriented* assignment models (Spiess and Florian, 1989) allow us to obtain only average line flows, which are mainly useful for off-line applications in strategic transportation planning. Conversely, aiming to predict, for example, on-board loads of single transit vehicles, the *run-oriented* approach is the only suitable way within network assignment methods and which is why it is the only approach considered below.

Run-oriented assignment models can be differently classified into *analytical* and *simulation-based* models (Gentile and Noekel, 2016). *Analytical run-oriented* models have been developed for assessment applications using equilibrium models (among the most recent papers see Ren and Lam, 2007; Sumalee et al., 2009), some of which apply a travel strategy approach (Hamdouch and Lawphongpanich, 2008; Rochau et al., 2011; Hamdouch et al., 2011, 2014). *Simulation-based or agent-based* models are more suitable for *dynamic* modelling (Cats et al., 2011) as they reproduce interactions over time among different agents: travellers, transit vehicles and sometimes also other vehicles sharing the right of way. These interactions can be reproduced in detail, moment by moment (micro-simulation models), or some interactions can be omitted or simplified, in relation to their application fields (mesoscopic models). Despite the greater level of output details of micro-simulation models, meso-simulation allows us to analyze large networks with respect to the limited networks that can be analyzed by the micro-simulation approach. Furthermore, mesoscopic models are easy to apply, computationally simpler and less time-consuming in obtaining the required output details (Cats et al., 2011; Gentile and Noekel, 2016). Aiming to predict on-board loads of single transit vehicles in real time, as required by APTS following the analyses reported in Chapter 1 (Nuzzolo, 2016), run-oriented simulation-based (or agent-based) mesoscopic assignment is the only suitable approach within assignment methods (Cats, 2011; Gentile and Noekel, 2016; Nuzzolo et al., 2016). Hence the run-oriented simulation-based assignment approach is the only one considered below.

As reported in Chapter 5 (Nuzzolo and Comi, 2016), transit path choice decision-making is influenced by several factors (e.g., knowledge of the transit network, service regularity, type of available information). One of the more complex situations is when travellers move on an unreliable network, where at certain nodes (*diversion* nodes) travel decisions have to be made according to random occurrences (e.g., bus arrivals at stops, on-board crowding). Even if a system predicting information on network states were available and travellers could be provided with such info, due to the uncertainty of such forecasts (Lu et al., 2015), choosing at departure time a specific path up to the destination may not be the best decision. Rather, with the aim of minimizing expected travel costs, an *optimal travel strategy* should be used, according to a predefined set of paths, i.e., *hyperpath* (Gallo et al., 1993). A strategy includes a set of *diversion nodes*, each with a set of *diversion links*, where travellers choose how to proceed to their destination, considering service occurrences which arise and using a given diversion choice rule. As reported in Chapter 5, to which readers are invited to refer for some concepts of travel strategies applied in this Chapter 6, optimal travel strategies are classified into:

- *objective or normative* strategies, when they allow the *objective minimum expected experienced travel cost* to be attained;
- *subjective or descriptive* optimal travel strategies, when, due to cognitive issues, a *subjective minimum expected experienced cost* is obtained by a traveller.

While normative strategies should be used in transit trip planning with a view to suggesting an objective optimal travel strategy, descriptive travel strategies should be applied in transit simulation in order to reproduce the traveller's actual behavior. Recent research, using data collected through new ticketing technologies, confirms that on unreliable transit networks with diversion nodes, travellers tend to apply descriptive strategies and that the kind of strategy depends on the information that is available and on user socio-economic characteristics (Kurauchi et al., 2012; Schmocker et al., 2013; Fonzone et al., 2013).

Traditionally, strategy-based path choice models are applied in *frequency-based* or *line-oriented* assignment models, developing different optimal-strategy models according to the different diversion rules that travellers apply to reach their destination. For example, Spiess and Florian (1989) proposed the following simple rule: 'take whichever bus comes first, belonging to the line attractive set', which is chosen among the available lines and which allows travellers to reach their destinations at minimum expected travel time. Stochastic strategy-based models have also been proposed: Nguyen et al. (1998) proposed a logit model for simulating hyperpath choices, while Florian and Constantin (2012) modified their choices in strategy transit assignments by including short walks to access attractive transit paths. Schmocker et al. (2013) presented a discrete choice model for hyperpath choices with explicit choice set generation based on Swait's (2001) formulation. Ma and Fukuda (2015) developed a hyperpath-based Network-GEV model to analyze route choice under travel time uncertainty.

In the context considered herein of real-time run-oriented simulation-based mesoscopic assignment models of unreliable transit systems with predictive info, path choice modelling requires a dynamic strategy-based approach which follows the within-day and day-to-day evolution of network performance and predictive info supplied. Literature relative to transit simulation-based mesoscopic assignment models is very limited and those applied in the transit simulation frameworks MILATRANS (Wahba and Shalaby, 2009, 2014), BUSMEZZO (Cats, 2011, 2014) and DYBUSRT (Nuzzolo et al., 2016) are mentioned. In order to consider service network unreliability, all the above simulation tools reproduce, explicitly or implicitly, the chosen path as the result of a travel strategy, but they present different modelling features with respect to model specification, as will be explored in the next part of this chapter.

The main objective of this chapter is to present a new dynamic transit path choice modelling with a descriptive travel strategy approach, suitable for real-time run-oriented simulation-based mesoscopic assignment models of unreliable transit systems, which provide predictive info on service characteristics. The presented path choice model allows us to overcome some behavioral hypotheses and model specifications of the previous recalled frameworks, as detailed on the next sections.

Chapter 6 is organized as follows: Section 6.2 gives an overview of the notation used throughout the chapter, while Section 6.3 analyzes the general behavioral assumption framework underlying the presented dynamic path choice modelling. In Section 6.4, the general formulation of the path choice model is reported. Finally, concluding remarks and research prospects for this topic are given in Section 6.5.

6.2 List of Notation

The notations used during this chapter chiefly comprise the following:

- Γ_{od}^{τ}, the set of all available paths (*master hyperpath*), considered by the traveller according to his/her past experience, connecting the origin-destination pair *od* at time τ;

- $K_{il,h}^{\tau}$, the *h*-th hyperpath at time τ with the root in the diversion link *il*;

- $S_{il}^{\tau,t}$, the travel strategy experienced at time τ of day *t* consisting of travelling on a sub-hyperpath $K_{il,h}^{\tau,t}$ according to a given diversion rule χ;

- $G_{il,h}$, the set of elementary paths *k* belonging to hyperpath $K_{il,h}^{\tau,t}$;

- $TU_{K_{il}^{\tau,t},\chi}^{\tau,t}$, the subjective experienced utility of strategy $S_{il}^{\tau,t}$;

- TU_{k}, subjective experienced utility of path *k* within hyperpath $K_{il,h}^{\tau,t}$;

- $TX_{k,j}^{\tau,t}$, the *j*-th experienced path attribute;

- $TU_{K_{il,h}^{\tau,t}}^{\tau,t}$, the experienced utility of hyperpath $K_{il,h}^{\tau,t}$;

- $SS*[K_{il}^{\tau,t}, \chi]$, the strategy with the subjective maximum expected experienced utility (subjective optimal strategy);

- $ETU_{K_{il}^{\tau,t},\chi}^{*\tau,t}$, the subjective maximum expected experienced utility of strategy $S_{il}^{\tau,t}$;

- $K_{il}^{*\tau,t}$, the hyperpath with the maximum expected experienced utility (optimal sub-hyperpath);

- $ETU_{K_{il}^{\tau,t}}^{*,\tau,t}$, the maximum expected experienced utility of strategy hyperpath K_{il}^{τ};

- $CS_{il}^{\tau,t}$, the choice set of hyperpaths with the root in the diversion link *il* at time τ of day *t*;

- $K_{il}^{*\tau,t}$, the best run hyperpath among all hyperpaths belonging to $CS_{il}^{\tau,t}$;

- $CS_{i}^{*\tau,t}$, the choice set of the best hyperpaths $K_{il}^{*\tau,t}$ at time τ of day *t*;

- $K_{i}^{**\tau,t}$, the best run hyperpath among all hyperpaths belonging to $CS_{i}^{*\tau,t}$;

- $AU_{K_{il,h}^{\tau}}^{\tau,t}$, the anticipated utility of hyperpath $K_{il,h}^{\tau}$ at time τ of day *t*;

- $AU_{k}^{\tau,t}$, the anticipated utility of path *k* belonging to hyperpath $K_{il,h}^{\tau}$ at time τ of day *t*;

- $P[k/K_{il,h}^{\tau}]$, the percentage use of path k within hyperpath $K_{ij,h}$;
- $TX_{k,j}^{\tau,t}$, the attribute value of path k experienced at time τ of day t;
- $PX_{k,j}^{\tau,t}$, the attribute value of path k perceived at time τ of day t;
- $AX_{k}^{\tau,t}$, the attribute value anticipated at time τ of day t,
- $FX_{k}^{\tau,t}$, the attribute value forecasted by info system at time τ of day t;
- $PX_{k}^{\tau,t}$, the attribute value perceived at time τ of day t;
- $AU_{K_{il}^{*\tau,t}}^{\tau,t}$, the value of anticipated utility of hyperpath $K_{il}^{*\tau,t}$;
- $AU_{K_{il}^{*}}^{\tau,t}\tau_{,t}$, the value of anticipated utility of hyperpaths $K_{il}^{*\tau,t}$;
- $L_{i}^{\tau,t}$, the set of all the diversion links from diversion node i;
- $p[il]$, the choice probability of diversion link il belonging to set $L_{i}^{\tau,t}$.

6.3 General Behavioral Assumption Framework

In this section, a general framework of path choice behavior in the context of an unreliable service network with diversion nodes and real-time information is presented. The generation of travel strategies and the diversion choice mechanism at different types of diversion nodes are then explored.

6.3.1 General Assumptions

The behavioral assumptions framework is defined here in the context of an unreliable or stochastic service network with diversion nodes (i.e., nodes where decisions are carried out according to occurrences of random events) and of frequent travellers (i.e., people who often travel on a given origin-destination pair) equipped with mobile devices that allow them to access real-time individual predictive information. Given the origin-destination (O-D) pair *od* and a departure time on day *t*, such information concerns predicted run arrival times at stops, as well as routes and relative predicted characteristics (i.e., travel time components and on-board crowding) from the traveller's current position to destination.

As reported in Section 6.1, even if a real-time predictive information system is available, decisions are made in the context of uncertainty, because travellers move on a stochastic or unreliable network and the predicted values could differ from the values that will be experienced. Thus, travellers are assumed not to choose, at origin, an entire path up to the destination, but to choose a *travel strategy* with adaptive behavior during the trip.

6.3.2 Strategies, Hyperpaths and Diversion Rules

A strategy entails a sequence of traveller's choices carried out at diversion nodes, combining transit services status, real-time predictive information and previous network experience. The first choice is at origin and includes the departure time and the first boarding stop. The others refer to choices during the trip at given points of the network (i.e., diversion nodes), like pedestrian nodes and stops, where travellers can make adaptive *en-route* decisions according to random occurrences and available updated predictive info.

A strategy can be topologically represented through a hyperpath, characterized by diversion nodes at the decision points, where the *diversion link* choices come with a *decision or diversion rule*. Given that line or run-oriented approaches can be used to represent service networks, the hyperpath can be defined in terms of a line-oriented or run-oriented graph (Fig. 6.1).

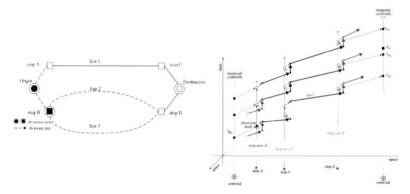

Figure 6.1 Example of line (left side) and run (right side) hyperpath.

It should be noted that different strategies, with different diversion nodes and/or diversion links (and therefore, different hyperpaths) and/or diversion rules can be considered and used on the same network by a traveller for the same origin-destination pair. For example, as presented in the introductive section, Spiess and Florian (1989) assume that travellers use the rule to board the first arriving line of a set of attractive lines, which allows them to reach their destinations at minimum expected travel time. Alternative and more complex rules are possible if, for example, information about the time until the next arrival of the attractive lines is available (Noekel and Weckek, 2009). Gentile et al. (2005) consider the availability of such information at stops and travellers are assumed to board the line of the attractive line set, which offers the best combination of displayed waiting time and expected travel time to the destination. The diversion rule can also include how to combine network experience and available info. Cats (2011) in BUSMEZZO and Nuzzolo et al. (2016) in DYBUSRT hypothesize, for transit systems with individual predictive info, that travellers choose to board the arriving line or to wait according to the maximum anticipated utility up to destination of each alternative where the anticipated utility combines experience and individual predictive info. Thus a strategy is defined when the related hyperpath and diversion rule are given. An *optimal strategy* is a combination of hyperpath and rule which minimizes the expected travel cost.

In the following, some hypotheses are reported on the traveller's process of determining the optimal strategy.

6.3.3 Master Hyperpaths

The travellers are assumed not to consider the whole set of all the available paths connecting the O-D pair at time τ (for example, between 7.25 and 7.35 a.m.), but they consider a sub-set, which, based on their past experience of the network service functioning as frequent users on that O-D pair *od* and in relation to a diversion rule used at its diversion nodes, has the minimum expected travel cost compared with all other possible sub-sets. This sub-set of paths is termed the *master hyperpath* Γ_{od}^{τ}.

6.3.4 Subjective Experienced Utilities and Optimal Master Hyperpath

A master hyperpath Γ_{od}^{τ} includes several diversion nodes i and then several sub-hyperpaths $K_{i'}$ from diversion node i up to destination. For example, the master hyperpath Γ_{od}^{τ} in Fig. 6.1 (left side) includes three elementary paths, seven alternative sub-hyperpaths at node *origin*, one sub-hyperpath at node A and three sub-hyperpaths at node B.

For a given diversion node i and time τ and in relation to each diversion link il from node i, each traveller (as a frequent user on the given O-D pair *od* at time τ) is assumed to identify and experience several strategies with sub-hyperpaths $K_{il,h}^{\tau}$, including transit lines. Let us assume that at time τ of day t, the traveller experiences a travel strategy $S_{il}^{\tau,t}$ consisting of travelling, according to a given diversion rule χ, on a sub-hyperpath $K_{il,h}^{\tau,t}$, starting from node i up to the destination, including the diversion link il and with $G_{il,h}$ elementary paths k. At the end of the path, the traveller associates to this strategy (i.e., hyperpath plus rule) a *subjective experienced (tried) utility* $TU_{K_{il}^{\tau,t},\chi}^{\tau,t}$, equal to the experienced utilities TU_k of path k within $K_{il,h}^{\tau,t}$, used as a result of the realizations of service stochastic events and of the choices with the adopted diversion rule χ. It is also assumed that the experienced path utility is expressed by a linear combination of the components TX of the experienced path attribute vector **TX**:

$$TU_k^{\tau,t} = \sum_j \beta_j \cdot TX_{k,j}^{\tau,t}$$

(6.1)

Over a long period, due to the stochasticity of the service network, different paths of strategy S are experienced and the experienced hyperpath utility $TU_{K_{il,h}^{\tau,t}}^{\tau,t}$ is a random variable. In order to select the optimal strategy which minimizes the travel cost, alternative available strategies (i.e., hyperpaths and decision rules) have to be considered and, according to decision theory (Von Neumann and Morgenstern, 1947), compared in terms of *subjective expected experienced path costs or utilities ETU*, which are obtained as a result of the learning process reported above.

Therefore, when several alternative strategies are available for the diversion link il, the subjective optimal strategy $SS*[K_{il}^{\tau,t}, \chi]$, and the optimal sub-hyperpath $K_{il}^{*\tau,t}$ are respectively the strategy with the subjective maximum expected experienced utilities $ETU_{K_{il}^{\tau,t}, \chi}^{*\tau,t}$ and the relative hyperpath with the maximum expected experienced utilities $ETU_{K_{il}^{\tau,t}}^{*\tau,t}$.

It is further assumed that, given a hyperpath, the traveller experiments with different diversion rules χ in order to find the optimal pair of hyperpath and diversion rule, that is the *subjective absolute optimal strategy*.

If all the paths of the subjective optimal strategies of all the diversion links are merged, the *subjective optimal master hyperpath* Γ_{od}^{τ} is obtained.

DYBUSRT and BUSMEZZO use an *optimal master hyperpath* obtained by imposing restrictions (filters) on all the available paths, based on some exogenous evidence and following logical and behavioral constraints indicated by Bovy (2009). Logical constraints avoid loops, as well as successive boarding of the same run or the use of opposite lines and behavioral constraints eliminate unrealistic alternatives in terms of maximum values of attributes, such as number of transfers, transfer time, access and egress times and schedule delay. The residual paths are combined to obtain a master hyperpath Γ_{od}^{τ} from origin o to destination d, within which travellers use *run sub-hyperpaths*. In MILATRAS, travellers are assumed to consider a universal choice set, which is subsequently upgraded according to their experience using the transit system and to a 'mental model'.

6.3.5 Diversion Nodes and Dynamic Diachronic Run Hyperpaths

When a traveller arrives at diversion node i at time τ, the line sub-hyperpaths of the master hyperpath are upgraded with respect to the service state and the available predictive info, and the master hyperpath $\Gamma_{od}^{\tau,t}$ at time τ of day t is considered by the traveller. For example, info on disrupted lines allows such lines to be cancelled in the master hyperpath and info on arrival times of next lines allows adaptive sub-hyperpaths to be defined.

In order to carry out a diversion choice, a traveller considers sub-hyperpaths defined in terms of runs in a *space-time (diachronic) diagram* (see right part of Fig. 6.1), with space-time diversion nodes. Several diachronic sub-hyperpaths, in terms of runs represented in a time-space diagram, correspond to a master hyperpath, which is defined in terms of lines.

6.3.6 Diversion Link Choice Rule

When a traveller is at a spatial diversion node i (e.g., at a stop) at time τ of day t, in order to choose the best diversion link il, s(he) compares the *anticipated utilities* AU_{il} (defined in the following sub-section) up to the destination of the run sub-hyperpaths with runs of lines belonging to the path master choice set Γ_{od}^{τ}, including the alternative diversion links il and departing from node i up to the destination. The diversion link il belonging to the sub-hyperpath K with the maximum AU is then chosen.

For each diversion link il, more than one run hyperpath could be available, composing the hyperpath choice set $CS_{il}^{\tau,t}$. Therefore, in principle, a two-step procedure can be assumed. In the first step, the best run hyperpath $K_{il}^{*\tau,t}$ is chosen within $CS_{il}^{\tau,t}$; in the second step, the traveller, by comparing the best run hyperpaths $K_{il}^{*\tau,t}$ of the different links il (composing the choice set $CS_{i}^{\tau,t}$), chooses the best optimal run sub-hyperpath $K_{i}^{**\tau,t}$ departing from node i and hence the best diversion link il. Subsequently, the traveller moves to the next diversion node belonging to $K_{i}^{**\tau,t}$, where the choice procedure is again applied.

6.3.7 Anticipated Utility

The anticipated utility AU_K *of a run hyperpath* K_{il} takes into account the current state of the service network, past experience and the available predictive info. At time τ of day t, the anticipated utilities $AU_{K_{il,h}^{\tau,t}}^{\tau,t}$ of a run hyperpath $K_{il,h}$ made of $G_{il,h}$ elementary paths k, starting from node i up to the destination and including the diversion link il, can be written as a function of the *anticipated utilities of its paths k* as follows:

$$AU_{K_{il,h}^{\tau,t}}^{\tau,t} = \sum_{k=1}^{G_{il,h}} p\left[k / K_{il,h}^{\tau}\right] \cdot AU_k^{\tau,t} \tag{6.2}$$

where $p[k/K_{il,h}^{\tau}]$ is the use percentage of path k within hyperpath $K_{ij,h}^{\tau}$, perceived by the traveller, on the basis of the previous use, at time τ of the line master hyperpath and $AU_k^{\tau,t}$ is the anticipated utility of path k at time

τ of day t, which takes into account the past experienced utilities and the predictive info, if available.

The anticipated path utility AU_k is a function of the anticipated attribute vector *AX*, which considers the traveller's prior experiences and info, that are, in turn, functions of *forecasted* attributes *FX* and *perceived* attributes *PX*:

$$AU_k = AU_k(AX) = AU_k(FX; PX)$$

The j-th perceived attributes $PX_{k,j}$ at time τ of day t are the results of a learning process, which considers the traveller's prior experiences:

$$PX_{k,j}^{\tau,t} = \upsilon' \cdot TX_{k,j}^{\tau,t-1} + (1-\upsilon') \cdot PX_{k,j}^{\tau,t-1} \tag{6.3}$$

where:

- $TX_{k,j}^{\tau,t-1}$ is the attribute value of path k experienced at time τ of day $t-1$;
- $PX_{k,j}^{\tau,t-1}$ is the attribute value of path k perceived at time τ of day $t-1$;
- $v'(\in [0,1])$ is the weight given to attributes perceived on day $t-1$, depending on the memory process.

The anticipated attributes *AX* are hence obtained as:

$$AX_k^{\tau,t} = \xi \cdot FX_k^{\tau,t} + (1-\xi) \cdot PX_k^{\tau,t} \tag{6.4}$$

where:

- $AX_k^{\tau,t}$ is the attribute value, anticipated at time τ of day t;
- $FX_k^{\tau,t}$ is the attribute value forecasted by the info system at time τ of day t;
- $PX_k^{\tau,t}$ is the perceived attribute value at time τ of day t;
- $\xi(\in [0,1])$ is the weight given by the traveller to the information provided;
- $v(\in [0,1])$ is the weight given to attributes experienced on day t-1, depending on the memory process.

Note that it is assumed, in the learning process, travellers search for the optimal weight ξ that maximizes the subjective expected experienced utility.

Continuing in the example of Fig. 6.1 (left side), the traveller arriving at stop B at time τ of day t will choose between the two sub-hyperpaths incorporating a run of line 2 and/or 3, respectively, by comparing the anticipated utilities associated to each of the hyperpaths.

The general approach outlined above also allows us to consider different hypotheses on how travellers' decisions are impacted by information and on type of choice: path or strategy-based. For example, travellers may not consider the information provided ($\xi = 0 \Rightarrow AX = PX$) or, as non-frequent travellers on that origin-destination pair, they may consider only the info values provided ($\xi = 1 \Rightarrow AX = FX$).

Cats (2011) in BUSMEZZO and Nuzzolo et al. (2016) in DYBUSRT hypothesize that travellers choose to board the arriving line or to wait according to the maximum anticipated utility up to destination of each alternative. By contrast, MILATRAS, as presented in Chapter 5 (Nuzzolo and Comi, 2016), uses bounded rationality choice in a reinforcement learning algorithm, and thus it seems more suitable for use in an equilibrium-oriented simulation, rather than to simulate transit networks in a real-time context. Therefore, the path choice modelling approach of MILATRAS is not considered in the following behavioral assumptions and in the section of model formulation analysis.

6.3.8 At-origin and At-stop Diversion Choice

Travellers' choices at spatial diversion nodes can be classified into *at-origin* and *en-route*. The latter, in turn, could be made at each pedestrian node and at each boarding stop. Travellers might also change their alighting stops, depending on incoming real-time information (RTI) or on-board factors, such as whether they could obtain a seat (Schmöcker et al., 2011). For the sake of simplicity, given that pedestrian and on-board diversion rules are similar to those of origin, travellers' choices only at origin and at boarding stops are analyzed below.

At-origin diversion rule. According to the availability of real-time information provided to travellers moving on the O-D pair *od* with target time τ_{TT}, at origin the traveller is assumed:

(i) to consider the diversion links of the master hyperpath departing from origin *o*;

(ii) to generate, for each of the above diversion links, a choice set of run hyperpaths according to the master hyperpath and available predictive information;

(iii) to associate an *anticipated* utility $AU_{K_{ol,h}^{\tau,t}}^{\tau,t}$ to each hyperpath $K_{ol,h}^{\tau,t}$ ($\forall K_{ol,h}^{\tau,t} \in CS_{ol}^{\tau,t}$) within the set $CS_{ol}^{\tau,t}$ of all hyperpaths with the root in the same diversion link *ol*;

(iv) to compare the anticipated utilities $AU_{K_{ol,h}^{\tau,t}}^{\tau,t}$ of these hyperpaths;

(v) to consider for each diversion link the optimal hyperpath $K_{ol}^{*\tau,t}$ with the maximum anticipated utility

$$AU_{K_{ol}^{*\tau,t}}^{\tau,t} = max_{K_{ol,h}^{\tau,t} \in CS_{ol}^{\tau,t}} \left\{ AU_{K_{ol,h}^{\tau,t}}^{\tau,t} \right\};$$

(vi) to compare the optimal hyperpaths of each diversion link ol from the diversion node o and belonging to the optimal hyperpath set $CS_o^{*\tau,t}$;

(vii) to use the diversion link ol included in the absolute optimal hyperpath $K_{ol}^{**\tau,t}$ with the maximum anticipated utility:

$$AU_{K_{ol}^{**\tau,t}}^{\tau,t} = max_{K_{ol}^{*\tau,t} \in CS_o^{*\tau,t}} \left\{ AU_{K_{ol}^{*\tau,t}}^{\tau,t} \right\};$$

At boarding stop diversion rule: At time τ, when the traveller is at the (first boarding or at an interchange) stop i, and a run r belonging to the hyperpath arrives (as depicted in Fig. 6.2 below), s(he) is assumed:

(i) to consider two diversion links: the diversion link il_r to board run r, and the diversion link il_w to wait;

(ii) to generate, for each of the above diversion links, a choice set of run sub-hyperpaths according to the line master hyperpath and available predictive information;

(iii) to associate *an anticipated* utility to each sub-hyperpath with the root in the above two diversion links;

(v) to compare the anticipated utilities of these sub-hyperpaths $AU_{K_{ilr,h}^{\tau,t}}^{\tau,t}$ and $AU_{K_{ilw,h}^{\tau,t}}^{\tau,t}$;

(v) to consider the optimal sub-hyperpaths $K_{ilr}^{*\tau,t}$ and $K_{ilw}^{*\tau,t}$ with the maximum anticipated utilities among the sub-hyperpaths with the same root in the diversion link il_r and il_w, respectively;

(vi) to board the run r if

$$AU_{K_{il_r}^{*\tau,t}}^{\tau,t} > AU_{K_{ilw}^{*\tau,t}}^{\tau,t},$$

that is the anticipated utility $AU_{K_{ilr}^{*\tau,t}}^{\tau,t}$ associated to the optimal hyperpath $K_{ilr}^{*\tau,t}$ incorporating run r is greater than the anticipated utility $AU_{K_{ilw}^{*\tau,t}}^{\tau,t}$ of the optimal hyperpath $K_{ilw}^{*\tau,t}$ associated to waiting link il_w, which includes all other hyperpaths $K_{ilr'}$, incorporating runs r' not yet arriving. If the traveller does not board run r, the process is re-applied when the next run arrives. It should be noted that when the traveller is at a stop, in scarcely congested transit networks and common bus line situations, the first arriving run dominates all others.

6.3.9 Non-expected Utility

As introduced in Chapter 5, there is much empirical evidence against expected utility EU, e.g., Allais and Ellsberg paradoxes (Kahneman and Tversky 1979). Therefore, based on these experimental results, *non-expected* oriented utility theories have been developed (Starmer, 2000; de Palma et al., 2008; Ramos et al., 2014), of which the most widespread are

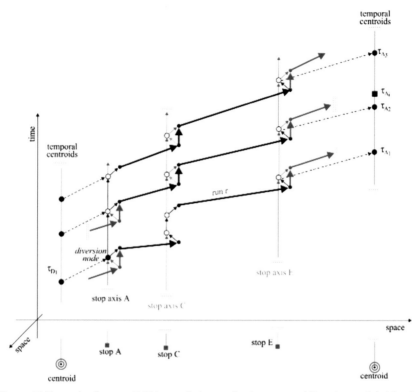

Figure 6.2 Example of sequential binary choice mechanism at stop (diversion node) *A* for the arriving run *r* (Nuzzolo et al., 2016).

prospect theory or PT (Kahneman and Tversky, 1979) and regret theory or RT (Bell, 1982). According to PT, decision makers frame possible outcomes as gains or losses based on a point of reference (e.g., status quo) and not according to final status as the classic interpretation of *EU* theory suggests. Besides, PT addresses non-linearity in the perception of probabilities. The theory has been widely used in the field of economics, but applications in transportation are relatively recent (Gao et al., 2010; Xu et al., 2011).

Similarly, regret theory (RT) has only recently been used in transportation (Chorus, 2012). RT is based on the fact that decision makers regret their decisions if they learn that, had they chosen differently, the outcome would have been better. In RT, choices are based on the anticipated performance (result) of the chosen alternative and on the possibility that the non-chosen alternative(s) outperform(s) the chosen one. In general, according to this theory, a user who has to choose either alternative a_i or alternative a_k would choose alternative a_i if and only if:

$$\sum_{j=1}^{m} P(\theta_j) \cdot R_{ik}(\theta_j) > 0 \qquad\qquad (6.5)$$

where $P(\theta_j)$ is the probability that status j occurs and $R_{ik}(\theta_j)$ is the regret of (or delight with) alternative i compared to alternative k, given the status θ_j. The papers by Ramos et al. (2014) and de Palma et al. (2008) are useful to obtain a complete overview of this topic and indications on how to use the non-expected approach in discrete choice modelling.

6.4 Path Choice Model Formulation

6.4.1 From Behavioral Assumptions to Model Formulation

From the previous behavioral assumptions, two types of run hyperpath anticipated utilities are involved in the link diversion choice. The first concerns the anticipated utility of optimal hyperpath K^* among the alternative hyperpaths available for the same diversion link il and composing the run hyperpath choice set CS_{il} of diversion link il. The second concerns the anticipated utility of the absolute best run hyperpath K^{**} among the optimal hyperpaths composing the optimal hyperpath choice set CS_i at diversion node i. The modeller knows neither the choice set of diversion links CS_i, nor the choice set CS_{il} of the hyperpaths of each link il (which depend on the line master hyperpath of the traveller), nor the hyperpath anticipated utility values used by travellers. Thus, the run hyperpath choice set modelling issue has to be tackled, as explored in the next section, where the anticipated utility values are assumed to be random variables, with the diversion link choice considered in terms of probabilities $p[il]$.

6.4.2 Existing Methods of Choice Set Modelling

Below three methods are recalled in order to tackle the problem that, in most cases, the choice sets (i.e., $CS_{il}^{\tau,t}$ and $CS_i^{\tau,t}$) considered by the decision maker are not known by the analyst. In the first approach, some analysts make the simplifying assumptions that the decision maker considers all potential alternatives in the universal choice set (i.e., set composed by all available options). However, there are many choice contexts where this is not a reasonable assumption. Examples are typically found in path choices where the number of alternatives evaluated is constrained by the individual's limited capacity for gathering and processing information.

In a different approach, the final observed traveller's choice is considered an outcome of two latent steps:

- choice set generation;
- choice of one alternative from the generated choice set, using a decision rule (e.g., utility maximisation).

In this two-stage framework, two methods can be applied. The *first two-stage method*, based on the theoretical framework proposed by Manski (1977), provides an explicit choice probability for each potential choice set. We need to enumerate all possible choice sets and compute probabilities of choosing different alternatives across all possible choice sets. Therefore, the probability $p[i]$ of choosing alternative i is explicitly modelled through the probability $p[C_m]$ of every possible choice set C_m, subset of the universal choice set C and the probability $p[i/C_m]$ of choosing an alternative i belonging to the choice set C_m:

$$p[i] = \sum_{C_m \subset C} p[i/C_m] \cdot p[C_m] \qquad (6.6)$$

Then, according to (6.6), for example, the probability $P[K_{il}^* = K_{il}]$, of choosing the run hyperpath K_{il} as the optimal K_{il}^* among those belonging to the different choice set $CS_{il,m} \subseteq CS_{il}$ can be expressed as:

$$p\left[K_{il}^* = K_{il} \right] = \sum_{CS_{il,m} \subset CS_{il}} p\left[K_{il}^* / CS_{il,m} \right] \cdot p\left[CS_{il,m} \right] \qquad (6.7)$$

where $p[K_{il}^*/CS_{il,m}]$ is the probability of the alternative hyperpath K_{il} being chosen given the choice set $CS_{il,m}$ (i.e., set of diversion links belonging to the set of hyperpaths considered by the traveller) and $p[CS_{il,m}]$ is the probability that the choice set $CS_{il,m}$ is considered among all non-empty subsets of the universal choice set CS_{il} (i.e., set of all run hyperpaths available from the diversion node where the traveller is).

Several models have been proposed for the specification and calibration of $p[CS_{il,m}]$. For example, Swait and Ben-Akiva (1987) provided a way to simulate $p[CS_{il,m}]$, incorporating random constraints in the discrete choice set generation. Ben-Akiva and Boccara (1995) proposed a latent choice-set model that incorporates in a single framework of choice set generation modelling the effects of stochastic constraints or elimination criteria and the influence of attitudes and perceptions on the choice set generation process. Swait (2001) generates the attractiveness of choice sets endogenously with the same attributes that determine the final choice. In the model, called 'Generation Logit', the choice set probabilities need make no use of exogenous information, but are determined on the basis of tastes and the choice context. The attribute thresholds are treated as exogenous variables.

An example of the application of Manski's method to transit networks is given by Schmocker et al. (2013) who specify and calibrate the hyperpath choice within the framework proposed by Swait (2001). They build a two-level model in which the upper level (choice set formation) is constrained by logit choice, while the lower level choice of a specific bus from the choice set is given by the frequency distribution of bus arrivals, assuming that travellers take the first. The model parameters are estimated using smart card data and the number of possible non-empty choice sets is quite limited.

In the Manski framework, enumeration of all possible choice sets can lead to an explosion as the number of alternatives in the universal choice set CS_{il} increases. For instance, even with just three alternatives in the universal choice set CS_{il}, the total number of possible latent choice sets is $(2^3 - 1 = 8)$ (assuming that there must be at least one alternative in the choice set). Although straightforward, it is not feasible to implement Manski's framework even for smaller universal choice sets.

Recently, modelling frameworks referred to as *'implicit choice set generation'* models were developed as an alternative to the Manski framework. Unlike Manski's model, the implicit choice set generation models work directly (in a single step) by making adjustments to the utility of alternatives in the universal choice set. These adjustments may be viewed as penalties that make an alternative less attractive if the probability of considering that alternative during decision making is low. The implicit choice set generation models in the literature vary according to the nature and functional form of the penalty imposed on the utilities (Cascetta and Papola, 2001; Martínez et al., 2009; Castro et al., 2013; Paleti, 2015).

In the *second two-stage method*, the choice set generation step is predominantly viewed as a non-compensatory mental process, while the second step of choosing from the considered set of options is assumed to be a compensatory one. In fact, experimental research suggests that in a complex choice situation (e.g., choosing among a large number of alternatives) decision makers adopt non-compensatory screening approaches (e.g., elimination-by-aspects, Twersky, 1972; Swait and Ben-Akiva, 1987; Pagliara and Timmermans, 2009; Zolfaghari et al., 2013) to reduce the number of alternatives to a smaller number before using a compensatory decision rule to make the final decision. The main approaches used refer to an attribute-based method where the decision maker is assumed to generate his/her choice set, by eliminating alternatives based on one or more constraints on the attributes of alternatives.

Despite these methodological developments, the methodology of route choice set generation is still open to debate.

The path choice models of the recalled simulation frameworks apply the above second two-stage method. As regards the choice sets CS_i and CS_{il}, in DYBUSRT, in order to simplify the procedure, the traveller is assumed to consider all diversion links il from node i of the master hyperpath, and for each diversion link il only the *inclusive run sub-hyperpath*, that is the sub-hyperpath of the master hyperpath. In BUSMEZZO, travellers choose in real time at each decision node (e.g., stops) considering the set of all feasible run paths, belonging to lines of the master hyperpath, from this node to destination.

6.4.3 Diversion Link Choice Probabilities

Once the master hyperpath is identified, at each diversion node, using hyperpath anticipated utilities, the traveller makes link diversion choices to reach the next diversion nodes, where the choice process is again performed.

In BUSMEZZO (Cats, 2014) path anticipated utilities are based on forecasted and experienced attributes and on logsum of utilities at downstream diversion nodes. In DYBUSRT, the anticipated run hyperpath utility is obtained through the anticipated path utilities of the run paths belonging to the hyperpath, weighted through the percentage of use in the previous simulation period. In accordance with the assumptions of the previous section, for each diversion link il, only the inclusive run hyperpath K_{il}^* is considered and thus the choice probability of the diversion link il (ol if the traveller is at origin) is obtained by assigning to each diversion link il ($\forall il \in L_i^{\tau,t}$), belonging to the set $L_i^{\tau,t}$ of all the diversion links from diversion node i, an anticipated utility $AU_{K_{il}^{*\tau,t}}^{\tau,t}$ equal to the utility of its inclusive run hyperpath K_{il}^*.

Therefore, the probability ($p[il]$) of using diversion link il is equal to the probability that the hyperpath $K_{il}^{*\tau,t}$ is considered the optimal one $K_i^{**\tau,t}$ among the optimal hyperpaths ($K_{il}^{*\tau,t}$; $\forall il \in L_i^{\tau,t}$) composing the choice set $CS_i^{*\tau,t}$ considered by the traveller:

$$p[il] = p\left[K_i^{**\tau,t} = K_{il}^{*\tau,t}\right] = prob\left[AU_{K_{il}^{*\tau,t}}^{\tau,t} > AU_{K_{il'}^{*\tau,t}}^{\tau,t}\right]$$
$$K_{il}^{*\tau,t} \neq K_{il'}^{*\tau,t}; \quad K_{il}^{*\tau,t}, K_{il'}^{*\tau,t} \in CS_i^{*\tau,t}$$

$$(6.8)$$

where $AU_{K_{il}^{*\tau,t}}^{\tau,t}$ and $AU_{K_{il'}^{*\tau,t}}^{\tau,t}$ are the values of anticipated utilities of hyperpaths $K_{il}^{*\tau,t}$ and $K_{il'}^{*\tau,t}$, respectively. Both in DYBUSRT and in BUSMEZZO, the probability expressed by eq. 6.8 is computed by a logit random utility model.

An application of the reported path choice model within the mesoscopic assignment included in DYBUSRT was carried out in Comi et al. (2016). The aim of the application was: (a) to compare the performance of the newly proposed model with that previously included in the former version of DYBUS; (b) to assess how different contributions of expected and forecasted utilities affect expected values of experienced utilities; (c) to analyze how these effects are influenced by stochasticity of services and forecasting precision. The results indicate that the optimal weights used for combining expected and forecasted hyperpath utilities can be estimated and that these weights strongly depend on features of the transit system.

6.5 Conclusions and the Road Ahead

The rapid evolution of information technology and telematics has led to the swift development of real-time information systems that assist travellers along their journeys. Opportunities are being opened up to set up new methods and models to support operations control decision making and to assess performances and impacts of RTI both from the travellers' and operators' perspective.

In this Chapter, a behavioral assumption and a model formulation framework were presented for transit path choice modelling with the descriptive travel strategy approach, for unreliable transit systems providing predictive info on service characteristics. Such modelling is suitable for real-time run-oriented simulation-based mesoscopic assignment models, which can support real-time predictive info and real-time operations control. The concepts of dynamic link diversion choice rule and of anticipated utility, as a combination of experienced and forecasted travel attributes, were introduced in a travel strategy behavioral context. Path choice model formulation, with the relative hyperpath choice set generation issue, was also explored. The new framework seems to be able to represent system dynamics both in terms of transit services and traveller choices. Test applications to real-size test networks showed the ability of this modelling framework to simulate the effects of providing real-time individual predictive information and to analyze how expected and forecasted utilities are influenced by stochasticity of services and forecasting precision.

The path choice models of three mesoscopic transit simulation systems presented in the literature, namely MILATRAS, BUSMEZZO and DYBUSRT, were also revisited and analyzed from the point of view of the above framework. In order to consider service network unreliability, all these simulation tools reproduce, explicitly or implicitly, the chosen path as the result of a travel strategy, but they present different modelling features with respect to model specification.

In terms of future perspectives, research should aim to develop more effective methods and models for path choice set generation. Besides, the presented modelling framework requires quite complex algorithms and several model parameters. Nevertheless, the new opportunities due to the availability of a large quantity of data derived from automated data collecting and bi-directional communication between travellers and information centers should allow model parameter estimation and upgrading to be more easily achieved in the near future.

Keywords: transit simulation models; unreliable transit networks; dynamic strategy-based path choice models; individual predictive traveller information; descriptive transit path choice models.

References

Bell, D. E. 1982. Regret in decision making under uncertainty. *In*: Operations Research 30(5): 961–981.

Ben-Akiva, M. E. and B. Boccara. 1995. Discrete choice models with latent choice sets. International Journal of Research in Marketing 12: 9–24.

Bovy, P. H. L. 2009. On modelling route choice sets in transportation networks: A synthesis. Transport Reviews: A Transnational Transdisciplinary Journal 29(1), 43–68 Taylor & Francis.

Cascetta, E. and A. Papola. 2001. Random utility models with implicit availability/perception of choice alternatives for the simulation of travel demand. Transport. Res. Part C 9(4): 249–263.

Castro, M., F. Martínez and M. A. Munizaga. 2013. Estimation of a constrained multinomial logit model. Transportation 40(3): 563–581.

Cats, O. 2014. An agent-based approach for modeling real-time travel information in transit systems. pp. 744–749. *In*: Procedia Computer Science 32, Elsevier.

Cats, O. 2011. Dynamic Modelling of Transit Operations and Passenger Decisions. Ph.D. thesis, KTH, School of Architecture and the Built Environment (ABE), Transport Science, Traffic and Logistics, Lindstedtsvägen 26, KTH, Stockholm.

Cats, O., H. N. Koutsopoulos, W. Burghout and T. Toledo. 2011. Effect of real-time transit information on dynamic path choice of passengers. Transportation Research Record: Journal of the Transportation Research Board, No. 2217: 46–54.

Chorus, C. G. 2012. Regret theory-based route choices and traffic equilibria. pp. 291–305. *In*: Transportmetrica 8(4), DOI:10.1080/18128602.2010.498391, Taylor and Francis.

Comi, A., A. Nuzzolo, U. Crisalli and L. Rosati. 2016. A new generation of individual realtime transit information systems. *In*: A. Nuzzolo and W.H.K. Lam (eds.). Modelling Intelligent Multi-modal Transit Systems, CRC Press.

de Palma, A., M. Ben-Akiva, D. Brownstone, C. Holt, T. Magnac, D. McFadden, P. Moffatt, N. Picard, K. Train, P. Wakker and J. Walker. 2008. Risk, uncertainty and discrete choice models. *In*: Market Letters 19: 269–285.

Florian, M. and I. Constantin. 2012. A note on logit choices in strategy transit assignment. EURO J. Transport. Logistics 1: 28–46.

Fonzone, A., J. D. Schmöcker, F. Kurauchi, S., Hassan, M. 2013. Strategy choice in transit networks. *In*: Proceedings of the Eastern Asia Society for Transportation Studies, Vol. 9.

Gallo, G., G. Longo, S. Nguyen and S. Pallottino. 1993. Directed hypergraphs and applications. Discrete Applied Mathematics 42: 177–201.

Gao, S., E. Frejinger and M. Ben-Akiva. 2010. Adaptive route choices in risky traffic networks: A prospect theory approach. *In*: Transportation Research Part C 18: 727–740.

Gentile, G. and K. Noekel. 2016. Modelling Public Transport Passenger Flows in the Era of Intelligent Transport Systems. Springer International Publishing.

Gentile, G., S. Nguyen and S. Pallottino. 2005. Route choice on transit networks with online information. Transportation Science 39(3): 289–297.

Hamdouch, Y. and S. Lawphongpanich. 2008. Schedule-based transit assignment model with travel strategies and capacity constraints. Transportation Research Part B: Methodological 42(7-8): 663–684.

Hamdouch, Y., H. W. Ho, A. Sumalee and G. Wang. 2011. Schedule-based transit assignment model with vehicle capacity and seat availability. pp. 1805–1830. *In*: Transportation Research Part B: Methodological 45(10).

Hamdouch, Y., W. Y. Szeto and Y. Jiang. 2014. A new schedule-based transit assignment model with travel strategies and supply uncertainties. Transportation Research Part B: Methodological 67: 35–67.

Kahneman, D. and A. Tversky. 1979. Prospect theory: An analysis of decision under risk. *In*: Econometrica 47(2): 263–291.

Kurauchi, F., J. D. Schmöcker, A. Fonzone, S. M. H. Hemdan, H. Shimamoto and M. G. H. Bell. 2012. Estimation of weights of times and transfers for hyperpath travellers. *In*: Transportation Research Records 2284: 89–99.

Lu, Y., F. C. Pereira, R. Seshadri, A. O'Sullivan, C. Antoniou and M. Be-Akiva. 2015. DynaMIT2.0: Architecture design and preliminary results on real-time data fusion for traffic prediction and crisis management. pp. 2250–2255. *In*: IEEE 18th International Conference on Intelligent Transportation Systems.

Ma, J. and D. Fukuda. 2015. A hyperpath-based network generalized extreme-value model for route choice under uncertainties. Transportation Research Part C 59: 19–31.

Manski, C. F. 1977. The structure of random utility models. Theory Decision 8(3): 229–254.

Martínez, F., F. Aguila and R. Hurtubia. 2009. The constrained multinomial logit: A semi-compensatory choice model. Transportation Research Part B 43(3): 365–377.

Nguyen, S., S. Pallottino and M. Gendreau. 1998. Implicit enumeration of hyperpaths in logit model for transit networks. *In*: Transportation Science 32(1): 54–64.

Noekel, K. and S. Weckcker. 2009. Boarding and alighting in frequency-based transit assignment. Proceedings of TRB 2009, Annual Meeting, Washington DC, USA.

Nuzzolo, A. and A. Comi. 2016. Real-time modelling of normative travel strategies on unreliable dynamic transit networks: A framework analysis. *In*: A. Nuzzolo and W. H. K. Lam (eds.). Modelling Intelligent Multi-modal Transit Systems, CRC Press.

Nuzzolo, A. 2016. Introduction to modelling multimodal transit systems in an ITS context. *In*: A. Nuzzolo and W. H. K. Lam (eds.). Modelling Intelligent Multi-modal Transit Systems, CRC Press.

Nuzzolo, A., U. Crisalli, L. Rosati and A. Comi. 2016. A mesoscopic transit assignment model including real-time predictive information on crowding. *In*: Journal of Intelligent Transportation Systems: Technology, Planning, and Operations, DOI: 10.1080/15472450.2016.1164047, Taylor & Francis.

Pagliara, F. and H. J. P. Timmermans. 2009. Choice set generation in spatial contexts: A review. Transportation Letters: The International Journal of Transportation Research 1: 181–196.

Paleti, R. 2015. Implicit choice set generation in discrete choice models: Application to household auto ownership decisions. Transportation Research Part B 80: 132–149.

Ramos, G. M., W. Daamen and S. Hoogendoorn. 2014. A state-of-the-art review: Developments in utility theory, prospect theory and regret theory to investigate travellers' behaviour in situations involving travel time uncertainty. *In*: Transport Reviews: A Transnational Transdisciplinary Journal 34(1): 46–67, DOI: 10.1080/01441647.2013.856356.

Ren, H. and W. H. K. Lam. 2007. Modeling transit passenger travel behaviors in congested network with en-route transit information systems. Journal of the Eastern Asia Society for Transportation Studies 7: 670–685.

Rochau, N., K. Nökel and M. G. H. Bell. 2011. A strategy-based transit assignment model for unreliable network. *In*: Proceedings of European Transport Conference.

Schmöcker, J. D., H. Shimamoto and F. Kurauchi. 2013. Generation and calibration of transit hyper-paths. Transportation Research C 36: 406–418.

Schmöcker, J. D., A. Fonzone, H. Shimamoto, F. Kurauchi and M. G. H. Bell. 2011. Frequency-based transit assignment considering seat capacities. *In*: Transportation Research Part B. 45(2): 392–408.

Spiess, H. and M. Florian. 1989. Optimal strategies. A new assignment model for transit networks. Transportation Research. Part B: Methodological 23B(2): 83–102.

Starmer, C. 2000. Developments in non-expected utility theory: The hunt for a descriptive theory of choice under risk. *In*: Journal of Economic Literature 38(2): 332–382.

Sumalee, A., Z. J. Tanm and W. H. K. Lam. 2009. Dynamic stochastic transit assignment with explicit seat allocation model. Transportation Research Part B 43, Elsevier, 895–912.

Swait, J. 2001. Choice set generation within the generalized extreme value family of discrete choice. Transportation Research 25B: 643–666.

Swait, J. and M. Ben-Akiva. 1987. Incorporating random constraints in discrete models of choice set generation. Transportation Research B 21: 91–102.

Tversky, A. 1972. Elimination by aspects: a theory of choice. Psychol. Rev. 79: 281.

Von Neumann, J. and O. Morgenstern. 1947. Theory of Games and Economic Behavior. Princeton: Princeton University Press.

Wahba, M. and A. Shalaby. 2009. MILATRAS. A new modelling framework for the transit assignment problem. pp. 171–194. *In*: N. H. M. Wilson and A. Nuzzolo (eds.). Schedule-Based Modeling of Transportation Networks: Theory and Applications, Kluwer Academic Publisher.

Wahba, M. and A. Shalaby. 2014. Learning-based framework for transit assignment modeling under information provision. pp. 397–417. *In*: Transportation 41(2), Springer.

Xu, H., J. Zhou and W. Xu. 2011. A decision-making rule for modeling travelers' route choice behavior based on cumulative prospect theory. *In*: Transportation Research Part C 19: 218–228.

Zolfaghari, A., A. Sivakumar and J. Polak. 2013. Simplified probabilistic choice set formation models in a residential location choice context. The Journal of Choice Modelling 9: 3–13.

Time-dependent Shortest Hyperpaths for Dynamic Routing on Transit Networks

Guido Gentile

ABSTRACT

This chapter introduces a unified framework for the computation of shortest hyperpaths on a public transport network with time-varying performance, which ii due to road traffic and passenger congestion features like service frequency, running times, waiting times, on-board comfort and line regularity change noticeably during the day, especially in urban systems.

Even where a fixed timetable exists, it may be unknown to the passenger or be not satisfied in practice. So, from the user perspective, the transit service results in a mix of schedule-based and frequency-based lines. Passengers will then optimize their route choice based on the available information about service performance adopting a strategic behavior with n-trip diversions in reaction to random events.

The classical static case, where the time dimension is neglected, as well as the case of simple paths, where no strategic behavior

Dipartimento di Ingegneria Civile, Edile e Ambientale, University of Rome 'La Sapienza', Via Eudossiana, 18 – 00184 Rome, Italy.
E-mail: guido.gentile@uniroma1.it

is considered, are seen as two particular instances of a more general dynamic routing problem. Both discrete and continuous representation of time are considered here. Two solution approaches are presented, namely; user trajectories and temporal layers. Extensions to departure (or arrival) time choice and to multimodal networks (e.g., with park and ride) are also provided. The proposed methodology can be applied in the context of dynamic transit assignment, as well as in the context of point-to-point navigation for passenger trips.

7.1 Introduction

This chapter on dynamic routing presents the formulation of a general search problem for shortest (hyper)paths on networks with time-varying arc performance (travel times and generalized costs), as well as the specific case of transit networks with discontinuous services, random headways and strategic passenger behavior.

The focus is on the theoretical context of the proposed methods, while the algorithms are presented with the aim of describing the solution approach. Therefore, numerical tests and experiments are out of scope. Moreover, no conclusion section is provided. The main achievements are summarized at the end of this introduction in a list of original contributions.

7.1.1 Motivations

The vast majority of the literature regarding shortest path search on transport networks addresses the classical static case, where the problem is to find the cheapest route connecting an origin node to a destination node on a graph with given arc costs that are constant in time.

As such, there are not so many contributions on time-dependent shortest paths (more papers are available on dynamic shortest path, that usually refers to re-computation after arc cost updates on a static network). This version of the problem addresses a more realistic situation where network performances vary during the day. In this case, every arc is characterized by two distinguished temporal profiles (i.e., functions of time)—one for the generalized cost and the other for the travel time. Besides being a major component of generalized costs travel times are indeed crucial to properly concatenate the entry instant of the user on each arc of a path and then evaluate the disutility perceived during that trip.

In this chapter we will highlight the time-dependent shortest path problem (which does not have an immediate definition) into the dynamic routing problem (DRP) with multiple roots and targets. We want to emphasize the fact that many more factors than just a cost/distance attribute shall be taken into account to provide an optimal solution for a passenger trip on an urban network; among which are included random events and strategic behavior.

We will show how the DRP can be addressed in principle through specific applications of static solution approaches to a space-time network. However, there are several questions that make this problem deserve attention and consideration, especially if we refer to large transit networks with thousands of lines and runs, where efficiency becomes a relevant issue. For example, journey planners must provide the solution to a passenger request in a few seconds.

In particular, the arc temporal profiles for waiting at stops may be characterized by sharp variations and discontinuities. Think of the coexistence of frequency-based and schedule-based services, where a bus delay of few minutes can hinder catching a favorable train connection and make the passenger wait for several additional minutes until the next run, or even make him reroute to another line.

7.1.2 Classical Algorithms for Static Networks

Before introducing dynamic routing, we briefly review the most established methods for shortest path computation when arc costs are constant in time.

Consider a directed graph (N, A) with nodes N numbered from 1 to $|N|$ whose generic arc $ij \in A$ has a cost c_{ij}. The aim of a (static) shortest path algorithm is to find the sequence of arcs with the minimum total cost that connects one node $i \in N$ to another node $j \in N$. Most methods iteratively improve an estimation of this cost, denoted w_{ij}, until the optimal value is reached. Therefore, they address the dual problem (which has a unique solution) and don't return (directly) by the shortest path itself (which may be not univocal).

Shortest tree methods find in the same procedure all shortest paths from origin (or root or source) $o \in N$ to all possible destinations (or targets) $d \in N$ by repeatedly applying the following update formula for each arc $ij \in A$ until no further improvement is achieved:

$$w_{oj} \leftarrow Min(w_{oj}, w_{oi} + c_{ij}). \tag{7.1}$$

This local condition, called Bellman optimality (Bellman, 1957), ensures that the global solution is found. The route cost w_{oi} of each node $i \in N$ is initialized to infinity, while w_{oo} is set to zero.

Most methods apply the above scheme. The existing algorithms then differ in the visiting order of arcs for possible update.

Dijkstra's algorithm selects the node with minimum route cost that has not yet been processed and performs the Bellman update (7.1) for all its outgoing arcs. If arcs have positive costs, this ensures a Label Setting algorithm, where nodes are processed only once because the route cost of the updated node j is higher than that of the processed node i. The original algorithm (Dijkstra, 1959) keeps the nodes with finite cost that have to be processed in a simple set, each time extracting that with minimum cost; it then runs in $O(|N|^2)$. The implementation based on a priority queue with a Fibonacci heap (Fredman and Tarjan, 1984) runs in $O(|A|+|N| \cdot Log(|N|))$; this is asymptotically the fastest known shortest tree algorithm for arbitrary directed graphs with unbounded non-negative costs.

In the particular case of acyclic graphs, by processing the nodes in any topological order yields a Label Setting algorithm that runs in $O(|A|)$.

By contrast, Bellman-Ford-Moore algorithm (Bellman, 1958; Ford, 1956; Moore, 1959) simply applies the update formula (7.1) for all the arcs and does this $|N|$-1 times (in the worst case); it runs then in $O(|A| \cdot |N|)$. This is a label correcting algorithm, where arcs are visited more than once (unlike Dijkstra); this allows some negative arc costs, but with no negative cycles. The method performance can be substantially improved in practice (although not in the worst case) by reducing the number of updates within each iteration. In particular, until node i is improved again, there is no need to visit its outgoing arcs (Yen, 1970). So we can introduce a FIFO list of nodes to be processed. A variant of this method is the LDEQUE algorithm (Pape, 1974), where the improved node is inserted in front of the list if it was already processed and (as normal) in the back, otherwise. This version runs though exponentially in $O(|N| \cdot 2^{|N|})$. In a second version of the same algorithm, called L2QUE (Gallo and Pallottino, 1988) and running in $O(|A| \cdot |N|^2)$, the queue is disjoined in two FIFO lists— one for already processed nodes, and one for first labelled nodes; nodes are extracted preferably from the first list if it is not empty.

Despite the higher complexity, label correcting usually outperform label setting algorithms on transport networks. However, the label setting algorithms are most suited for optimal strategies on transit networks (see Section 7.3.5). Moreover, if arc costs are bounded and positive, it is possible to implement label setting through a list with n buckets (see Section 7.4.2) running in $O(|A| + n)$.

Different to tree-based methods, the Floyd-Warshall algorithm (Floyd, 1962; Warshall, 1962) finds the cost of the shortest path between all pairs of nodes running in $O(|N|^3)$. Let w_{ijk} denote the minimum cost from $i \in N$ to $j \in N$ using only the first k nodes as intermediate points along the way. Given w_{ijk} for each node pair $ij \in N \times N$, our goal is to find $w_{ij\,k+1}$. This could be given by either a path that only uses the first k intermediate nodes or a path that goes from i to $k+1$ and then from $k+1$ to j. Therefore, it is:

$$w_{ij\,k+1} \leftarrow Min(w_{ijk}, w_{i\,k+1\,k} + w_{k+1\,j\,k}) \ \forall ij \in N \times N. \qquad (7.2)$$

The algorithm is initialized with:

$$w_{ij0} \leftarrow c_{ij}, \text{ if } ij \in A, w_{ij0} \leftarrow \infty, \text{ otherwise } \forall ij \in N \times N, \tag{7.3}$$

and the process continues iterating (7.2) for $k = 1, 2, \ldots, |N|$, until we have found the minimum cost w_{ij} for all pairs $ij \in N \times N$ using any intermediate node: $w_{ij} = w_{ij|N|}$.

Taking the above paradigms as a basis, the latest methodological efforts on the computation of shortest paths are largely devoted to the acceleration of point-to-point cost estimation applied to journey planners for continental road networks (Bast et al., 2015).

7.1.3 State-of-the-art on Algorithms for Dynamic Networks

Most of the scientific literature on time-dependent shortest paths focusses on *discrete models*, where it is assumed that the time dimension is a finite set of instants. Consistently, the exit time from each arc evaluated at one of these instants shall fall exactly in a later one of such instants.

In this case, the time-dependent shortest paths problem can be addressed by introducing the so-called *diachronic graph*, where each node of the transport network is replicated into a vertex for every instant and an edge connects the vertex relative to the initial node of each arc at every possible entry instant with the vertex of its final node at the corresponding exit instant. Waiting at nodes can be allowed or not; in the former case waiting edges are introduced to connect the consecutive vertices in time of the same node. The dynamic concatenation of travel times is thus ensured by the topology of the diachronic graph, while any generalized cost can be associated to each edge. The dynamic problem is then reduced to a static one and any shortest path algorithm can be directly applied on the diachronic graph.

Moreover, the resulting *space-time network* has no cycles and a chronological ordering of its vertices provides a topological order. This makes the problem even easier than in the static case.

Exploiting this property Pallottino and Scutellà (1998) proposed to find a shortest tree from an origin vertex (given a node and a departure time) whose label is initialized to zero, by performing a chronological visit where only the labelled vertices are processed. To this end, one can record the nodes for each (later) temporal layer whose cost is updated (from infinite to finite). The shortest path to a given destination can be found ex-post by finding the minimum cost vertex for that node.

Often in dynamic traffic assignment the time-dependent shortest path problem is solved for every departure time from each origin towards the same destination. For this case, Chabini (1998) proposed an algorithm which solves all such instances jointly in an optimal running time. The method finds the next edge from each vertex to reach the destination at a later time with the minimum cost (the vertex label). This is achieved by processing vertices in reverse chronological order. The Bellman update is in this case still applied to the outgoing edges but in a reverse way, with the aim of improving the label of the processed vertex, by comparing it with the sum of the edge cost and the label of its head. The label of destination vertices is initialized to zero.

The major drawback of the methods for discrete models based on the paradigm of the diachronic graph is the need to handle multiple labels for each node, one for every instant. Moreover, a large number of instants with short intervals is required to reproduce the arc travel times with suitable accuracy (usually one second, in urban contexts). This results in simple but time-consuming procedures, especially in the case of journey planning, where the computation is requested only for one (or few) departure instant.

In *continuous models*, temporal profiles for travel times and generalized cost are associated to each arc of a directed graph that represents only the space topology of the transport network. However, exact solutions (Orda and Rom, 1990) are still infeasible for large networks or require further assumptions. For example, Ding et al. (2008) propose a Dijkstra-like method for finding the least travel time to reach a destination node, departing from an origin node at any time of an interval.

In Gentile and Meschini (2004) we explore the possibility of allowing some approximation when operating on the label temporal profiles, with the aim of achieving fast heuristics. Specifically, we deal with piecewise linear functions of time defined on a pre-specified set of instants with long intervals (several minutes) and keep their form during the procedure, accepting the resulting (small) distortions. This approach is further developed in this chapter.

A shortest path is said to satisfy the concatenation property if its two sub-paths from the origin to any intermediate node and from this node to the destination are themselves the shortest paths. In general, this does not hold in the dynamic context referring to the transport network (it does on the space-time network). However, if the FIFO rule is satisfied for each arc, i.e., no overtaking can occur on links, then the concatenation property holds true on the transport network with respect to travel times.

In this case, indeed, arriving earlier at an intermediate node not only maximizes the available opportunities to follow on towards the destination, but also minimizes the cost of reaching that node. We can then consider the simpler problem to find a time-dependent fastest path (which cannot contain space cycles), i.e., a path with the least travel time, directly on the transport network, without introducing the space-time network. Based on the FIFO rule, any classical algorithm with only one label for each node can be applied in order to find the fastest trajectory. This special case is attracting considerable attention since a long time, especially for highway applications (Kaufmann and Smith, 1993; Ziliaskopoulos and Mahmassani, 1993; Sung et al., 2000) and transit applications (Cooke and Halsey, 1966; Dreyfus, 1969; De Palma et al., 1993; Nachtigall, 1995).

This specific problem has recently been the object of further research to implement some of the acceleration methods that are available for static networks (Baum et al., 2015).

7.1.4 Dynamic Strategies on Transit Networks

In this chapter we address the specific case of public transport networks, where the service provided by transit lines is intrinsically discontinuous in time.

Two behavioral approaches are considered to represent how passengers wait for transit lines at a stop—frequency-based and scheduled-based (for the dynamic case, see: Meschini et al., 2007; Papola et al., 2009). In the first case, the arrival of transit carriers at the stop are so frequent or so irregular/unreliable that the passengers see no advantage in timing their arrival to the stop with the runs of the line schedule, which are instead considered in the second case.

The presence of random recurrent events on transit networks induces passengers to adopt a strategic behavior. Passenger congestion (Bellei et al., 2000) tends to equilibrate path costs and thus further increases the convenience of this approach to route choice. Consider, for instance, the line headway distribution at stops of frequency-based services, the fail-to-seat probability to find a free chair in a crowded vehicle, or the fail-to-board probability of a passenger mingling on a crowded platform to get on the next arriving run.

As a result, the route to travel from the origin to the destination chosen by a rational decision-maker cannot anymore be represented on the graph by a simple path, but becomes a hyperpath, with diversion nodes and arc probabilities (Gallo et al., 1993). Nevertheless, in a given day, the passenger will actually use a certain path among those included in the hyperpath, depending on the outcome of the events.

Moreover, if early information is available in real-time on these events (e.g., through smartphone applications for passenger navigation), the traveller can anticipate diversions and the spectrum of the optimal strategy may be enlarged to include paths that are rarely used in practice, but may turn useful in particular situations (Noekel and Wekeck, 2009).

The computation of shortest hyperpaths on dynamic networks is still considered an open problem (Bell et al., 2012; Trozzi et al., 2013). Here, we will address two specific issues that arise in this context:

- conditional exit times of hyperarc branches need to be explicitly calculated;
- the impossibility of choosing the arrival time conflicts with the need of calculating hyperpaths backward from the destination.

On dynamic networks, where travel time matters, it is not sufficient to calculate a combined cost for the hyperarc (the local representation of a strategy) and a diversion probability for each one of its branches. We need also to calculate the conditional time of each branch for the concatenation of the dynamic route choice (Gentile and Noekel, 2016). Indeed, when events induce to take a particular diversion arc its (conditional) exit time for a given entry time is usually not the same as of other branches.

Take the case of a passenger waiting at the stop for the first one of two lines with (deterministic) headways of 5 minutes and 50 minutes, respectively. When the passenger boards the second line, it means that the first line is not arrived yet. This has a linearly decreasing probability with the wait time that goes to 0 after 5 minutes (with a triangular shape). Thus, the conditional wait time for the second line is 5/3 minutes. When the passenger boards the first line, it means that the second line is not arrived yet. This has a linearly decreasing probability with the wait time that goes to 0 after 50 minutes (and is then almost flat during the first 5 minutes). Thus, the conditional wait time for the first line is slightly lower than 5/2 minutes.

When using frequency-based lines, a passenger can be sure only of his departure time from the origin, while the arrival time at the destination is not univocal and has a (non-trivial) random distribution. Given the stochastic nature of transit services, he can then choose only the average arrival time. Moreover, when adopting a strategic behavior, each path of the chosen hyperpath has its own average travel time, which implies that the distribution of arrival times is in general not monomodal.

Nonetheless, computation of the shortest hyperpaths shall be done starting from the destination and proceeding backward in space towards the origin, because a forward hyperarc can be used to update the label of its tail given the labels of its heads, but not the contrary (the head labels could not be updated with probability equal to one).

Because, as it was shown above, the conditional travel times are different among the branches of a same hyperarc, in dynamic hyperpath computations we can update in principle the label (cost to destination) of the tail at a given instant based on the labels of the heads at various instants.

7.1.5 The Coexistence of Frequency-based and Schedule-based Services

If the headway distributions have tails (like in the classical exponential case), there is no upper bound to the passenger waiting time at the stop. This hinders the possibility of travelling on a network in the presence of both schedule-based and frequency-based services, because it becomes impossible for the passenger to properly use the connections between lines.

So we shall then assume that the waiting time distributions have an upper bound, i.e., there exists a maximum headway. A distribution with an infinite tail can be modified to meet the above requirement by dividing the original probability density function by the complement of the original distribution function calculated at the desired maximum headway and truncate it after this value. A way of choosing the maximum headway can be that of assuming it to be equal to twice the expected waiting time of the original distribution, like for regular lines. This is what happens if the headway stochasticity derives from disturbances along the line up to the vehicle pairing, but the departures from the terminal are regular, with the maximum waiting time being roughly equal to twice the headway, i.e., twice the expected waiting time of the exponential distribution that has the highest irregularity (with a variation coefficient equal to one).

Moreover, it is wrong to consider the average waiting time for the concatenation of dynamic routing, because wait times longer than the average can bring to missing connections. In practice, delays have not only a local consequence in terms of disutility, but also a global effect in terms of the opportunities actually available to complete the current trip. How should the cost of a given route (path or hyperpath) be calculated in the presence of random travel times?

In this case, the exact dynamic calculation of the expected path (or hyperpath) cost would be very cumbersome. In theory, we should consider the joint distribution of all events and for each possible outcome of different events (e.g., the wait time for each attractive line at each stop of the route) we should compute the cost and time of the used path (depending on which lines arrive first at the stops), fully considering the dynamic concatenation. This computation for each joint outcome may result in large cost differences in the presence of scheduled services because the passenger may have to wait for a later run of a line (perhaps the next day, leading to a modelling paradox).

Consider the case of a bus plus train connection. The bus can arrive after 10, 20 or 30 minutes. For convenience, in this exercise let's assume a discrete distribution, where each outcome has the equal probability of 1/3. Given the departure time of the passenger, let's assume that in the first two cases he is able to get a given run of the train, otherwise in the third case he will have to wait additional 30 minutes for the next run. Assuming that in the second case the passenger enjoys a perfect coincidence (i.e., the train headway is 40 minutes), then the total wait time for the three distinguished outcomes of the bus wait are: 20 minutes (10 plus 10), 20 minutes (20 plus 0), and 60 minutes (30 plus 30). Therefore the exact expected value of the total wait time is 33 minutes. It would be wrong to consider an average of 20 minutes wait time for the bus and assume that this allows him/her to catch for sure the early run of the train, with a total wait time of 20 minutes.

Note that in the presence of time-varying performances, the need of computing the expected travel times of a dynamic route by separately considering each possible joint outcome of the random events applies also when all services are frequency-based.

The calculation of the exact expected cost may become a computational nightmare and lead to a combinatorial problem, which is practically impossible to solve on real networks for trips with several transfers. Moreover, this is probably not what the passenger can conceive. We then have to opt for a heuristic, possibly based on some behavioral consideration.

Note that it is possible to build up other examples where the approximation of using the average travel time for the dynamic concatenation leads to a pessimistic evaluation (whereas in the proposed case it is optimistic). However, considering the lesson of the Prospect Theory, where it is shown how people are much more sensitive to losses than to gains, we prefer to adopt a risk adverse approach.

A totally risk adverse behavior would consider only the worst case in the evaluation. For our example, the total wait time is in this case equal to 60 minutes (30 plus 30).

We believe that such an approach is too conservative or pessimistic and leads to travel cost estimations that are too high. Therefore, we propose here a suitable compromise: the generalized cost of the arc (or hyperarc) is calculated as usual, considering its expected value and not its maximum value, plus only a share of the difference; but the (conditional) travel time of each (diversion) arc is calculated considering its upper bound. In other words, the passenger is conservative when considering the dynamic concatenation of travel times. This way he will never miss the connections, but he is more risk prone in the overall evaluation of the solution (risk adverse components can in any case be introduced on arcs to penalize variances). For our example, the total wait time is equal to 50 minutes (20 plus 30).

This also allows the possibility to strengthen the importance of transit service regularity because in the proposed framework, it plays a crucial role for the concatenation of dynamic route choice and not only for disutilities. Basically, poor regularity leads to losing connection opportunities, besides increasing the expected waiting times.

Recent studies on the robustness of transport networks take explicitly into account travel times uncertainties (Yu and Yang, 1998; Montemanni and Gambardella, 2004). In this chapter a heuristic approach is adopted with the aim of tackling this issue within a more general framework.

7.1.6 Contributions

This chapter presents the many innovative aspects in the domain of dynamic routing:

- a unified framework for several dynamic routing problems and algorithms;
- a deep analysis of strategies and hyperpaths in the framework dynamic networks;
- a new approach to random events with distinction between travel times used for performances and exit times used for concatenation;
- several implementations, including the temporal-layer and user-trajectory approaches, derived from one general algorithm;
- a new multi-label algorithm is proposed which minimizes the computational effort for journey planning under FIFO assumptions in the presence of tolls and fares;
- an application to the case of coexistent frequency-based and schedule-based services.

The specific original contribution of this chapter focuses mainly on the graph visiting mechanisms in shortest path problems, with specific attention given to the case of time-varying arc performance. In particular we will outline a general framework for dynamic routing and two main algorithm specifications—one suited to address route choice in dynamic traffic or transit assignment, where we are interested in solving the shortest tree problem towards one destination from all possible origins and for all possible departure times at once (temporal-layer approach); another suited for point-to-point navigation, where we have to solve the fastest path problem (generalized costs are proportional to travel times) for one arrival time only (user-trajectory approach).

Moreover, a new algorithm that shares features from both approaches is proposed, and it requires handling a Pareto frontier of non-dominated points for each node in terms of generalized costs and travel times.

Contributions include the extension of these three specifications to:

- the case of continuous time modelling;
- the case of hyperpath and strategies;
- the case of intermodal routing;
- the case of departure or arrival time choice.

The chapter is organized as follows. In Section 2, the variables of the proposed general framework for routing computation are introduced, including the space-time network to represent the time-varying performances and the concept of hyperarcs to represent strategies in travel behavior. In Section 3, the dynamic routing problem is formulated and the basic solution algorithm is presented together with its extensions for continuous time modelling, departure time choice representation, intermodal routing and hyperpaths. In Section 4, the some suitable solution approaches are presented, including algorithm implementation details. In Section 5, a specialization of the new multi-label algorithm is applied to journey planning on transit networks with a mix of schedule-based and frequency-based lines; this way we show the applicability of the proposed methodological framework to the case of discontinuous services.

7.1.7 Future Research

The approach proposed in this chapter to deal with the randomness of travel times is based on risk adverseness. In essence, passengers anticipate departures, so that they are (almost) sure to catch transfers, consistently with the correctness of the assumption to consider twice the expected time when calculating the dynamic concatenation of events. Clearly, in the case of headway distributions with tails (e.g., exponential headways), they cannot be 100 per cent certain.

This leads to the idea of stochastic shortest paths (Samaranayake et al., 2012), where the objective is to maximize the probability that a passenger arrives on time at a destination given a departure time and a time budget. This problem has not been yet addressed for transit networks in coexistence with frequency-based and schedule-based lines, and will thus be the object of future research.

7.2 A Mathematical Framework for Dynamic Routing and Strategies

In this section, a unified framework for dynamic routing computation is presented without specific reference to transit networks. A space-time network is built up from the base transport network to represent the time of the day. Continuous time models as well as hyperpath-based models, that allow to consider strategic behavior, will then be presented as an extension of this discrete framework.

7.2.1 Topology

The transport network is represented here through a directed graph (N, A), where N is the set of the *nodes* and $A \subseteq N \times N$ is the set of the *arcs*. Thus, the generic arc $a \in A$ is an ordered couple of nodes. Its initial node, referred to as the *tail*, is denoted $a[-] = a^- \in N$, while its final node, referred to as the *head*, is denoted $a[+] = a^+ \in N$. The set of arcs exiting from a given node $i \in N$ is referred to as its *forward star* and denoted $i^+ = \{a \in A: a^- = i\}$, while the set of arcs entering into a given node $i \in N$ is referred to as its *backward star* and denoted $i^- = \{a \in A: a^+ = i\}$.

The time dimension is represented here as a chronologically ordered list T of discrete *instants*, with integer indices from 0 to η, which cover a bounded *simulation period* (e.g., all seconds in one day). The *clock time* of the generic instant $t \in T$ is denoted $\tau_t \in \Re$, and $T = \{\tau_t \in \Re: t \in T\}$ is their set.

The generic interval $t \in T\text{-}\{0\}$ spans from time τ_{t-1} to time τ_t.

7.2.2 Performance

We associate to each arc and instant three main performance attributes—the travel time, the exit time and the generalized cost. Indeed, in dynamic networks it is important to know how long it takes to travel along an arc and not only how much disutility is paid by the user from the tail to head.

No time or cost is suffered by the users to traverse a node. The representation of turn delays and permissions requires then to explode the node topology (or to adopt a dual graph). But this is not important in the case of public transport where the proposed framework will be applied.

In dynamic models, for a user entering the generic arc $a \in A$ at instant $t \in T$,

- t_{at} – the travel time;
- $\theta_{at} \in T$ – the exit time;
- c_{at} – the generalized cost.

In the presence of random events, as explained in the introduction, we assume that t_{at} derives from a neutral evaluation of the arc travel time (e.g., its expected value), while the exit time θ_{at} derives from a risk adverse evaluation of the arc travel time (e.g., its upper bound). Then, θ_{at} differs from $(\tau_t + t_{at})$ and in general, we have:

$$\theta_{at} \geq \tau_t + t_{at}. \tag{7.4}$$

To allow the representation of discontinuous transport services (e.g., schedule-based lines) and to manage the boundary of the simulation period, the above variables are not defined for each instant $t \in T$. We assume in general that each arc $a \in A$ offers a connection from its tail, leaving only at some:

- $T_a \subseteq T$ *availability instants.*

By convention, it is: $t_{at} = \theta_{at} = c_{at} = \infty$ if $t \notin T_a$.

If the service is *discontinuous*, T_a is a specific sub-set of instants (e.g., the departure times of runs); if the service is *continuous*, T_a is every possible instant such that the corresponding exit time falls within the simulation period.

It is useful to introduce the following *index function*:

- $\theta(a, t) = e \in T : \theta_{at} = \tau_e$, if $t \in T_a$; $\theta (a, t) = \eta + 1$ otherwise (by convention).

We assume the fundamental property that travel times are strictly positive:

$$\theta_{at} > \tau_t \text{ and } \theta(a, t) > t, \forall t \in T_a. \tag{7.5}$$

Waiting is allowed only at a possibly empty subset of nodes $N^{wait} \subseteq N$. For each *waiting node* $i \in N^{wait}$ and instant $t \in T$:

- c_{it} the *waiting cost* at node i from time τ_t unit the next instant $\tau_{t+1} \in T$; $c_{i\eta} = \infty$.

The above costs usually depend on the class of user making the trip. Without loss of generality, we can assume:

$$c_{at} = \gamma^{vot} \cdot \gamma_a \cdot t_{at} + c_{at}{}^{nt} + \gamma^{risk} \cdot \gamma^{vot} \cdot (\theta_{at} - \tau_t - t_{at}) \tag{7.6}$$

$$c_{it} = \gamma^{vot} \cdot \gamma_i \cdot (\tau_{t+1} - \tau_t) \tag{7.7}$$

All the following parameters are class (or user) specific:

- γ^{vot} – is the *value of time;*
- γ_a – is the *discomfort coefficient* of arc a;
- γ_i – is the *discomfort coefficient* of node i;
- $c_{at}{}^{nt}$ – is the *non-temporal cost* of arc a (e.g., the monetary cost);
- γ^{risk} – is the *risk adverse coefficient.*

The passenger then suffers only a share $\gamma^{risk} \in [0, 1]$ of the additional travel time he has to consider as a margin to achieve concatenation without risks.

The reference to the class is omitted, under the assumption that the dynamic routing is computed for a given user considering his specific costs.

7.2.3 The Space-time Network

We will analyze the problem of dynamic routing in a *space-time* network represented through a directed graph (V, E), where $V \subseteq N \times T$ is the set of *vertices* and $E \subseteq V \times V$ is the set of *edges*. The space-time network is also called *diachronic graph*. Note that this approach is mainly useful for presentation of the methodology; instead, in the implementation of the algorithms, we will avoid to introduce explicitly the space-time network for the sake of efficiency.

The generic vertex $(i, t) \in V$ is thus an ordered couple of a node $i \in N$ and an instant $t \in T$. It can have at most one exiting edge for each arc of the node forward star, plus one edge that represents the wait at the node. A user that enters a given arc $a \in i^+$ at a certain instant $t \in T_a$ will be at its head a^+ at one following instant $\theta(a, t)$. A user that is waiting in a given node $i \in N^{wait}$ at a certain instant $t \in T\text{-}\{\eta\}$ will still be at node i at the next instant $t + 1$.

Therefore, the space-time network is built-up as follows:

- a *travelling edge* connects the vertex (a^-, t) to the vertex $(a^+, \theta(a, t))$, for every arc $a \in A$ and instant $t \in T_a$, with cost c_{at}; it is also identified for short by (a, t);
- a *waiting edge* connects the vertex (i, t) to the vertex $(i, t + 1)$, for every node $i \in N^{wait}$ and instant $t \in T\text{-}\{\eta\}$ with cost c_{it}; it is also identified for short by (i, t).

In practice all isolated vertices, if any, can be eliminated from the space-time network, so that in general it is $V \subseteq N \times T$.

The resulting graph is acyclic by construction under the assumption of positive travel times. Note that this important property does not hold true in the static case, unless we consider only a suitable subset of all arcs (e.g., the efficient arcs that bring the user closer to its destination with respect to some cost metric). In this sense, the dynamic case is simpler than the static case. Moreover, because the space-time network is acyclic, in the dynamic case the presence of negative costs is not a problematic issue (no loop exists where the user can enjoy minus infinite costs).

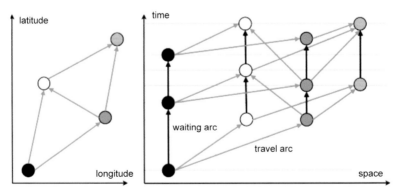

Figure 7.1 Generic space-time network, or diachronic graph, and the corresponding base network. Waiting arcs are depicted in black, travel arcs are depicted in grey.

7.2.4 The Concept of Topological Order

On any acyclic graph it is possible to establish a *topological order* of its nodes (of its vertices, in the case of the space-time network). By definition:

- the topological order of a node is lower than the topological order of the nodes that can be reached through its forward star and higher than the topological order of the nodes that can reach it through its backward star.

When travelling on an acyclic graph, the topological order of the passed nodes always increases.

Visiting the arcs of an acyclic graph in a (*front*) topological order means that, for every node, the arcs of its forward star will be visited only when all the arcs of its backward star have already been visited. When visiting it in a *reverse* topological order, the opposite is true.

In *forward visits*, for each node the arcs of its forward star are visited consecutively. In *backward visits*, for each node the arcs of its backward star are visited consecutively.

Once a topological order is identified, we can then define the following four types of graph visits:

- front forward visit;
- front backward visit;
- reverse forward visit;
- reverse backward visit.

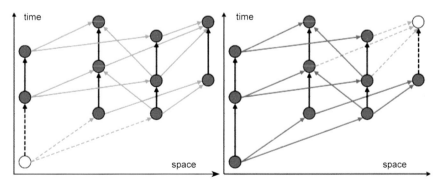

Figure 7.2 Forward (front) and backward (reverse) visits.

Thus, without loss of generality, we will concentrate on reverse visits as they allow also for shortest hyperpath computation, as explained later on.

Given an acyclic graph, in general, there are many possible numbering of its nodes that give a topological order. One way to obtain a reverse topological order of the nodes is to:

- Initialize
  - a label for each node with the cardinality of its forward star, and
  - set-up a queue of nodes having a null label;
- extract iteratively the first node of the queue, then
  - for each arc of its backward star subtract one to the label of the initial node, and
  - insert it in the queue if the label becomes zero;
- the order of extraction from the queue is the desired result.

However, for the space-time network there is a much simpler way—the reverse *chronological order* of the vertices is a reverse topological order.

Another topological order which turns useful is that based on the *minimum cost* of the nodes to reach the destination. This is clearly not known in advance, but can be constructed iteratively during the shortest path procedure, if a label-setting approach is adopted (as in the Dijkstra algorithm).

A generalization of this approach, that allows for origin-oriented reverse searches from a given destination, is achieved by using a reverse topological order based on increasing node *cost priority*. For example in the A* (a-star) algorithm, the cost priority of a node is defined as the sum of the (current minimum) expected cost from the node to the destination and a lower bound of the cost from the origin to the node. The real benefits of the A* algorithm wrt Dijkstra depend on how close the lower bound is to the actual cost. Actually the better the estimation, the smaller the number of visited arcs and the faster the search.

7.2.5 Path Costs in Presence of Random Arc Performance

A *path k* is a sequence or arcs connecting an *origin* node $k^- \in N$ to a (different) *destination* node $k^+ \in N$. In the static case, the path cost c_k may be then calculated by recursion based on the *successor* arc $i_k^+ \in A$ exiting from the current node $i \in N$, starting from the origin $i = k^-$ where $c_{k=0}$, until the destination is reached, i.e., $i = k^+$:

$$a \leftarrow i_k^+ \ ; \ c_k \leftarrow c_k + c_a \ ; \ i \leftarrow a^+. \tag{7.8}$$

Random performance perturbations (mainly in terms of travel time) due to probabilistic recurrent events (occurring systematically on the network) increase the variance of route cost. For a user who follows a fixed path determined *a priori* before starting the trip this is clearly a disadvantage as it leads to less reliable travel times. This can be reflected in additional arc costs.

In dynamic networks, where arcs have time-varying performance parameters, the effect of random perturbations is not limited to variance increase. Due to the concatenation of travel times, the expected value of the path cost is also affected, as shown in the example of the introduction.

The expected cost of path k for a user starting his trip at time $t \in T$ is here calculated through the following modified recursion:

$$a \leftarrow i_k^+ \; ; \; c_k \leftarrow c_k + c_{at} \; ; \; i \leftarrow a^+ \; ; \; t \leftarrow \theta(a, t) \qquad (7.9)$$

It must be stressed that this is just an approximation: the dynamic concatenation is accomplished on the base of the exit time $\theta(a, t)$, which derives (by assumption) from the upper bound of the arc travel time (this way no connection is lost), while the generalized cost c_{at} derives (by assumption) from the expected value of the arc travel time t_{at} through Eq. 7.6, where only a share of the additional travel time is taken into account. Instead, the exact expected value is obtained by considering each possible joint outcome of the random variables and then calculating the corresponding path cost through dynamic concatenation. The result is the average of the above path costs weighted by the probability of the corresponding joint outcome.

If the path is defined on the space-time network, the problem is reduced to static and the graph topology expresses also the dynamic concatenation. The resulting cost is analogous to that provided by 7.9; the same approximation is introduced, but is less evident.

7.2.6 Modelling Strategies Through Hyperarcs

Events can, however, become an opportunity of cost minimization for a user who adopts a strategic behavior. This implies modification of en-trip his/her route at diversion nodes where information about the outcomes of events that are preventively seen as random unknowns become available.

A travel *strategy* is a set of routing rules to reach the destination from the origin. It is chosen a priory before starting the trip, while en-trip decisions are rather a predefined reaction to events.

The classic example is the possibility of passengers boarding at the stop the first arriving carrier which serves any transit line of an *attractive* set, instead of waiting for a single line.

The chosen route is not anymore a simple path; it includes nested diversions, each associated with event probabilities and mergings. Sub-paths can be identified between such nodes. Strategies are represented mathematically (and topologically) as hyperpaths, while local diversions are represented through hyperarcs, which are formally introduced as follows.

A *hyperarc* $\breve{a} \subseteq i^+$ is a nonempty set of (diversion) arcs, called its *branches*, exiting from a same *diversion node* $i \in N^{div} \subseteq N$. The one initial node of a hyperarc is referred to as its *tail* and denoted $\breve{a}^- \in N$. The set of its final nodes, referred to as its *head* and denoted $\breve{a}^+ \subseteq N$, is given by the heads of the hyperarc branches: $\breve{a}^+ = \{a^+ \in N: a \in \breve{a}\}$.

Not all possible subsets of a diversion forward star constitute a hyperarc. The rules to identify the actual hyperarcs are specific to each application. In general, let H be the set of hyperarcs.

The *availability instants* of a hyperarc $\breve{a} \in H$ are denoted $T_{\breve{a}} \subseteq T$. Usually, it is:

$$t \in T_{\breve{a}} \Leftarrow t \in T_a, \forall a \in \breve{a}. \tag{7.10}$$

In dynamic models, for users entering each branch $a \in \breve{a}$ of the generic hyperarc $\breve{a} \in H$ at instant $t \in T_{\breve{a}}$ it is seen:

- $p_{a|\breve{a}t}$ — the *diversion probability*;
- $t_{a|\breve{a}t}$ — the *conditional travel time*;
- $\theta_{a|\breve{a}t} \in T$ — the *conditional exit time*;
- $\theta(a|\breve{a}, t)$ — the *index function* $\theta(a|\breve{a}, t) = e \in T: \theta_{a|\breve{a}t} = \tau_{e'}$ if $t \in T_{\breve{a}}$ or $\theta(a|\breve{a}, t) = \eta + 1$ otherwise.

Again, we assume that the conditional exit time $\theta_{a|\breve{a}t}$ derives from a risk adverse evaluation of the branch travel times (e.g., its upper bound), while the conditional travel time $t_{a|\breve{a}t}$ derives from a neutral evaluation of the branch travel time (e.g., the expected value). In general, we have:

$$\theta_{a|\breve{a}t} \geq \tau_t + t_{a|\breve{a}t}. \tag{7.11}$$

Then, the *combined travel time* $t_{\breve{a}t}$ of the hyperarc is given by:

$$t_{\breve{a}t} = \sum_{a \in \breve{a}} t_{a|\breve{a}t} \cdot p_{a|\breve{a}t}. \tag{7.12}$$

It is the main driver of the *combined travel cost* $c_{\breve{a}t}$, which can be expressed as:

$$c_{\breve{a}t} = \gamma^{vot} \cdot \gamma_a \cdot t_{\breve{a}t} + \sum_{a \in \breve{a}} \gamma^{risk} \cdot \gamma^{vot} \cdot (\theta_{a|\breve{a}t} - \tau_t - t_{a|\breve{a}t}) \cdot p_{a|\breve{a}t'} \tag{7.13}$$

under the assumption that no non-temporal cost is associated with hyperarcs.

The diversion probabilities are not the result of a local route choice (which instead occur among hyperarcs), while they reflect (on average) random events happening on the network.

In the classical case of optimal strategies (Spiess and Florian, 1989), the diversion probability represents the probability of boarding a given service, which is equal to the frequency of the line divided by the cumulative frequency of all the attractive lines, while the combined travel time is equal to the inverse of the cumulative frequency. These results are correct under the assumption of independent exponential headways. In this case, the conditional travel times are equal to the combined travel time, though this is not a general result, as shown in the introduction for the case of deterministic headways. The conditional exit times will instead be assumed equal to the entry time plus twice the headway. This is a suitable compromise for safety while the distribution is actually unbounded.

In some models the diversion probabilities of branches and the corresponding conditional travel times are destination specific and can be computed only during the solution of the routing problem. This is, for example, the case when the waiting times for the lines serving the stop are displayed in real-time and the passenger can locally optimize his route according to this information (e.g., Gentile et al., 2005).

In Fig. 7.3, there are 7 possible hyperarcs exiting from the diversion node $i \in N^{div}$, i.e., all the possible combinations of diversion arcs a, b and c: $\{a\}, \{b\}, \{c\}, \{a, b\}, \{a, c\}, \{b, c\}, \{a, b, c\}$, but among them only one hyperarc, i.e., $i_k^+ = \breve{a} = \{a, b\}$, can belong to a given hyperpath k.

It is intended that exiting from a diversion node, no diversion arc can be used per se in a hyperpath but only hyperarcs can be. It is possible to clearly define a singleton hyperarc made of only one diversion arc.

To include hyperarcs, the space-time network is modified as follows:

- all travelling edges $\{(a, t) \in E: a^- \in N^{div}\}$ whose arc a exits from a diversion node are eliminated from E;
- a *hyperedge* is introduced to connect the vertex $(\breve{a}^-, t)$ to the vertices $(a^+, \theta(a \,|\, \breve{a}, t))$ $\forall a \in \breve{a}$, for every hyperarc $\breve{a} \in H$ and instant $t \in T_{\breve{a}}$; it is also identified for short by $(\breve{a}, t)$. The set of such hyperedges is denoted $\breve{E}$.

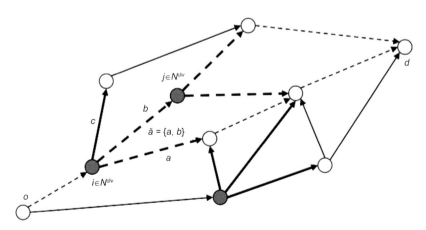

Figure 7.3 Example of a hyperpath k from origin $o = k^-$ to destination $d = k^+$. The hyperpath is depicted with broken lines. The diversion nodes are in grey. The bold lines are diversion arcs.

7.2.7 The Cost of Hyperpaths

An *hyperpath* $k = (N_k, A_k \subseteq N_k \times N_k)$ is an acyclic sub-graph of (N, A) connecting an *origin* node $k^- \in N$ to a (different) *destination* node $k^+ \in N$ with exactly one *successor* arc or hyperarc i_k^+ exiting from each one $i \in N_k$ of its nodes, except for the destination that has none, and one or more *predecessor* arc or hyperarc branch i_k^- entering each one of its nodes, except for the origin that has none.

The hyperpath cost shall be computed through a reverse forward visit by fixing recursively the expected cost to reach the destination of each node in reverse topological order. Front visits would require either handling a new sub-route for each further diversion (because in the dynamic case, merging nodes are reached at different instants), which is cumbersome and inefficient (leading to combinatorial complexity), or a symmetrical definition of hyperarcs (with several tail nodes and a single head), which is contrary to the causality rule (the random event happens first and only then on the base of its outcome the optimal en-trip option can be selected).

In the static case, given the successor hyperarc $\check{a} = i_k^+$ of the diversion node $i \in N_k \cap N^{div}$, the *expected cost* to destination w_i is given by:

$$w_i \leftarrow c_{\check{a}} + \sum_{a \in \check{a}} w_{a[+]} \cdot p_{a|\check{a}'} \tag{7.14}$$

while the following hold for the successor arc $a = i_k^+$ of the other nodes $i \in N_k$ with $i \notin N^{div}$:

$$w_i \leftarrow c_a + w_{a[+]}. \tag{7.15}$$

The cost of hyperpath k is equal to the expected cost of its origin node k^-.

The dynamic extension of hyperpaths is tricky because to consistently define its cost we have to specify the departure time from its origin or the arrival time to its destination. In the first case, we would be induced to undertake a front visit but we have already excluded this option for various reasons. In the second case, a question arises—how can the user decide the arrival time at destination in the presence of probabilistic diversions where different sub-paths included in the hyperpath may have different travel times (not counting the presence of random events, for which we consider the risk adverse behavior)? Indeed, he cannot.

Two further options are then proposed:

- Option 1. Forward cost computation for one tail instant;
- Option 2. Backward cost computation for different head instants.

The first option is to calculate jointly the hyperpath cost for all the possible departure instants from the origin. For all the possible arrival instants at the destination, the expected cost is fixed at 0. A reverse forward visit of the hyperpath nodes is then performed starting from the destination. Equations 7.14 and 7.15 are extended as follows and repeatedly applied to the successor $\breve{a} = i_k^+$ or $a = i_k^+$ of node i for each instant t in any order:

$$w_{it} \leftarrow c_{\breve{a}t} + \sum_{a \in \breve{a}} w_{a[+] \, \theta(a \mid \breve{a}, \, t)} \cdot p_{a \mid \breve{a}t}, \qquad (7.16)$$

$$w_{it} \leftarrow c_{\breve{a}t} + w_{a[+] \, \theta(a, \, t)}. \qquad (7.17)$$

The cost of hyperpath k is equal to the expected cost of users departing from the origin k^- at the desired instant t. The route resulting on the space-time network is not strictly speaking a hyperpath because by following different paths the users reach the destination at different instants.

This approach is convenient in assignment models, where the computation is required for all possible departure times; as mentioned, the corresponding arrival times are not univocal.

The second option also implies performing a reverse forward visit of the hyperpath nodes. But in this case, starting from one given arrival instant at the destination, only one vertex per node is considered. This leads to large efficiency improvement at the price of some complications and approximations. The instant t_i for which the expected cost of node i is evaluated is the highest possible under the constraint that the index $\theta(a \mid \breve{a}, \, t_i)$ of the conditional exit time for each branch a of its successive hyperarc $\breve{a} = i_k^+$ is not higher than the instant $t_{a[+]}$ for which the head expected cost $w_{a[+]}$ is evaluated:

$$t_i \leftarrow Max(t: \theta(a \mid i_k^+, \, t) \leq t_{a[+]}, \, \forall a \in i_k^+). \qquad (7.18)$$

Also in this case, the result is not a hyperpath on the space-time network. This approach is convenient in routing models, where the computation is required for a given arrival time.

This solution corresponds to a risk adverse approach in terms of travel times such that the user is sure to reach the destination within the given arrival time. The expected cost w_i is then obtained by applying the following equation for $t = t_i$ and $\breve{a} = i_k^+$:

$$w_i \leftarrow c_{\breve{a}t} + \sum_{a \in \breve{a}} w_{a[+]} \cdot p_{a \mid \breve{a}t} + \sum_{a \in \breve{a}} \gamma^{risk} \cdot \gamma^{vot} \cdot (\tau_{ta[+]} - \theta_{a \mid \breve{a}t}) \cdot p_{a \mid \breve{a}t}. \qquad (7.19)$$

We are then assuming that the lost time from $\theta_{a \mid \breve{a} t}$ to $\tau_{ta[+]}$ which is spent at the head of the branch $a \in \breve{a}$ has the same unitary disutility of the time margin for risk adverseness from $\tau_{ti} + t_{a \mid \breve{a} t}$ to $\theta_{a \mid \breve{a} t}$.

In practice, the passenger cannot and will not wait at the head of the branch, but instead will use the service as soon as it is available, which is earlier than $\tau_{ta[+]}$ (actually, earlier also than $\theta_{a \mid \breve{a} t}$). He will then suffer an expected cost which is different from $w_{a[+]}$ (but we take the latter as a suitable approximation) and will probably arrive at the destination earlier than the desired arrival time.

Finally, it's worth noting that the cost $c_{ăt}$ and the probability $p_{a|ăt}$ can in practice be calculated (with suitable approximation) based on attributes of the branch $a \in ă$ that are evaluated at instant $t_{a[+]}$ instead of that at instant t_j.

In the above options, waiting at nodes (not to be confused with waiting arcs of transit networks) has not been considered, else the departure or arrival times and the space topology of the hyperpath are not enough to evaluate its cost (this is true also for a simple path).

7.2.8 Extension to Continuous Time Modelling

Modelling time as a continuous variable allows a coherent representation of continuous transport services, such as roads or frequency-based transit lines. To this end we can:

- remove the assumption that the exit times belong to the finite set of instant clocks;
- remove the assumption of strictly positive travel times;
- introduce a function of time $\theta_a(\tau) \geq \tau$ yielding the exit time for users entering arc $a \in A$ at time τ;
- introduce a function of time $c_a(\tau)$ yielding the cost for users entering arc $a \in A$ at time τ.

In practice of these functions we know only the value at some clocks τ_t of the time discretization $t \in T_a : \theta_a(\tau_t) = \theta_{at}$, $c_a(\tau_t) = c_{at}$. The rest of their plot may be reconstructed, where possible, under some assumption, e.g., through the following piecewise linear interpolation; for $\tau_{t-1} < \tau \leq \tau_t$:

$$\theta_a(\tau) = \theta_{a\,t-1} + (\tau - \tau_{t-1})/(\tau_t - \tau_{t-1}) \cdot (\theta_{at} - \theta_{a\,t-1}), \qquad (7.20)$$

if $t \in T_a$ and $t - 1 \in T_a$; $\theta_a(\tau) = \infty$, otherwise

$$c_a(\tau) = c_{a\,t-1} + (\tau - \tau_{t-1})/(\tau_t - \tau_{t-1}) \cdot (c_{at} - c_{a\,t-1}), \qquad (7.21)$$

if $t \in T_a$ and $t - 1 \in T_a$; $c_a(\tau) = \infty$, otherwise

The index function assumes the following meaning:

- $\theta(a, t) = e : \tau_{e-1} < \theta_{at} \leq \tau_{e'}$ if $t \in T_a$ or $\theta(a, t) = \eta + 1$ otherwise (by convention).

In the solution algorithms for continuous time models it is relevant to distinguish the two cases of *long time intervals* and *short time intervals*. The latter satisfies for each arc $a \in A$ and instant $t \in T_a$ the following rule:

$$\theta_{at} > \tau_{t+1} \text{ and } \theta(a, t) > t + 1. \qquad (7.22)$$

7.3 Formulation and Solution of the Dynamic Routing Problem

For the sake of simplicity, we will refer to the case of simple path searches on the space-time network, leaving continuous modelling and the computation of optimal hyperpaths as extensions of this base problem.

7.3.1 Route Search with Roots and Targets

In this chapter we consider the specialization of the time-dependent shortest path problem and which we will call the *dynamic routing problem* (DRP). Each search u is in this case identified by two subsets of vertices on the space time network:

- the *roots* (or sources) $R \subseteq V$ and
- the *targets* $S \subseteq V$.

The search aims at finding the costliest connection on the space-time network between *any* root and *every* target. Each root and target is characterized by an *additional cost*, which sums up to the additive cost of the route edges on the space-time network, denoted respectively:

- c_r – additional cost of root $r \in R$;
- c_s – additional cost of target $s \in S$.

There are two possible search *directions*:

- *front* searches (forward in time and space), where origins are associated with roots and destinations are associate with targets;
- *reverse* searches (backward in time and space), where the opposite is true.

The computation of shortest hyperpaths is possible only through reverse searches. Usually, there are fewer roots than targets.
Noticeable example of searches are:

- *front path* (point-to-point) where the root $R = (o, t)$ identifies a specific *origin* node $o \in N$ and a specific *departure instant* $t \in T$, while the target $S = \{(d, h): h > t\}$ identifies a different *destination* node $d \in N$ at any later arrival instant $h \in T$;
- *reverse path* (point-to-point), where the root $R = (d, h)$ identifies a specific destination node $d \in N$ and a specific *arrival instant* $h \in T$, while the target $S = \{(o, t): t < h\}$ identifies a different origin node $o \in N$ at any earlier departure instant $t \in T$;
- *front tree* (one-to-all), where the root $R = \{(o, t): t \in T\}$ identifies a specific *origin* node $o \in N$ and all possible departure instants $t \in T$, while the target $S = V$ is made by all possible destination nodes $d \in N$ at any arrival instant $h \in T$;

- *reverse tree* (all-to-one), where the root $R = \{(d, h): h \in T\}$ identifies a specific destination node $d \in N$ and all possible arrival instants $h \in T$, while the target $S = V$ is made by all possible origin nodes $o \in N$ at any departure instant $t \in T$.

In all the above searches additional root and target costs are typically null.

Paths are typical searches in journey planning, while trees are typical searches in assignment algorithms. But the *DRP* allows for quite general searches, such as, for example, the user wants to arrive at the destination not later than a given arrival time and accepts to arrive with some additional cost at the maximum 30 minutes earlier, leaving the origin at any departure time.

A generic search may not have a feasible solution: the existence of a feasible connection depends on the input (root and target) of the requested search and on the available services.

7.3.2 General Algorithm

Now we will refer to reverse searches because in the field of transportation users naturally aim at destinations. It's worth noticing once more that in a reverse search, the root is where the users are directed, while the target is where they come from. The names are maybe counterintuitive but are justified by the solution mechanism, which starts from the destination root and explores the space-time network backward in space and time (wrt the trajectory of the user) with the aim of reaching the origin(s) target.

All users leaving node $i \in N$ at time $t \in T$ are directed towards the same root $R \subseteq V$, i.e., they aim at any of such vertices with the objective of minimizing their travel cost. Under the assumption of additive costs, they will make the same route choice, i.e., will follow their trip using the same successive edge, regardless of their origin and departure time. As a consequence, the result of any reverse search is a family of separate shortest sub-trees, or *forest*, on the space-time network with each one rooted at some $r \in R$. A symmetrical result holds for front search. Thus, for each vertex, that is part of the solution (the forest), the routing algorithm shall identify:

- the successive edge;
- the sub-tree root;
- the route cost to such root.

The search starts at the roots and ends when no target can be improved. A post processing is required to identify the solution, that is, to follow the sequence of successive edges, starting from every target and arriving to any root.

To find a shortest forest on the space-time network implies the computation of several separate shortest trees, one for each root $r \in R$ that is performed during the same routing procedure u.

Route searches are typically solved through an iterative procedure where a current forest is repeatedly modified until a solution is (possibly) found.

To completely define such an algorithm, we have to establish:

- how to represent the current forest;
- how to initialize the current forest;
- how to modify the current forest at each step;
- how to verify whether the current forest is shortest (or no solution exists for this search).

In the following, without loss of generality, we will refer to a reverse search.

For what concerns point 1 (representation), the following variables are used to deal with the current forest:

- s_{it} the (one) *successive edge* of vertex $(i, t) \in V$, which defines the current forest; by convention: $s_{it} = a \in i^+$ will denote the travelling edge $(a, t) \in E$; $s_{it} = 0$ will denote the waiting edge $(i, t) \in E$; $s_{it} = -1$ will indicate that the vertex at hand is a root, i.e., $r = (i, t) \in R$;
- w_{it} the *route cost* to reach any root $r \in R$ from vertex $(i, t) \in V$ travelling on the current forest (also referred to as vertex *label*);
- $r_{it} \in R$ the *closest root* reached by users leaving vertex $(i, t) \in V$ travelling on the current forest; $r_{it} = (d, h)$ also identifies the destination $d \in N$ and the arrival time τ_h.
- u_{it} the *last search* that included (visited) vertex $(i, t) \in V$ in the current forest; $u_{it} = u$ indicates that (i, t) is part of the current forest.

For what concerns the fourth point (that is, verification), based on the Bellman theorem, we know that a given (current) forest (defined by the successive edge of each vertex) is shortest if and only if no edge can (locally) improve the route cost of its tail, i.e., if the following holds true:

$$w_{a[-]t} \le c_{at} + w_{a[+]\,\theta(a,\,t)}, \quad \forall a \in A, \forall t \in T_a; \tag{7.23}$$

$$w_{it} \le c_{it} + w_{i\,t+1}, \quad \forall i \in N^{wait}, \forall t + 1 \in T. \tag{7.24}$$

For what concerns the third point (modification), a valid shortest forest algorithm can be obtained by visiting the edges of the space-time network in reverse topological order and executing for each edge the following instructions:

$$\textbf{if } w_{a[-]t} > c_{at} + w_{a[+]\,\theta(a,\,t)} \textbf{ then } w_{a[-]t} \leftarrow c_{at} + w_{a[+]\,\theta(a,\,t)}, \, s_{it} \leftarrow a, \tag{7.25}$$

for a travelling edge $(a, t) \in E$;

if $w_{it} > c_{it} + w_{i\,t+1}$ **then** $w_{it} \leftarrow c_{it} + w_{i\,t+1}$, $s_{it} \leftarrow 0$,   (7.26)
for a waiting edge $(i, t) \in E$.

For what concerns the second point (initialization), we set the route cost of each root equal to the root cost (that is typically null) and the route cost of each other vertex equal to infinity:

$$w_{dh} \leftarrow c_r\,, \forall r = (d, h) \in R; w_{it} \leftarrow \infty, \forall(i, t) \in V\text{-}R.$$ (7.27)

The latter task can be avoided by assuming implicitly an infinite cost for all vertices that are not part of the current forest, i.e., all vertices that have not been visited (yet) in the current search u:

$$u_{it} < u \Rightarrow w_{it} = \infty.$$ (7.28)

Summing up, the basic algorithm to solve the dynamic routing problem through a reverse search is presented below.

Algorithm 1. General algorithm to solve the dynamic routing problem through a reverse visit.

function GDRP(R, c_R, S, c_S)
  $u \leftarrow u + 1$ *set a new search
  $w_{dh} \leftarrow c_r\,, s_{dh} \leftarrow -1, r_{dh} \leftarrow r, u_{dh} \leftarrow u, \forall\, r = (d, h) \in R$ *initialization
  for each $e \in E$ in reverse topological order
    if $e = (i, t)$ **then**   *e is a waiting edge
      $w \leftarrow c_{it} + w_{i\,t+1}$ *does the tentative label w
      if $u_{it} < u$ **or** $w_{it} > w$ **then**   *improve the current solution?
        $w_{it} \leftarrow w, s_{it} \leftarrow 0, r_{it} \leftarrow r_{i\,t+1}, u_{it} \leftarrow u$ *if yes, update the solution
      end if
    end if
    if $e = (a, t)$ **then**   *e is a travelling edge
      $i \leftarrow a[-]$ *i is the tail of its arc
      $w \leftarrow c_{at} + w_{a[+]\,\Theta(a, t)}$ *does tentative label w
      if $u_{it} < u$ **or** $w_{it} > w$ **then**   *improved current the solution?
        $w_{it} \leftarrow w, s_{it} \leftarrow a, r_{it} \leftarrow r_{a[+]\,\Theta(a, t)}, u_{it} \leftarrow u$ *if yes, update the solution
      end if
    end if
  next e
  $w_{ot} \leftarrow w_{ot} + c_s, \forall s = (o, t) \in S$ *add target costs
end function

In principle, we have to process all the edges before we can identify the shortest forest. In practice, during the search we can skip:

- all edges that have an earlier initial vertex than any target, because by construction they cannot be used to connected targets to roots;
- all edges that have a later final vertex than any root, because by construction they cannot be used to connected targets to roots;

- all edges whose final vertex is not yet been visited, because the latter has an infinite route cost so that the Bellman verification is by definition satisfied;
- all remaining edges after the forward star of all targets has been processed, because by construction no further improvement is possible for targets.

For this reason it is useful to introduce a list of edges (or vertices) to be visited that shall be updated during the search.

In some applications (e.g., journey planning) we are also interested in identifying the best possible target s^* and the corresponding path k^*. To this end we can extend Algorithm 1 as follows:

$s^* = (o^*, t^*) \leftarrow \text{argmin}\{w_{ot} : s = (o, t) \in S\}$ *the best target may not be unique
$k^* \leftarrow \varnothing, i \leftarrow o^*, t \leftarrow t^*$ *initialize the best path from the best target
do until $s_{it} = -1$ *repeat until a root is found
  if $s_{it} = 0$ **then**   *successive edge is waiting
    $k^* \leftarrow k^* \cup (i, t)$ *add the edge to the best path
    $t \leftarrow t + 1$ *update the time index
  else   *successive edge is travelling
    $a \leftarrow s_{it}$ *update the arc index
    $k^* \leftarrow k^* \cup (a, t)$ *add the edge to the best path
    $i \leftarrow a[+]$ *update the node index
    $t \leftarrow \theta(a, t)$ *update the time index
  end if
loop

At the end of this process, it should be: $(i, t) = r_{o^* t^*}$.

7.3.3 Extension to Departure and Arrival Time Choice

The choice of departure or arrival instants can be easily included in the generic search by defining proper additional costs to roots or targets, depending on the search direction (front or reverse).

For example, we can define a front search with a *desired departure time* τ, where:

- the root $R = \{(o, t): \tau - t^{ant} \leq \tau_t \leq \tau + t^{del}\}$ identifies a specific origin node $o \in N$ and a choice set of departure instants $t \in T$ whose clock times τ_t fall in a given interval;
- the target $S = \{(d, h): h \in T\}$ identifies a specific destination node $d \in N$ at any arrival instant $h \in T$;
- additional target costs are null, while additional root cost c_r for each $r = (o, t) \in R$ are the result of a linear disutility (more complex functions can be defined) for early or late departure wrt the desired time τ:

$$c_r = \gamma^{del} \cdot \gamma^{pot} \cdot (\tau_t - \tau) \text{ , if } \tau_t \geq \tau, \qquad (7.29)$$

$$c_r = \gamma^{ant} \cdot \gamma^{vot} \cdot (\tau - \tau_t) \text{ , if } \tau_t \leq \tau \text{ .} \tag{7.30}$$

In this case, the choice model is characterized by the following class specific parameters:

- t^{ant} – the *maximum anticipation*
- t^{del} – the *maximum delay*
- γ^{del} – the *delay coefficient*
- γ^{ant} – the *anticipation coefficient*

We can also define a front search with a *desired arrival time* τ, where:

- the root $R = \{(o, t): t \in T\}$ identifies a specific origin node $o \in N$ at any departure instant $t \in T$;
- the target $S = \{(d, h): \tau - t^{ant} \leq \tau_h \leq \tau + t^{del}\}$ identifies a specific destination node $d \in N$ and a choice set of arrival instants $h \in T$ whose clock times τ_h fall in a given interval;
- the above equations can be used to define the target additional costs.

For reverse searches, additional costs for roots are defined for desired arrival time and additional costs for targets are defined for desired departure time.

7.3.4 Extension to Intermodal Routing

Intermodal routing is based on the possibility of travelling from an origin to a destination, using a sequence of (different) modes instead of a single mode. Mode changes can take place only at a subset of *intermodal nodes* $N^{int} \subseteq N$, such as: interports, harbors, airports, stations, parking, terminals. Each mode has a dedicated partition of arcs in the transport network.

A typical example is Park And Ride. A commuter drives on the road network from his/her origin (say, home) toward a parking facility, where he/she can leave his/her individual vehicle (car, motorcycle, bicycle) at a reasonable price and start his/her trip on the public transport network (usually exploiting fast mass transit) toward his/her actual destination (say, work). This allows avoidance of severe traffic congestion to access the city as well as the problem of finding a (costly) parking slot downtown. On return from work to home, the sequence of modes is inverted.

Another example is RORO (Roll-On/Roll-Off). A commercial vehicle drives on the road network from its origin (say, production site) towards a harbor, where it can board a RORO ship (a vessels designed to carry wheeled cargo) that will take it closer to its actual destination (say, consumption site). This allows avoidance of cost of transshipment as well as the problems related to limitations in continuous driving hours. After alighting at the final harbor, the last leg of the trip to the destination is again on the road network.

An intermodal search requires the additional identification (besides direction, roots, targets and their additional costs) of a mode sequence M.

The set of intermodal nodes $P(M[m], M[m+1]) \subseteq N^{int}$ that allow transfer from the m-th mode to $(m + 1)$-th mode is generated automatically consistently with rules based on the characteristics of the search (for example, the distance between the origin and the node or between the destination and the node).

Intermodal routing requires $n = |M|$ subsequent route searches, one for each mode in M. The additional root costs play a crucial role in the computation of intermodal routes. Hereon we list the sequence of searches for a front search, where roots are origins and targets are destinations.

- The first search will seek for routes on the first mode $M[1]$ from any root in $R_1 = R$ to the each intermodal node in $S_1 = P(M[1], M[2])$. Additional root costs for R_1 are typically null.
- The second search will seek for routes on the second mode $M[2]$ from any intermodal node in $R_2 = P(M[1], M[2])$ to each intermodal node in $S_2 = P(M[2], M[3])$. Additional root costs for R_2 are equal to the route costs calculated in the previous search for S_1.
- The m-th search will seek for routes on the m-th mode $M[m]$ from any intermodal node in $R_m = P(M[m - 1], M[m])$ to each intermodal node in $S_m = P(M[m], M[m + 1])$. Additional root costs for R_m are equal to the route costs calculated in the previous search for S_{m-1}.
- The last search will seek for paths on the last mode $M[n]$ from any intermodal node $R_n = P(M[n - 1], M[n])$ to each target in $S_n = S$. Additional root costs for R_n are equal to the route costs calculated in the previous search for S_{n-1}. Additional target costs for S_n are typically negligible.

A reverse search requires to invert the sequence of modes.

7.3.5 Extension to Strategic Behavior and the Greedy Approach

The introduction of hyperedges at diversion nodes in the space-time network does not modify the structure of the general algorithm for dynamic routing. As mentioned, only reverse searches are possible.

For what concerns the representation of the hyperforest solution, the following variables change:

- s_{it} – if $i \in N^{div}$ is a diversion node, can be either null, if the successor of vertex (i, t) is the waiting edge, or it can be a vector containing the branches of a hyperarc $ă \in H$, if the successor is a hyperedge;
- w_{it} – is now referred also as the *expected cost*, because the presence of random events allows to know the value of the route cost only in probabilistic terms;

- $r_{it} \in R$ – the closest root is now undefined, because the user can probabilistically reach several vertices of the root.

The Bellman relations 7.23–7.24 can be extended as follows:

$$w_{\check{a}[-]t} \leq c_{\check{a}t} + \sum_{a \in \check{a}} w_{\check{a}[+]\,\theta(a|\check{a},t)} \cdot p_{a|\check{a}t} = w, \; \forall \check{a} \in H, \; \forall t \in T_{\check{a}}. \tag{7.31}$$

The modification 7.25–7.26 is then extended as follows:

if $w_{\check{a}[-]t} > w$ **then** $w_{\check{a}[-]t} \leftarrow w$, $s_{it} \leftarrow \check{a}$, for a hyperedge $(\check{a}, t) \in \check{E}$. $\tag{7.32}$

The reverse tree search can be seen as an application of the Option 1 presented in Section 7.2.7, where the dynamic concatenation is fixed by the structure of the space-time network.

Reverse path searches require definition of a suitable set of arrival times, possibly penalized for anticipation and delay wrt a desired instant; else the users have to wait somewhere on the network in order to utilize hyperarcs whose branches have different expected times to reach the destination.

Typically, it is not convenient to consider explicitly all possible hyperarcs exiting from a diversion node $i \in N^{div}$ because their amount is equal to the combinations of any number of arcs in the forward star i^+ (e.g., if $|i^+| = 3$ there are 7 hyperarcs, if $|i^+| = 4$ there are 15 hyperarcs). Moreover, handling the successor of a diversion node as a vector is not straightforward.

By looking at the specific phenomenon underlined by the hyperarc, it is usually possible to simplify the model. For example, for fail-to-sit and fail-to-board probabilities at transit stops (Gentile, 2015), there is just one hyperarc exiting the diversion node with no possibility of waiting. The potential issue in this case is the need for processing the tail only once both the heads are labelled. Indeed, the expected cost of the tail can be lower than the expected cost of the failing head. This requires attention if the topological order of the visit is based on costs.

In the case of waiting for frequency-based services, the model can be simplified by considering the *greedy* approach, introduced by Nguyen and Pallottino (1988) for the static case. More specifically, the arcs exiting the diversion node i are ordered based on the expected cost of their head node: only the hyperarcs constituted by the first 1, 2, 3, … arcs of the forward star i^+ are iteratively evaluated until w_{it} is consequently reduced by applying (7.32). No waiting edge is considered from this diversion node because waiting is represented by specific arcs of the transit network.

In the case of information at the stop (Gentile, 2015), the greedy approach can exclude a line whose expected cost once boarded is worse than the expected cost at the stop, but that can turn useful in the unlucky case where all the other lines are very late. Despite of some cases, the greedy approach provides only a sub-optimal strategy—it guarantees the acyclicity of the solution route.

7.4 Algorithm Implementations

In this section, we present in detail five different specifications of the general algorithm in terms of discrete or continuous time modelling, visiting order of edges, targets and roots:

- the discrete temporal-layer approach, particularly suited for transit assignment models, especially if services are discontinuous and schedules are provided;
- the continuous temporal-layer approach, particularly suited for traffic assignment models, where the road network includes short arcs, and performances are described as piece-wise linear functions defined on long time intervals;
- the discrete and the continuous user-trajectory approach, particularly suited for shortest (fastest) path searches;
- the multi-label approach, particularly suited for point-to-point journey planning with tolls and fares.

All the proposed approaches can be casted for front research as well, but their extension to strategies is infeasible, as mentioned earlier.

7.4.1 Temporal-Layer Approach

The temporal-layer (TL) approach for discrete time models is based on the following assumptions:

- reverse chronological order of each vertex and forward visit of its exiting edges;
- root made-up by all instants for a given destination $d \in N$;
- target is all vertices;
- no additional costs.

Because each root has the same destination node d, it can here be indicated with its instant, say t.

Algorithm 2. The reverse temporal-layer algorithm for discrete time modelling.

function DLYR(d)
  $u \leftarrow u + 1$ *set a new search
  for $t = \eta$ **to** 0 **step** -1 *in reverse chronological order
    $w_{dt} \leftarrow 0$, $s_{dt} \leftarrow -1$, $r_{dt} \leftarrow t$, $u_{dt} \leftarrow u$ *initialize the layer with the destination
    for each $i \in N-\{d\}$ *for each other node in no order
      if $i \in N^{wait}$ **and** $t + 1 \leq \eta$ **and** $u_{i\,t+1} = u$ **then**   *if the waiting edge is feasible
        $w_{it} \leftarrow c_{it} + w_{i\,t+1}$, $s_{it} \leftarrow 0$, $r_{it} \leftarrow r_{i\,t+1}$, $u_{it} \leftarrow u$ *initialize node solution with waiting

```
        end if
        for each a∈i⁺                          *for each arc of the forward star
          j ← a[+]                             *set the arc head
          e ← θ(a, t)                          *set the exit time index
          if e ≤ η and u_je = u then           *if the travelling edge is feasible
            w ← c_at + w_je                     *does the tentative label w
            if u_it < u or w_it > w then         *improve the current solution?
              w_it ← w , s_it ← a , r_it ← r_je , u_it ← u   *if yes, update the node solution
            end if
          end if
*A      next a
      next i
    next t
end function
```

The extension to strategies is immediate. Option 1 can be applied by visiting the hyperedges of the node forward star after instruction *A as follows:

```
for each ă∈H : i = a[−]                     *for each hyperarc of the forward star
  w ← c_ăt                                   *initialize the tentative label w
  for each a∈ă                               *for each branch of the hyperarc
    j ← a[+]                                 *set the branch head
    e ← θ(a | ă, t)                          *set the conditional exit time index
    if e ≤ η and u_je = u then               *if the travelling branch is feasible
      w ← w + w_je · p_a|ăt                   *update the tentative label w
    else
      w ← ∞                                  *otherwise, set the tentative label to infinity
      exit for
    end if
  next a
  if u_it < u or w_it > w then               *does w improve the current solution?
    w_it ← w , s_it ← ă, u_it ← u            *if yes, update the node solution
  end if
next ă
```

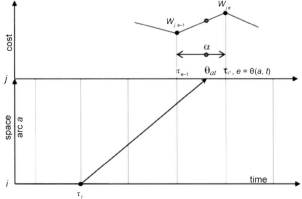

Figure 7.4 Interpolation of costs in the continuous version of the temporal-layer algorithm.

In the case of continuous time modelling, the instruction $w \leftarrow c_{at} + w_{je}$ can be replaced with:

$$w \leftarrow c_{at} + (1 - \alpha) \cdot w_{je-1} + \alpha \cdot w_{je}, \text{ where } \alpha = (\theta_{at} - \tau_{e-1})/(\tau_e - \tau_{e-1}). \qquad (7.33)$$

The extension is immediate for short time intervals, when the assumption 7.22 holds true, because intra-layer improvements are impossible.

In the contrary case of long time intervals, when the exit time from an arc of a user entering it at the beginning of an interval can fall within the same interval, a few things shall be modified.

- A ordered list B of nodes to be visited in each temporal layer t based on the node label w_{it} is introduced. The list is initialized with the destination. Nodes are then extracted in order of cost. In practice, though, B can be implemented as a simple set and the costliest node can be identified before each extraction.
- After a node is extracted, its backward star is visited. So we have a reverse backward visit.
- An ordered list O of nodes already extracted from B, which identifies the reverse topological order of visit inside each temporal layer t is introduced. In practice though, O can be implemented as a simple set to check whether a node has been extracted yet.
- Travelling edges are checked for Bellman only if their tail has not been extracted yet in the current temporal layer. This is to avoid absorbing cycles that could occur in presence of decreasing node costs. Indeed, list B is ordered in terms of the node cost evaluated at the time of the layer, while the tail cost updates may consider a lower head cost at a later layer.
- The closest root represents now the arrival time.

Algorithm 3. The reverse temporal-layer algorithm for continuous time modelling.

```
function CLYR(d)
  u ← u + 1                              *set a new search
  for t = η to 0 step −1                 *in reverse chronological order
    w_dt ← 0 , s_dt ← −1, r_dt ← τ_t , u_dt ← u  *initialize the layer with the destination
    B ← {d} , O ← ∅                      *initialize B and O
    do until B = ∅                       *repeat until B is empty
      j ← ArgMin(w_it : i∈B)             *identify the costliest node in B
      B ← B−{j} , O ← O∪{j}              *extract it from B and put it in O
      if j∈N^wait and t + 1 ≤ η and u_{j t+1} = u then  *if the waiting edge is feasible
        w ← c_jt + w_{j t+1}             *does the tentative label w
        if u_jt < u or w_jt > w then     *improve the current solution?
          w_jt ← w , s_jt ← 0 , r_jt ← r_{j t+1} , u_jt ← u  *if yes, update the node solution
        end if
      end if
```

```
        for each a∈j⁻                          *for each arc of the backward star
          i ← a[−]                             *set the arc tail
*A        if i∉O then                          *if the tail has not been extracted yet
            e ← θ(a, t)                        *set the exit time interval
            if e ≤ η and u_je = u then         *if the edge is feasible
              α ← (θ_at − τ_e-1)/(τ_e − τ_e-1) *interpolation
              w ← c_at + (1 − α) · w_je-1 + α · w_je   *does the tentative label w
              if u_it < u or w_it > w then     *improve the current solution?
                w_it ← w , s_it ← a , u_it ← u *if yes, update the tail solution
                r_it ← (1 − α) · r_je-1 + α · r_je  *and interpolate the arrival time
                B ← B∪{i}                      *if not present, insert the tail in B
              end if
            end if
          end if
        next a
      next i
    next t
end function
```

If we force $\theta(a, \eta) = \eta$ for each arc $a \in A$, then we may have results for all vertices, even at late layers.

The extension to hyperarcs can be done through the greedy approach. This way we can elude to introduce explicitly the hyperarc, set as this is built-up recursively by considering the first best 1, 2, 3, … head expected cost to destination. To this end, the following code is to be inserted after instruction *A:

```
if i∈N^div and u_it = u then                   *if the tail is a labelled diversion node
  ă ← s_it ∪{a}                                *build-up the hyperarc from the successor
  w ← c_ăt                                      *initialize the tentative label w
  r ← 0                                         *initialize the tentative arrival time r
  for each b∈ă                                  *for each branch of the hyperarc
    k ← b[+]                                    *set the branch head
    e ← θ(b|ă, t)                               *set its conditional exit time index
    if e ≤ η and u_ke = u then                  *if the travelling branch is feasible
      α ← (θ_b|ăt − τ_e-1)/(τ_e − τ_e-1)        *interpolation
      w ← w + ((1 − α) · w_ke-1 + α · w_ke) · p_b|ăt   *update the tentative label w
      r ← r + ((1 − α) · r_ke-1 + α · r_ke) · p_b|ăt   *update the tentative arrival time r
    else
      w ← ∞                                     *otherwise, set the tentative label to infinity
      exit for
    end if
  next b
  if w_it > w then                              *does w improve the current solution?
    w_it ← w , r_it ← r , s_it ← ă              *if yes, update the tail solution
    B ← B∪{i}                                   *if not present, insert the tail in B
  end if
else
```

In Section 7.5.3 we will see how in the particular case of waiting for the first arriving carrier of an attractive set of lines with exponentially distributed independent headways, there is no need for exploring all the branches of the new hyperarc and the tentative solution can be updated considering only the new line.

7.4.2 User-Trajectory Approach

In dynamic networks it may be convenient to lose time with the aim of travelling later on certain arcs at a lesser cost. Think, for example, about a higher train frequency in the morning peak, or a limited traffic zone that reopens in the evening. If waiting at nodes is not allowed or is costly, time decreasing costs may even lead to best routes that contain (space) cycles.

The implementation of the general algorithm results if considerably simplified for each arc $a \in A$ and instant $t \in T$:

- travel and waiting costs coincide (or are proportional) with travel and waiting times,

$$c_{at} = \gamma^{vot} \cdot (\theta_{at} - \tau_t), \tag{7.34}$$

$$c_{it} = \gamma^{vot} \cdot (\tau_{t+1} - \tau_t), \tag{7.35}$$

- the First In First Out (FIFO) rule on holds true,

$$t < e \Rightarrow \theta_{at} < \theta_{ae}. \tag{7.36}$$

On these basis, to delay the entrance on an arc is never convenient.

Under these assumptions, the concatenation rule holds for each path. Thinking of front searches, the cheapest way of reaching a node departing from a given origin at a given time is also the fastest one, i.e., the earliest one and thus the one offering most connection opportunities to follow the trip.

In this case, one label for each node is enough. Then we can drop the time index from the notation.

In reverse searches, we look for the fastest route a user leaving node $i \in N$ can take to reach a given destination $d \in N$ in a given instant $h \in T$, i.e., we aim at the latest instant the user can depart from node i to reach destination d in instant h. The node label is then chosen consistently:

- $t_i \in T$ the latest instant when a user leaves node $i \in N$ to reach the root (d, h)

The User-Trajectory (UT) approach for discrete time models is based on the following assumptions:

- topological order based on cost priority of each vertex and backward visit of its entering edges;

- the root is a vertex, identifying one destination node $d \in N$ and one arrival time instant $h \in T$;
- target set made-up by all earlier instants for a given origin node $o \in N$;
- no additional costs.

The cost priority of node $i \in N$ is defined as follows:

$$y_i = w_i + \lambda \cdot c_{oi}, \tag{7.37}$$

where we denoted:

- y_i – the cost priority of node $i \in N$ (low y_i means high priority);
- λ – A* (a-star) parameter (0 = Dijkstra, 1 = A-Star, > 1 Best First);
- c_{oi} – estimated cost to travel from node $o \in N$ to node $i \in N$ (lower bound of path costs).

The cost priority allows implementation of classical shortest path algorithms, ranging from Dijkstra (Dijkstra, 1959), for $\lambda = 0$, to A-Star (Hart et al., 1968), for ($\lambda = 1$); for $\lambda > 1$, we obtain a Best First search (Pearl, 1984). The proof correctness and the justification of these algorithms goes beyond the scope of our chapter, whose aim is extension of such approaches in the dynamic framework.

The topological order based on cost priority requires to introduce an ordered list of nodes B: each time the priority of a node is reduced, we need to update its position in the list. A convenient way to implement the ordered list is to introduce a reasonable number n of adjacent buckets (Dial, 1969) covering a fixed cost priority interval $[y^0, y^1]$. When a node $i \in N$ is inserted in the list, to find the proper bucket $b \in [0, n]$ is an immediate operation requiring only an integer division:

$$b = (y_i - y^0) \backslash ((y^1 - y^0)/n). \tag{7.38}$$

Then the ordering operations involve only the nodes (if any) falling in that bucket. To reduce the time wasted in scanning empty buckets one can apply a multi-level bucket list (Denardo and Fox, 1979).

To keep the list ordered in case of node updates we can extract the node from the list and reinsert it in the proper bucket.

For backward visits, if the topology of the space-time network is not explicitly defined and only the temporal profiles of the arc performance (i.e., θ_{at} and c_{at}) are defined, it is useful to introduce the following inverse of the index function, which yields the latest connection on arc $a \in A$ for a given exit instant $e \in T$:

- $\theta^{-1}(a, e) = Max(-1; t \in T_a : \theta_{at} \le \tau_e)$.

Algorithm 4. The reverse user-trajectory algorithm for discrete time modelling.

function DTRJ(o, d, h)

$u \leftarrow u + 1$	*set a new search
$t_d \leftarrow h$, $s_d \leftarrow -h$, $u_d \leftarrow u$	*initialize the destination solution
$B \leftarrow \{d\}$	*initialize the bucket list B with the destination
do until $B = \varnothing$	*repeat until B is empty
$j \leftarrow ArgMin\{w_i + \alpha \cdot c_{oi} : i \in B\}$	*identify the node with lowest cost priority in B
if $j = o$ **then stop**	*terminate when the origin is extracted
$B \leftarrow B\text{-}\{j\}$	*extract the node from the list
for each $a \in j^-$	*for each arc of the backward star
$i \leftarrow a[-]$	*set the arc tail
$t \leftarrow \theta^{-1}(a, t_j)$	*set the latest entry time index
if $t \geq 0$ **and** $(\theta(a, t) = t_j$ **or** $\theta(a, t+1)$	*entry index is defined
$\leq \eta$ **or** $a[+] \in N^{wait})$ **then**	
if $u_i < u$ **or** $t_i < t$ **then**	*does it improve the current solution?
$t_i \leftarrow t$, $s_i \leftarrow a$, $u_i \leftarrow u$	*if yes, update the tail solution
$B \leftarrow B \cup \{i\}$	*if not present, insert the node tail in B
end if	
end if	
next a	
loop	

end function

Once the entry time index t is set, at least one of the following three conditions must hold to consider the backward arc a:

- the exit index falls exactly at the latest instant of the head;
- the service is locally continuous;
- waiting in the head of the arc is allowed.

If the algorithm terminates because list B is empty, then no solution may be found.

In the case of continuous time modelling, the following further changes and variables are required:

- the root identifies one destination node $d \in N$ and one arrival time $\tau_0 \leq \tau \leq \tau_\eta$;
- z_i is the latest time when a user leaves node $i \in N$ to reach the destination d at time τ;
- t_i becomes the interval that contains z_i.

Algorithm 5. The reverse User-Trajectory algorithm for continuous time modelling.

function CTRJ(o, d, τ)

$u \leftarrow u + 1$	*set a new search
$h \leftarrow 0$, **do until** $\tau_h \geq \tau : h \leftarrow h+1$ **loop**	*find the arrival time interval
$z_d \leftarrow \tau$, $t_d \leftarrow h$, $s_d \leftarrow -1$, $u_d \leftarrow u$	*initialize the destination solution
$B \leftarrow \{d\}$	*initialize the bucket list with the destination
do until $B = \varnothing$	*repeat until the list is empty
$j \leftarrow ArgMin\{w_i + \lambda \cdot c_{oi} : i \in B\}$	*identify the node with lowest cost priority in B
if $j = o$ **then stop**	*terminate when the origin is extracted

```
    B ← B-{j}                                *extract the node from the list
      for each a ∈ j⁻                        *for each arc of the backward star
*A    i ← a[-]                               *set the arc tail
      (t, z) ← θ⁻¹(a, tⱼ, zⱼ)                *set the latest entry time z to exit before time zⱼ
      if t > 0 then                          *if the entry time interval is defined
        if uᵢ < u or zᵢ < z then             *does the tentative label z improve the solution?
          zᵢ ← z , tᵢ ← t , sᵢ ← a , uᵢ ← u  *if yes, update the tail solution
          B ← B∪{i}                          *if not present, insert the tail in B
        end if
      end if
    next a
  loop
end function

function θ⁻¹(a, e, ω)
  t ← θ⁻¹(a, e)+1 , do until t = 0 or θ_{a t-1}      *find the entry time interval
  ≤ ω : t ← t − 1 loop
  if t > 0 and (θ(a, t) ≤ η or a[+]∈N^{wait}) then   *if the entry time interval is defined
    if θ(a, t) ≤ η then                              *is the service locally continuous?
      α ← (ω − θ_{a t-1})/(θ_{at} − θ_{a t-1})       *if yes, interpolate
    else
      α ← 0                                          *otherwise, take the initial instant
    end if
    z ← (1 − α) · τ_{t-1} + α · τ_t                  *set the tentative label z
  end if
  return (t, z)
end function
```

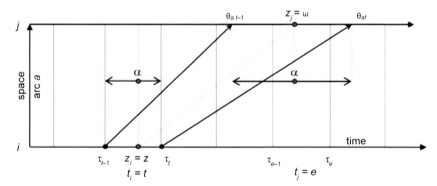

Figure 7.5 Interpolation of the leaving time in the continuous version of the user-trajectory algorithm.

The above entry time function yields the latest entry time z to exit arc a before a given time ω. It requires as an input also the index of the interval e containing ω and provides as an output also the index of the interval t containing z.

We assume that in presence of service discontinuity, where $t − 1 \in T_a$ but $t \notin T_a$, if wait at the head of arc a is allowed then the passenger will leave arc a at instant $t − 1$ (i.e., $\alpha = 0$).

The extension to strategies can be achieved through the greedy approach. This time Option 2 is applied and the tentative leaving time is the earliest inverse of the conditional exit times among all branches of the current hyperarc. To this end, the following is to be inserted after instruction *A:

if $i \in N^{div}$ **and** $u_i = u$ **then**	*if the tail is a labelled diversion node
$\breve{a} \leftarrow s_i \cup \{a\}$	*build-up the hyperarc from the successor
$(t^*, z^*) \leftarrow (0, \infty)$	*initialize the tentative label z^*
for each $b \in \breve{a}$	*for each branch of the hyperarc
$k \leftarrow b[+]$	*set the branch head
$(t, z) \leftarrow \theta^{-1}(b \mid \breve{a}, t_k, z_k)$	*set the latest entry time z to exit before time z_k
if $t > 0$ **then**	*if the travelling branch is feasible
if $z < z^*$ **then** $(t^*, z^*) \leftarrow (t, z)$	*save the earliest entry time
else	
$z^* \leftarrow -\infty$	*otherwise, set the tentative label to minus infinity
exit for	
end if	
next b	
if $z_i < z^*$ **then**	*does z^* improve the current solution?
$z_i \leftarrow z^*$, $t_i \leftarrow t^*$, $s_i \leftarrow \breve{a}$	*if yes, update the tail solution
$B \leftarrow B \cup \{i\}$	*if not present, insert the tail in B
end if	
else	

7.4.3 The Multi-Label Algorithm

The main advantage of the temporal-layer approach is its correctness. The main disadvantage is its weak efficiency. If an assignment problem is to be solved to one destination from all origins and for each departure time, then we have to explore (almost) all vertices and edges of the space-time network anyhow. But in a journey planning search from o to d for a given arrival time τ, this is computationally heavy wrt the complexity of a similar path search on a static network.

The main advantage of the User-Trajectory approach is its efficiency. The main disadvantage is its weak correctness. The version proposed in Section 7.4.2 based on cost priority order is specifically conceived for journey planning. But if generalized costs are not proportional to travel time, e.g., in the presence of relevant tolls and fares, then the algorithm provides only a suboptimal solution on a dynamic network.

Is there a routing method which combines correctness and efficiency for dynamic journey planning?

In this section we propose a new Multi-Label algorithm that has the same structure of the User-Trajectory algorithm but handles several points (or vertices, in its discrete version) for each node. More specifically, two assumptions are made:

- only the non-dominated points of a node in terms of travel time and generalized cost are kept in a list and considered for further exploration;
- the generalized costs follow a FIFO rule like the travel times, so that it is never convenient to wait (when possible) for a later service on the same arc (and to anticipate an exit for reverse searches).

As Fig. 7.6 shows, given a set of points relative to a given node, only a subset of them is dominating, or Pareto optimal. Point 1 (depicted in red) has the best route cost to destination, Point 2 (depicted in green) has the best route time to destination. These two points are the pillar of the Pareto frontier. Points 3 and 4 (depicted in blue) are not dominated and they belong to the Pareto frontier. All the other points are dominated as they belong to the influence area (first quadrant) of one or more Pareto point.

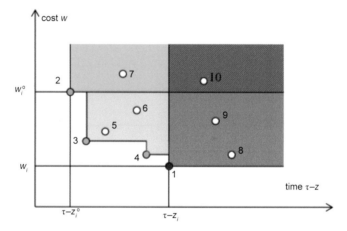

Figure 7.6 Set of dominating and dominated points for one node i.

The need for keeping several labels (points) for the two objectives (best cost and best time) and not only for the main objective (best cost) depends on the fact that a better time to destination for a given node can imply more opportunities (e.g., connections) to achieve a better cost for other nodes in the search.

Let's take an example. In Fig. 7.7, we can see that if we keep only the best solution in terms of route cost from node 3, that is to use arcs (3, 2) and (2, 1) at a minimum generalized cost of 9, then the solution for node 4 has a cost of 14. Instead, if we consider also the direct route (3, 1) which has a higher cost of 10 but allows for a later connection, then the minimum cost from node 4 is 13.

The cost FIFO assumption is made to avoid the consideration of more convenient earlier connections (for reverse searches, that is what we

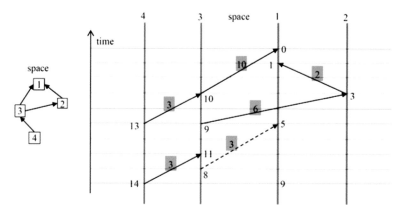

Figure 7.7 Example of network with discontinuous connections. The numbers next to edges are costs. The numbers next to vertices are minimum route costs to destination 1 for users arriving at the root instant. The cost of waiting for one interval is 1.

consider here; for forward searches, it would be later connections) where waiting is allowed at a node. This way, only the latest available connection for a given arc (of the backward star) is to be explored.

If this assumption is not made, we have to explore all the earlier connections until the waiting cost at the head is higher than the cost of the latest connection; if waiting is not very costly, we might have to explore lots of connections backward in time in the hope that a cheap connection is found.

Formally, the time FIFO (7.36) holds while the cost FIFO for arc $a \in A$ and instant $t \in T_a$ is stated as follows:

$$c_{at} \leq \gamma^{vot} \cdot \gamma_i \cdot (\tau_e - \tau_t) + c_{ae} \; \forall e \in T_a \geq t, \tag{7.39}$$

$$c_{at} \leq \gamma^{vot} \cdot \gamma_j \cdot (\theta_{at} - \theta_{ae}) + c_{ae} \; \forall e \in T_a \leq t, \tag{7.40}$$

where the waiting costs are given by (7.7).

To illustrate the cost FIFO rule, let's refer again to the network of Fig. 7.7. The dashed edge of arc (3, 1) does not satisfy FIFO cost. In fact it is convenient to anticipate the exit as this would provide a route cost of 8 for node 3. In this case, we have to explore all the connection of arc (3, 1) exiting later than the earliest vertex which has a cost of 9 (equal to the current solution for node 3).

Below the multi-label algorithm is presented for continuous time models. To handle the Pareto points, the following labels and variables are required for each node $i \in N$:

- w_i – the minimum route cost from i to reach destination d at time τ;
- z_i – the time when the user leaves i in the above least cost solution;
- z_i° – the latest time when a user can leave i to reach destination d time τ;

- w_i° – the route cost from i to reach destination d in the above latest solution;
- X_i – the set of Pareto points $(\tau - z, w)$ of node i;
- w_{ix} – the route cost from point $x \in X_i$ of node i;
- z_{ix} – the latest instant from point $x \in X_i$ of node i;
- t_{ix} – the index of the interval that contains z_{ix};
- s_{ix} – successive point from point $x \in X_i$ of node i to reach destination d at time τ.

Algorithm 6. The multi-label algorithm for continuous time modelling.

function CMLB(o, d, τ)	
$u \leftarrow u + 1$	*set a new search
$h \leftarrow 0$, **do until** $\tau_h \geq \tau : h \leftarrow h+1$ **loop**	*find the arrival time interval
NDOM($d, \tau, 0$) , $x \leftarrow (0, 0)$	*set the root point
$w_{dx} \leftarrow 0$, $z_{dx} \leftarrow \tau$, $t_{dx} \leftarrow h$, $s_{dx} \leftarrow -1$	*initialize the destination solution
$X_d \leftarrow X_d \cup \{x\}$	*insert the root point in the frontier of d
$B \leftarrow \{(d, x)\}$	*initialize the bucket list with the destination
do until $B = \varnothing$	*repeat until the list is empty
$(j, q) \leftarrow ArgMin\{w_{ix} + \lambda \cdot c_{oi} : (i, x)\in B\}$	*identify the node point with lowest priority in B
if $j = o$ **then stop**	*terminate when the origin is extracted
$B \leftarrow B-\{(j, q)\}$	*extract the node point from the list
for each $a\in j^-$	*for each arc of the backward star
*A $i \leftarrow a[-]$	*set the arc tail
$(t, z) \leftarrow \theta^{-1}(a, t_{jq}, z_{jq})$	*set the latest entry time to exit before z_{jq}
if $t \geq 0$ **then**	*if the entry time is defined
$w \leftarrow c_a(z) + w_{jq} + \gamma^{risk} \cdot \gamma^{vot} \cdot (z_{jq} - \theta_a(z))$	*set the tentative route cost
*B **if** NDOM(i, z, w) **then**	*is the tentative point x for node i
$x \leftarrow (\tau - z, w)$	*not dominated by the current solution?
$w_{ix} \leftarrow w$, $z_{ix} \leftarrow z$, $t_{ix} \leftarrow t$	*update the tail solution
$s_{ix} \leftarrow (j, q)$	*register the successive point
$X_i \leftarrow X_i \cup \{x\}$	*if not present, insert x in the frontier of i
$B \leftarrow B\cup\{(i, x)\}$	*if not present, insert the tail point in B
end if	
end if	
next a	
loop	
end function	

The successor of a node point is another node point and not an arc as in the other algorithms. This allows full definition of the dynamic concatenation of the resulting route in presence of multiple labels for each node.

To evaluate efficiently the functions $c_a(z)$ and $\theta_a(z)$ defined in Eqs. 7.20 and 7.21, it is necessary to know also the time index $t \in T$ that contains z. To this end, the latest entry time to exit arc a before z_{jq} is evaluated through the function $\theta^{-1}(a, t_{jq}, z_{jq})$ of Algorithm 5, which requires as an input the interval containing the leaving time z_{jq}. All these operations are further simplified if a discrete model is considered.

The main feature of the proposed algorithm is the introduction of Pareto points. The following function, not only defines if a tentative point $(\tau - z, w)$ for node i is dominated or not, but in the latter case the frontier and list B are cleaned up from dominated points and its defining labels and variables are updates.

If a point is not dominated, then in the main function the solution from the node is updated by inserting the new point in the frontier and its successive arc is recorded. Finally, the node point is added to list B to be later examined for further improvements.

function NDOM(i, z, w)

ψ = FALSE	*initialize the result
if $u_i < u$ **then**	*if node i has not been visited yet, initialize
$w_i \leftarrow \infty$, $z_i \leftarrow -\infty$	*the labels of the least cost solution (point 1)
$w_i^{\circ} \leftarrow \infty$, $z_i^{\circ} \leftarrow -\infty$	*the labels of the latest solution (point 2)
$X_i \leftarrow \varnothing$	*the set of frontier points
$u_i \leftarrow u$	*mark i as visited
end if	
if $w \leq w_i$ **then**	*tentative point below point 1
if $z \geq z_i$ **then**	*tentative point left of point 1
for each $x \in X_i$	*for each point of the frontier
if $z \geq z_{ix}$ **then**	*if it is dominated by the tentative point
$X_i \leftarrow X_i - \{x\}$	*remove the point from the frontier
$B \leftarrow B - \{(i, x)\}$	*if present, remove the node point from B
end if	
next x	
end if	
$w_i \leftarrow w$, $z_i \leftarrow z$	*update the labels of the least cost solution
ψ = TRUE	*the tentative point is not dominated
end if	
if $z \geq z_i^{\circ}$ **then**	*tentative point left of point 2
if $w \leq w_i^{\circ}$ **then**	*tentative point below point 2
for each $x \in X_i$	*for each point of the frontier
if $w \leq w_{ix}$ **then**	*if it is dominated by the tentative point
$X_i \leftarrow X_i - \{x\}$	*remove the point from the frontier
$B \leftarrow B - \{(i, x)\}$	*if present, remove the node point from B
end if	
next x	
end if	
$w_i^{\circ} \leftarrow w$, $z_i^{\circ} \leftarrow z$	*update the labels of the latest solution
ψ = TRUE	*the tentative point is not dominated
end if	
if $w_i < w < w_i^{\circ}$ **and** $z_i^{\circ} < z < z_i$ **then**	*tentative point between point 1 and point 2
for each $x \in X_i - \{1, 2\}$	*for each other point of the frontier
if $z_{ix} \geq z$ **and** $w_{ix} \leq w$ **then**	*if its influence area contains the tentative point
ψ = FALSE	*the tentative point is dominated
return ψ	*the procedure stops and returns FALSE
end if	
next x	
ψ = TRUE	*the tentative point is not dominated

```
    for each x∈X_i – {1, 2}              *for each other point of the frontier
      if z_{ix} < z and w_{ix} > w then   *if it is dominated by the tentative point
        X_i ← X_i – {x}                   *remove the point from the frontier
        B ← B – {(i, x)}                  *if present, remove the node point from B
      end if
    next x
  end if
  return ψ
end function
```

It is probably evident to the expert reader that the complexity of the Multi-Label algorithm is NP-hard because the number of points making up the non-dominated Pareto set can grow indefinitely in theory. However, some simple countermeasure can be adopted to prevent untractability in practice. For example, we could define elementary temporal intervals (e.g., of one minute) with no need to consider transfers and assume that for each of these intervals only the point with the minimum cost is kept in the Pareto set. As an extreme alternative, only the least cost solution (point 1) and the latest solution (point 2) are considered. Clearly, this leads to a heuristic.

7.5 Implementation for a Journey Planner

In this section we first introduce some notation to formalize the transit supply, then we propose a new routing algorithm that allows consideration in the same framework scheduled-based and frequency-based lines.

In case of journey planning, a passenger may requests for indications to make a trip on the transit network from an origin address (or the current position) to a destination address (or a Point Of Interest), possibly specifying a desired departure time or arrival time and his/her travel preferences. The latter will affect the result of the search through the coefficients that multiply the relevant arc (or turn) attributes to yield the generalized cost. However, they are expressed as a bundle (e.g., find the fastest route or the cheapest one) and/or through a limited set of options (avoid walking or stairs).

7.5.1 The Transit Network

In this section, the (spatial) topology of the transit network is defined through stops and lines, to represent (static) frequency-based services. Schedule-based services are then defined through dynamic attributes, such as the timetable that represents single runs. Thus, for the sake of efficiency, no space-time network (or diachronic graph) will be explicitly introduced here.

The topology and the main attributes of the infrastructures, such as roads and rails, are described through the base network (N^{base}, $A^{base} \subseteq N^{base} \times N^{base}$). Each base node $i \in N^{base}$ has geographic coordinates and each base arc $a \in A^{base}$ is described by a polyline with intermediate points (also called shape).

A transit network consists of a set S of stops between which services operate. A stop $s \in S$ is indeed a unique location (with geographic coordinates) where passengers can board and/or alight from transit services (e.g., a platform or curbside). The defining characteristic of a stop is that transfers within a single stop take zero walking time. Each stop $s \in S$ is associated with a base node (e.g., the closest one) denoted $N_s^{base} \in N^{base}$.

Transit services are organized in a set L of lines. A line $\ell \in L$ serves in one direction an ordered set of stops with no repetitions; this stop sequence (or itinerary) is denoted $S_\ell \subseteq S$. Thus, circular lines and side-trips within lines are excluded to simplify the presentation. Stops which are passed without boarding and alighting are omitted here from the stop sequence. We ignore special cases where boarding may be allowed at a stop, but not alighting, or vice versa. The first stop of a line, denoted $S_\ell^- \in S_\ell$, and its last stop, denoted $S_\ell^+ \in S_\ell$, are also called line terminals. The successive stop of stop $s \in S_\ell - S_\ell^+$ is denoted $s_\ell^+ \in S_\ell$; the previous stop of stop $s \in S_\ell - S_\ell^-$ is denoted $s_\ell^- \in S_\ell$. The part of a line between one stop $s \in S_\ell - S_\ell^+$ and the successive one s_1^+ is called a line segment; by convention we refer to it as to its first stop, so that the line segment associated to the last stop is a dummy. Each line segment $s \in S_\ell - S_\ell^+$ is associated with an acyclic path on the base network, whose support arcs are denoted $A_{\ell s} \subseteq A^{base}$.

Clearly, a same transit stop and a same support arc can be shared by more lines. A transit trip consists in general of several phases:

- starting the trip from the origin, which is point in geographic coordinates (e.g., obtained by geocoding a given address);
- accessing a stop from the origin by walking;
- waiting at that stop for a vehicle, of a line or of a run, depending on whether the service is frequency-based or schedule-based;
- boarding the vehicle while it is dwelling at the stop;
- running (or riding) on-board the vehicle through a sequence of stops;
- alighting the vehicle while it is dwelling at another stop;
- (possibly) transferring between two stops by walking;
- (possibly) repeat the phases from waiting to transferring a certain number of times;
- finally, egress from a transit stop to the destination by walking;
- and arriving at the destination, which is point in geographic coordinates (e.g., obtained by geocoding a given address).

Each trip phase is represented by a sequence of arcs with a same type (which specifies the nature of the trip phase) on the transit network. The latter is composed by:

- the pedestrian network articulated in base nodes and base arcs, including connectors created on the fly to represent access from origin point and egress to destination point;
- the line network, with a sub-network for each transit line, articulated in boarding, running, dwelling and alighting arcs, plus the stops shared by several lines;
- intermodal arcs at each stop to access the line network from the pedestrian network.

Several layers of nodes are introduced, among which we can distinguish the following:

- the base (pedestrian) nodes N^{base};
- the stop nodes S; each stop node $s \in S$ is distinct from its base node $N_s^{base} \in N^{base}$ because the line network and the base network often derive from two independent data sources;
- the line nodes N_ℓ, with one layer for each line $\ell \in L$; $N^{line} = \cup_{\ell \in L} N_\ell$; $L_i = \ell$ is the line associated with line node $i \in N_\ell$, S_i is the stop associated with it ($L_i = S_i = 0$, if $i \in N^{line}$).

The key feature of transit models is the representation of waiting as a separate trip phase. To this aim, when building-up the graph supporting the routing algorithm the stop must be exploded into a set of arcs and nodes. In the scheme depicted in Fig. 7.8, two nodes for each stop of line $\ell \in L$ are then introduced so as to represent consistently dwelling and running:

- the arrival node $N_{\ell s}^{arr} \in N_\ell$, $\forall s \in S_\ell - S_\ell^-$;
- the departure node $N_{\ell s}^{dep} \in N_\ell$, $\forall s \in S_\ell - S_\ell^+$.

To build up the transit network the following types of arcs and hyperarcs are introduced:

- the base (walking) arcs A^{base};
- the stop arcs $A^{stop} = \{(N_s^{base}, s): \forall s \in S\}$;
- the running arcs $A^{run} = \{(N_{\ell s}^{dep}, N_{\ell s[+\ell]}^{arr}): \forall s \in S_\ell - S_\ell^+, \forall \ell \in L\}$;
- the dwelling arcs $A^{dwell} = \{(N_{\ell s}^{arr}, N_{\ell s}^{dep}): \forall s \in S_\ell - S_\ell^- - S_\ell^+, \forall \ell \in L\}$;
- the alighting arcs $A^{alight} = \{(N_{\ell s}^{arr}, s): \forall s \in S_\ell - S_\ell^-, \forall \ell \in L\}$;
- the waiting arcs $A^{wait} = \{(s, N_{\ell s}^{dep}): \forall s \in S_\ell - S_\ell^+, \forall \ell \in L\}$;
- the waiting hyperarcs $H^{wait} \subseteq \{\breve{a} \subseteq s^+: \forall s \in S\}$.

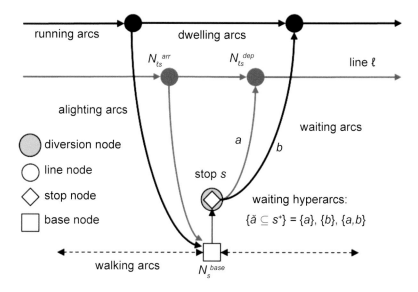

Figure 7.8 Topology of the transit network.

Consistently with the greedy algorithm, in the above scheme not necessarily all subsets of waiting arcs exiting from a given stop form a waiting hyperarc.

Each base arc $a \in A^{base}$ is characterized by the following features:

- l_a length of arc a;
- s_a^{walk} walking speed of arc a; the arc is walkable only if this is positive;
- s_a^{supp} speed of line segments supported by arc a (not including the stop time).

Note that the walking speed can also be seen as a characteristic of the user class. However, this would imply that travel times become class specific, which is inconsistent with the proposed model (7.6). However, different users (e.g., elderly people) can well have a different perception of walking times through the class specific discomfort coefficient.

Each line segment $s \in S_\ell - S_\ell^+$ of line $\ell \in L$ is characterized by the following features:

- $s_{\ell s}$ – commercial speed (including stop time);
- $c_{\ell s}^{kfee}$ – kilometric fee;
- $h_{\ell s}$ – expected headway; $f_{\ell s} = 1/h_{\ell s}$ is the frequency;
- $\sigma_{\ell s} \in [0,1]$ – irregularity or variation coefficient;
- $s_{ts} = 0$ deterministic headway, 1 exponential headway.

Each line $\ell \in L$ is characterized by the following additional features:

- c_ℓ^{bfee} – boarding fee;
- t_ℓ^{ab} – dwelling time available for alighting and boarding passengers;
- t_ℓ^{stop} – time lost to make a stop, including acceleration, door operation.

Each user class (or single user) is specified through its generalized cost coefficients:

- γ^{vot} – value of time;
- γ^{mfee} – fee multiplier;
- γ^{walk} – walking discomfort coefficient;
- γ^{wait} – waiting discomfort coefficient, including adverseness to variance;
- c^{tran} – perceived cost for each transfer.

Stops possess characteristics that can significantly reduce the high psychological burden of waiting, which is mainly due to the fact that passengers must pay attention and continuously check for carrier arrival whose instant is unknown. The provision of proper information (e.g., the arrival times of line vehicles) can play a crucial role, as well as the ergonomy of the stop in general (e.g., shelter, seating, air conditioning, entertainment, safety and security, protection from road noise and pollution). For longer waits, when the passenger can spend some time in the station before boarding a specific run without concerns about the departure time, the presence of other activities (e.g., shops) can have a relevant value.

Lines possess additional characteristics that may heavily impact comfort on-board, such as: seat ergonomy, vehicle style, air conditioning, cleanliness and silence inside carriers, safety and security on-board, additional services like (wifi) telecommunications, refreshments, entertainments. Another relevant (bundle) characteristic is the physical means of transport used to provide the service (e.g., bus, tram, metro, regional train, coach, high speed train, plane) on which passengers may have preferences.

Stops $s \in S$ and lines $\ell \in L$ are then characterized by qualitative Boolean attributes, that are relevant for the route choice of passengers, which make up two vectors $\mathbf{a}_s$ and $\mathbf{a}_\ell$, respectively. These (often very important) features are differently perceived by each user class and are connected with the travel or waiting time, that gives the exposure to discomfort. They can then be synthetized in:

- γ_s^{stop} – stop discomfort coefficient;
- γ_ℓ^{line} – line discomfort coefficient.

The following log-linear form can be assumed:

$$\gamma_s^{stop} = Exp(\sum_{c \in C^{cs}} \beta_c^{stop} \cdot a_{sc} / Exp(1)) \qquad (7.41)$$

$$\gamma_\ell^{line} = Exp(\sum_{c \in C^{cl}} \beta_c^{line} \cdot a_{\ell c} / Exp(1)) \qquad (7.42)$$

where β_c^{stop} and β_c^{line} are the class specific utility coefficients relative to stop and line attributes a_{sc} and $a_{\ell c}$, while C^{cs} and C^{cl} are the set of such attributes (if the sum is equal to 1, 2 and 3, the discomfort is approximately 10 per cent, 20 per cent and 30 per cent higher; if the sum is equal to –1, –2 and –3, the discomfort is approximately 10 per cent, 20 per cent and 30 per cent lower; if the sum is 0, the discomfort is unchanged).

7.5.2 Timetable and Dynamic Attributes

Each line $\ell \in L$ is served by an ordered set of runs, called run sequence and denoted R_ℓ. The set of all runs $R = \cup_{\ell \in L} R_\ell$ defines the whole transit service, while the line of run $r \in R$ is denoted $L_r \in L$.

A run $r \in R_\ell$ is constituted by one vehicle serving all stops on its line in order. The first run and the last run of line $\ell \in L$ are denoted, respectively, $R_\ell^- \in R_\ell$ and $R_\ell^+ \in R_\ell$. The successive run of run $r \in R_\ell - R_\ell^+$ is denoted $r^+ \in R_\ell$; the previous run of run $r \in R_\ell - R_\ell^-$ is denoted $r^- \in R_\ell$. As before, the part of a run between one stop $s \in S_\ell - S_\ell^+$ and the successive one s_ℓ^+ is called a run segment. Where in reality some runs of a line may be confined to only a sub-sequence of the stops, this can be represented as serving all stops of a separate, shorter, line.

We assume that each run $r \in R_\ell$ has a well-defined schedule, at least in the form of a working timetable defined by the operator (and maybe not communicated to passengers), where:

- τ_{rs} – is the arrival time for each stop $s \in S_\ell - S_\ell^-$;
- θ_{rs} – is the departure times for each stop $s \in S_\ell - S_\ell^+$;
- K_r – is the validity, i.e., a set of days when the run is operated.

The availability of a timetable is however not enough for a passenger to adopt a schedule-based behavior. Lines at stops are simulated as scheduled based if the following conditions jointly hold true:

- runs are specified, i.e., $R_\ell \neq \varnothing$;
- the expected headway is higher than a threshold, i.e., $h_{\ell s} > HMIN$ (say, 10 min.);
- the headway is deterministic, i.e., $\sigma_{\ell s} = 0$;

otherwise, they are simulated as frequency based. In the algorithm, to identify which one of the two passenger behaviors is considered, we introduce the following function, yielding if a given line $\ell \in L$ at a given stop $s \in S$ is or not treated as frequency based:

function ISFB(ℓ, s)

    return ($R_\ell = \varnothing$ **or** $h_{\ell s} \leq HMIN$ **or** $\sigma_{\ell s} > 0$)

end function

If $R_\ell \neq \varnothing$ the passenger will be routed by the algorithm on specific runs (even if he is not aware of this) and the dynamic attributes that are needed to calculate his perceived costs (line frequency, headway irregularity, running time) will be obtained for each stop or segment directly from the timetable (as in Section 5.4), with no need for introducing temporal profiles for these variables.

Therefore, a relevant feature of the proposed algorithm is run indexing with respect to absolute time. To enhance this operation, a periodization with intervals of equal duration $\Delta\tau$ (e.g., 30 min. or 1000 sec.) is introduced. The following variables are not an input but are calculated at initialization to improve the search engine efficiency:

- R_{it} – index of the first run serving line $\ell = L_i$ that arrives at the stop $s = S_i$ of node $i \in N_\ell$ before time $\tau_t = t \cdot \Delta\tau$, with $t \in [0, \eta]$; 0 means no preceding run (for the sake of simplicity, in the following we will use the run index to denote also the run object);
- $K_\ell = \cup_{r \in R\ell} K_r$ – the line validity.

7.5.3 Application of the Multi-Label Algorithm

Below we adapt the multi-label algorithm to compute shortest hyperpaths on transit networks with coexistent scheduled- and frequency-based lines. The proposed method is also implemented in the journey planner HyperPath (by PTV-SISTeMA), including parallelization (Attanasi et al., 2015) and Astar acceleration based on cluster precomputation (Maue et al., 2009).

Beside the arrival time τ, the procedure takes as input also the specific day k, so as to check for service validities in the arc performance function.

The service time of frequency-based lines is random. Waiting hyperarcs are processed through the greedy algorithm: given the new line, we try to include it in the attractive set of the existing solution(s), here represented by multiple frontier points. To compute the tentative route cost, the recursive update formula for exponential headways is adapted to comply with Option 2 (see Section 7.2.7), by including the cost of anticipating the leaving time of the current point, if $\delta > 0$, or of the new line, otherwise. If headways are not exponential, the benefit of higher regularity is reflected in a lower (upper bound of the) waiting time t provided by the arc performance function and thus is a higher equivalent frequency. The result is just a suitable approximation of more complex computations (Gentile et al., 2005). For example, the travel time of each branch of a hyperarc is its conditional waiting time; here we assume that these are all equal to the waiting time of the whole hyperarc (this is true for independent exponential headways, not necessarily for other headway distributions).

The algorithm requires definition of the following additional variables for the generic point $x \in X_i$ of node $i \in N$:

- r_{ix} – index of the run used at line node $i \in N^{line}$; 0 means that no specific run is used;
- f_{ix} – the (equivalent) cumulative frequency of the attractive set at stop $i \in S$.

Algorithm 7. The multi-label algorithm for coexistent frequency and scheduled-based lines.

function MLJP(o, d, τ, k)
  proceed like in CMLB until the instruction *A included
  * compute arc performances
  $\ell \leftarrow L_i$ *identify the line of the node
  if $i \in S$ **then** $s \leftarrow i$ **else** $s \leftarrow S_i$ *identify the stop of the node
  $(t, c, r) \leftarrow$ APF(a, r_{jq}, z_{jq}, k)   *evaluate the arc performance function
  if $t < \infty$ **then**   *if the entry time is defined
    * waiting hyperarcs
    if $a \in A^{wait}$ **and** ISFB(ℓ, s) **then**   *for frequency-based lines
      for each $p \in X_i : j \notin s_{ip}$ *for each point of the frontier not containing j

        $f \leftarrow f_{ip} + 1/t$ *set the tentative cumulative frequency
        $\delta \leftarrow z_{ip} + 1/f_{ip} - z_{jq}$ *δ compares the point leaving time once boarded

      if $\delta > 0$ **then**   *with the leaving time of the new line
        $w \leftarrow (w_{ip} \cdot f_{ip} + w_{jq}/t + \gamma^{risk} \cdot \gamma^{vot} \cdot \delta \cdot f_{ip})/f$ *set the tentative route cost
        $z \leftarrow z_{jq} - 1/f$ *set the entry time of the tentative point
      else
        $w \leftarrow (w_{ip} \cdot f_{ip} + w_{jq}/t - \gamma^{risk} \cdot \gamma^{vot} \cdot \delta/t)/f$ *set the tentative route cost
        $z \leftarrow z_{ip} + 1/f_{ip} - 1/f$ *set the entry time of the tentative point
      end if
      if NDOM(i, z, w) **then**   *is the tentative point x for node i
        $x \leftarrow (\tau - z, w)$ *not dominated by the current solution?
        $w_{ix} \leftarrow w, z_{ix} \leftarrow z$ *update the tail solution
        $r_{ix} \leftarrow r$ *set the tail run *1
        $f_{ix} \leftarrow f$ *set the cumulative frequency *2
        $s_{ix} \leftarrow s_{ip} \cup (j, q)$ *add the successive point to the solution
        $X_i \leftarrow X_i \cup \{x\}$ *if not present, insert the point in the frontier of i
        $B \leftarrow B \cup \{(i, x)\}$ *if not present, insert the tail point in B
      end if
      next x
    end if   *standard arc
    $z \leftarrow z_{jq} - t$ *set the tentative entry time
    $w \leftarrow w_{jq} + c$ *set the tentative route cost

if $a \in A^{wait}$ **and** ISFB(ℓ, s) **then** $f \leftarrow 1/t$ *initialize the cumulative frequency [*2]
    proceed like in CMLB from the instruction [*B] included
end function

    [*1] The run index substitutes the interval index of the CMLB.
    [*2] The cumulative frequency is to be saved together with the other point features.

7.5.4 Transit Arc Performance

The arc performance function yields the travel time t and the generalized cost c for given arc a and exit time z. Like for function $\theta^{-1}(a, e, \omega)$, to speed-up indexing the run r (which plays here the role of the time interval containing z) is taken as further input and provided as further output. As mentioned earlier, the run index represents a reference to calculate the dynamic attributes. In case no run is specified (i.e., $r = 0$), then the calculation of the attributes uses static information.

    Algorithm 7. The arc performance function.

function APF(a, r, z, k)
  $i \leftarrow a[-]$ *set the arc tail
  $j \leftarrow a[+]$ *set the arc head
  $\ell \leftarrow L_i$, **if** $\ell = 0$ **then** $\ell \leftarrow L_j$ *identify the line of the node
  $s \leftarrow S_i$, **if** $s = 0$ **then** $s \leftarrow S_j$ *identify the stop of the node
       *update the run index
  if $a \in A^{alight}$ **then**   *in case of alighting arc
    if $k \notin K_\ell$ **or** $R_\ell = \varnothing$ **then**   *if the line is not valid or has no run defined
      $r \leftarrow 0$ *the run is undefined
    else
      $e \leftarrow 1 + z \backslash \Delta\tau$ *let e be the interval containing z [*1]
      $r \leftarrow R_{ie}$ *start from the corresponding run index
      do until $r = 0$ **or** $(\tau_{rs} \leq z$ **and** $k \in K_r)$ *find the first valid run arriving before z
        $r \leftarrow r - 1$ *shift to the previous run
      loop
    end if
  else if $a \in A^{wait}$ **then**   *in case of waiting arc
    $r \leftarrow 0$ *reset the arc index of the stop
  end if

                              *calculate the travel time and the generalized cost
  if $a \in A^{run}$ **then**   *in case of running arc
    if $r > 0$ **then**   *a run is defined
      $t \leftarrow \tau_{r\,s+1} - \theta_{rs}$ *running time based on the timetable
    else if $s_{\ell s} > 0$ **then**   *the commercial speed of the line segment is
      $t \leftarrow (\Sigma_{b \in A\ell s}\, l_b)/s_{\ell s}$ defined
    else
      $t \leftarrow \Sigma_{b \in A\ell s}\, l_b / s_b^{supp} + t_1^{stop}$
    end if
    $c \leftarrow \gamma^{vot} \cdot \gamma_\ell^{line} \cdot t + \gamma^{mfee} \cdot (\Sigma_{b \in A\ell s}\, l_b) \cdot c_{\ell s}^{kfee}$
  else if $a \in A^{dwell}$ **then**   *in case of dwelling arc

> **if** $r > 0$ **then**   *a run is defined
> $t \leftarrow \theta_{rs} - \tau_{rs}$ *dwell time based on the timetable
> **else**
> $t \leftarrow t_{\ell}^{ab}$
> **end if**
> $c \leftarrow \gamma^{vot} \cdot \gamma_{\ell}^{line} \cdot t$
> **else if** $a \in A^{base}$ **then**   *in case of base arc
>   **if** $s_a^{walk} > 0$ **then**   *the arc is walkable
> $t \leftarrow l_a / s_a^{walk}$
>   **else**
> $t \leftarrow \infty$
>   **end if**
> $c \leftarrow \gamma^{vot} \cdot \gamma^{walk} \cdot t$
> **else if** $a \in A^{alight}$ **then**   *in case of alighting arc
>   **if** $r > 0$ **then**   *a run is defined
>   **if** ISFB(ℓ, s) **then**   *frequency-based approach
> $t \leftarrow 0$
>   **else**
> $t \leftarrow z - \tau_{rs}$ *waiting time based on the timetable *2
>   **end if**
>   **else if** $k \in K_{\ell}$ **and** $R_{\ell} = \varnothing$ **then**   *the line is valid
> $t \leftarrow 0$
>   **else**   *no valid run or line available
> $t \leftarrow \infty$
>   **end if**
> $c \leftarrow \gamma^{risk} \cdot \gamma^{vot} \cdot t + c^{tran} + c_{\ell}^{bfee}$
> **else if** $a \in A^{wait}$ **then**   *in case of waiting arc
>   **if** $r > 0$ **then**   *a run is defined
>   **if** ISFB(ℓ, s) **then**   *frequency based approach
> $h \leftarrow r + 1$, **do until** $h > |R_{\ell}|$ **or**   *find the first valid run after r
> $k \in K_h : h \leftarrow h + 1$ **loop**
>   **if** $h > |R_{\ell}|$ **then**   *no next run is found
> $t \leftarrow 0.5 \cdot h_{\ell s}$
>   **else**
> $t \leftarrow 0.5 \cdot (\theta_{hs} - \theta_{rs})$ *waiting time based on the timetable
>   **end if**
>   **else**
> $t \leftarrow 0$
>   **end if**
>   **else**
> $t \leftarrow 0.5 \cdot h_{\ell s}$
>   **end if**
> $t \leftarrow t \cdot (1 + \sigma_{\ell s}^2)$ *consider the line irregularity (null if scheduled-
> $c \leftarrow \gamma^{vot} \cdot \gamma^{wait} \cdot \gamma_s^{stop} \cdot t + \gamma^{risk} \cdot \gamma^{vot} \cdot t$ based)*3
> $t \leftarrow 2 \cdot t$ *the upper bound is twice the expected time *4
> **else if** $a \in A^{stop}$ **then**   *in case of stop arc
> $t \leftarrow 0$
> $c \leftarrow 0$
> **end if**
> **return** (t, c, r) *return the travel time and the generalized cost
> **end function**

Waiting time after alighting is strange *[2]. This is not what usually happens, as the passenger proceeds immediately towards the next stop, or towards the destination. In principle, he will wait there for the same amount of time, but it can also happen that he achieves boarding an earlier run. A concatenation of favorable events can bring him to the destination before the desired arrival time. Moreover, there is always some risk in travelling, so that people feel better to get closer to the destination earlier in order to have more opportunities for reacting to events. A similar consideration is also valid for the anticipation of the branches leaving time that is necessary to evaluate hyperarcs, as well as for the compensation of the risk adverse concatenation *[3].

On this basis, it is clear how $\gamma^{risk} \in [0, 1]$ incorporates in a bundle two different factors: 0 means that the passenger recognizes that on an average he will be able to reach the destination earlier than the desired arrival time and can fully exploit his time there; 1 means that the passenger is totally risk adverse or cannot exploit an earlier arrival at destination.

In any case, the passenger departs form the origin at a time that allows him to reach the destination within the desired arrival time, i.e., he is risk adverse wrt the time concatenation (e.g., scheduled connections). The above considerations regards only the generalized costs.

*[1] The integer division "\" is the base for run indexing operations.

*[3] For waiting arcs the generalized cost is proportional to the travel time. In fact, only in this case the cost updating formula of the greedy approach is valid.

*[4] In case of waiting arcs for frequency-based lines, t is an estimation of the travel time upper bound. Its assumption as twice the expected value is valid only for deterministic headways, otherwise it is just a viable approximation.

keywords: public transport assignment; multimodal journey planning; service perturbation and strategic behavior; coexistence of frequency-based and schedule-based lines.

References

Attanasi, A., E. Silvestri, P. Meschini and G. Gentile. 2015. Real world applications using parallel computing techniques in dynamic traffic assignment and shortest path search. IEEE 18th International Conference on Intelligent Transportation Systems, 316–321.

Bast, H., D. Delling, A. Goldberg, M. Müller-Hannemann, P. Sanders, D. Wagner and R. Werneck. 2015. Route Planning in Transportation Networks. Technical Report, Department of Informatics, Karlsruhe Institute of Technology, Karlsruhe, Germany.

Baum, M., J. Dibbelt and D. Wagner. 2015. Dynamic Time-dependent Route Planning in Road Networks with User Preferences. Technical Report, Department of Informatics, Karlsruhe Institute of Technology (KIT), Karlsruhe, Germany.

Bell, M. G. H., V. Trozzi, S. H. Hosseinloo, G. Gentile and A. Fonzone. 2012. Time-dependent hyperstar algorithm for robust vehicle navigation. Transportation Research Part A 46: 790–800.

Bellei, G., G. Gentile and N. Papola. 2000. Transit assignment with variable frequencies and congestion effects. Proceedings of the 8th Meeting of the EURO Working Group on Transportation, Rome, Italy.

Bellman, R. E. 1957. Dynamic Programming. Princeton University Press, Princeton, NJ, USA.

Bellman, R. E. 1958. On a routing problem. Quarterly of Applied Mathematics 16: 87–90.

Billi, C., G. Gentile, S. Nguyen and S. Pallottino. 2004. Rethinking the wait model at transit stops. Proceedings of TRISTAN V, Guadeloupe, French West Indies.

Chabini, I. 1998. Discrete dynamic shortest path problems in transportation applications: Complexity and algorithms with optimal run time. Transportation Research Record 1645: 170–175.

Cooke, L. and E. Halsey. 1966. The shortest route through a network with time-dependent intermodal transit times. Journal of Mathematical Analysis and Applications 14: 492–498.

Denardo, E. V. and B. L. Fox. 1979. Shortest-route methods: 1. reaching, pruning, and buckets. Operations Research 27: 161–186.

De Palma, A., P. Hansen and M. Labbé. 1993. Commuters' paths with penalties for early or late arrival times. Transportation Science 24: 276–286.

Dial, R. B. 1969. Algorithm 360: shortest-path forest with topological ordering. Communications of the ACM 12: 632–633.

Dijkstra, E. W. 1959. A note on two problems in connexion with graphs. Numerische Mathematik 1: 269–271.

Ding, B., Yu J. Xu and L. Qin. 2008. Finding time-dependent shortest paths over large graphs. Proceedings of the 11th International Conference on Extending Database Technology: Advances in Database Technology, 205–216.

Dreyfus, S. E. 1969. An appraisal of some shortest-path algorithms. Operations Research 17: 395–412.

Fredman, M. L. and R. E. Tarjan. 1984. Fibonacci heaps and their uses in improved network optimization algorithms. Proceedings of the 25th Annual Symposium on Foundations of Computer Science, 338–346.

Ford, L. R. 1956. Network Flow Theory. Paper P-923. Santa Monica, California: RAND Corporation.

Floyd, R. W. 1962. Algorithm 97: Shortest path. Communications of the ACM 5: 345.

Gallo, G., G. Longo, S. Nguyen and S. Pallottino. 1993. Directed hypergraphs and applications. Discrete Applied Mathematics 42: 177–201.

Gallo, G. and S. Pallottino. 1988. Shortest path algorithms. Annals of Operations Research 7: 3–79.

Gentile, G. and K. Noekel. 2016. Modelling Public Transport Passenger Flows in the Era of Intelligent Transport Systems: COST Action TU1004 (TransITS). Springer Tracts on Transportation and Traffic, 10, Springer, Switzerland.

Gentile, G., S. Nguyen and S. Pallottino. 2005. Route choice on transit networks with online information at stops. Transportation Science 39: 289–297.

Gentile, G. and L. Meschini. 2004. Fast Heuristics for Continuous Dynamic Shortest Paths and All-or-nothing Assignment. Presented at AIRO 2004, Lecce, Italy.

Gentile, G., K. Noekel, J-D. Schmoecker, V. Trozzi and E. Chandakas. 2016. The theory of transit assignment: demand and supply phenomena. pp. 387–486. *In*: G. Gentile and K. Noekel (eds.). Modelling Public Transport Passenger Flows in the Era of Intelligent Transport Systems: Cost Action Tu1004 (Transits). Springer Tracts On Transportation And Traffic 10, Springer, Switzerland, Doi: 10.1007/978-3-319-25082-3_7. Hart, P. E., N. J. Nilsson and B. Raphael. 1968. A formal basis for the heuristic determination of minimum cost paths. IEEE Transactions on Systems Science and Cybernetics 4: 100–107.

Kaufman, D. E. and R. L. Smith. 1993. Fastest Path in time dependent networks for intelligent vehicle-highway system applications. IVHS Journal 1: 1–11.

Maue, J., P. Sanders and D. Matijevic. 2009. Goal-directed shortest-path queries using precomputed cluster distances. ACM Journal of Experimental Algorithmic 14.3.2: 1–27.

Meschini, L., G. Gentile and N. Papola. 2007. A frequency based transit model for dynamic traffic assignment to multimodal networks. Transportation and Traffic Theory 2007 (Papers selected for presentation at ISTTT17, a peer reviewed series since 1959. R. E. Allsop, M. G. H. Bell and B. G. Heydecker (eds.), Elsevier, London, UK, 407–436.

Montemanni, R., and L. M. Gambardella. 2004. An exact algorithm for the robust shortest path problem with interval data. Computers & Operations Research 31: 1667–1680.

Moore, E. F. 1959. The shortest path through a maze. Proceedings of International Symposium Switching Theory. 1957. Part II: Harvard University Press, Cambridge, MA, USA, 285–292.

Nachtigall, K. 1995. Time-depending shortest path problems with applications to railway networks. European Journal of Operational Research 83: 154–166.

Nguyen, S. and S. Pallottino. 1988. Equilibrium traffic assignment for large-scale transit networks. European Journal of Operational Research 37: 176–186.

Noekel, K. and S. Wekeck. 2009. Boarding and alighting in frequency-based transit assignment. Transportation Research Record 2111: 60–67.

Orda, A. and R. Rom. 1990. Shortest-path and minimum-delay algorithms in network with time-dependent edge length. Journal of the ACM 37: 607–625.

Pallottino, S. and M. G. Scutellà. 1998. Shortest path algorithms in transportation models: Classical and innovative aspects. pp. 245–281. *In*: P. Marcotte and S. Nguyen (eds.). Equilibrium and Advanced Transportation Modelling, Kluwer Academic Publisher, Dordrecht, The Netherlands.

Papola, N., F. Filippi, G. Gentile and L. Meschini. 2009. Schedule-based transit assignment: A new dynamic equilibrium model with vehicle capacity constraints. pp. 145–171. *In*: N. Wilson and A. Nuzzolo (eds.). Schedule-Based Modelling of Transportation Networks. Theory and Applications, Springer.

Pape, U. 1974. Implementation and efficiency of Moore-algorithms for the shortest route problem. Mathematical Programming 7: 212–222.

Pearl, J. 1984. Heuristics: Intelligent Search Strategies for Computer Problem Solving. Addison-Wesley, 48.

Samaranayake, S., S. Blandin and A. Bayen. 2012. A tractable class of algorithms for reliable routing in stochastic networks. Transportation Research C 20: 199–21.

Spiess, H. and M. Florian. 1989. Optimal strategies: a new assignment model for transit networks. Transportation Research B 23: 83–102.

Sung, K., M. Bell, M. Seong and S. Park. 2000. Shortest paths in a network with time-dependent flow speeds. European Journal of Operational Research 121: 32–39.

Trozzi, V. G. Gentile, M. G. H. Bell and I. Kaparias. 2013. Dynamic user equilibrium in public transport networks with passenger congestion and hyperpaths. Transportation Research B 57: 266–285.

Yen, J. Y. 1970. An algorithm for finding shortest routes from all source nodes to a given destination in general networks. Quarterly of Applied Mathematics 27: 526–530.

Yu, G. and J. Yang. 1998. On the robust shortest path problem. Computers & Operations Research 25: 457–468.

Warshall, S. 1962. A theorem on Boolean matrices. Journal of the ACM 9: 11–12.

Ziliaskopoulos, A. K. and H. S. Mahmassani. 1993. Time-dependent shortest path algorithm for real-time intelligent vehicle highway system applications. Transportation Research Record 1408: 94–100.

8

Real-time Reverse Dynamic Assignment for Multiservice Transit Systems

Francesco Russo[a] and *Antonino Vitetta*[b,]*

ABSTRACT

In this chapter the Reverse Dynamic Assignment (RDA) is specified in terms of model and solution procedures for transit systems. In transportation systems output data of assignment models, considering path choice behavior models, are flows and costs given as input the parameters of demand and supply. The RDA reverses the classic assignment problem. The variables obtained from the observed flows and costs are the demand in terms of values and parameters and the parameters of some link cost attributes relative to the supply. The method is proposed in terms of optimization model and solution procedures. A section is dedicated to the dynamic assignment model, one of the constraints of the reverse assignment. A numerical application on a test system is reported. The aim of the numerical test is to demonstrate the method applicability in a fixed rolling horizon.

Dipartimento di Ingegneria dell'Informazione, delle Infrastrutture e dell'Energia Sostenibile, Università Mediterranea di Reggio Calabria, Italy.

[a] E-mail: francesco.russo@unirc.it
[b] E-mail: vitetta@unirc.it
* Corresponding author

In real applications, more detailed models can be used in relation to the available data, elaboration power and elapsed time available for the optimization.

8.1 Introduction

Considering the interaction between demand and supply in transportation systems, four themes are identified: (i) assignment, (ii) cost function calibration, (iii) demand update, and (iv) reverse assignment.

(I) *Assignment* (Fig. 8.1) gives as output flows and costs starting from the supply and demand models with fixed parameters. The core elements of assignment models are the path choice models. The state of the art relative to the assignment models, for transit systems, is reported in the previous chapters of this book.

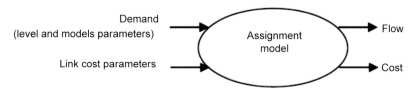

Figure 8.1 Assignment model.

In a transport system, the traffic assignment model can be defined as (in the paper the following Eq. 8.1 is described in Sections 8.2 and 8.3 and reported as Eq. 8.6):

$$q(\alpha, \beta, d) = P(\xi(\alpha), \beta) \cdot d \tag{8.1}$$

with

- q, the flow values vector;
- α the vector of the parameters relative to link cost attributes;
- β the vector of the parameters relative to expected user's utility;
- d, the demand values vector for transit systems;
- P, the assignment matrix;
- $\xi(\alpha)$, the time and cost attributes vector, function of α.

In case of equilibrium, the function ξ depends also on q.

The assignment problem is interpreted in terms of geometric representation in Fig. 8.2a, considering an O/D pair with only one perceived bus line that connects the O/D and a steady state system (considered also for the other cases in the following). The input and the output variables are reported separately in the figure. A demand function and a cost function relative to the trip (Russo and Vitetta, 2011b) are considered. The traffic assignment considers a pre-defined supply and demand functions; with some properties on the link cost and demand functions, the analytical solution is in the point of intersection between the supply and the demand functions and, in the case reported in Fig. 8.2a, it is a stable equilibrium point. On the left of the equilibrium point, the supply cost is lower than the cost associated with the demand and the system moves versus the equilibrium point, considering that the demand increases. On the right of the equilibrium point, the supply cost is higher than the cost associated with demand and the system moves versus the equilibrium point, considering that the demand decreases.

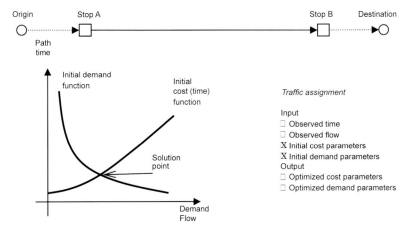

Figure 8.2a Geometric interpretation of the assignment problems.

(II) *Cost function calibration* from observed times gives as output the link cost parameters (α), given the demand parameters (β) and values (d), the functional form of the demand at route choice level (P) and supply functions (ξ).

A wide review of the calibration method for cost function parameters in transit systems is reported in Moreira-Matias et al. (2015). The parameter calibration in a transit network is studied also in Parveen et al., 2007. The transit arrival prediction time with AVL (Automatic Vehicle Locations) data is studied in Horbury, 1999; Wall and Dailey, 1999; Cathey and Dailey, 2003.

The cost function calibration problem is interpreted in terms of geometric representation in Fig. 8.2b (left). In these types of problems, the demand function is supposed fixed and is thus not calibrated. The cost function is calibrated considering the observed flows and costs. The method to calibrate the function is derived from different theoretical approaches (i.e., minimum least square, Kalman filter). The demand and the cost functions before the calibration are shown in the figure. After the calibration, in relation to the data and calibration method, a new cost function is obtained. Once the assignment problem with the new cost function is solved, considering that the demand function does not modify, the new solution is the intersection point between the new cost function and the initial demand function. If, in the minimization function of the calibration method, a measure of the distance between the initial and the observed cost is considered, the solution belongs to the feasible set reported in the figure with a bold line. Considering link cost calibration, the cost function can change but the demand function is a constraint; the solution is in a mono-dimensional set.

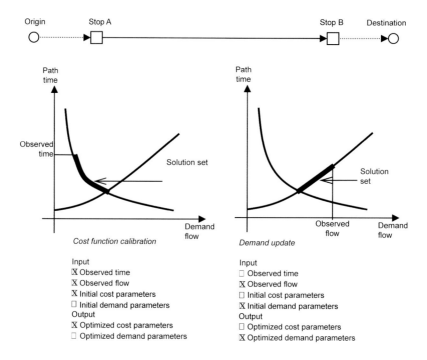

Figure 8.2b Geometric interpretation of cost function calibration and demand update problems.

(III) *Demand update* from observed flows gives as output the demand values (d) and/or demand parameters (β), considering given the cost parameters (α), the functional form of the demand at route choice level (P) and supply functions (ξ).

The demand update from observed flows (Cascetta and Nguyen, 1988) can be used for updating the demand values (i.e., with maximum entropy Van Zuylen and Willumsen, 1980; with Bayesian, Maher, 1983; with generalized least squares Cascetta, 1984; on two levels, Fisk, 1988; dynamic context Cascetta et al., 1993) and demand parameters (Cascetta and Russo, 1997). In relation to the transit system, the problem is studied in Nguyen et al., 1988.

The demand update is interpreted in terms of the geometric representation in Fig. 8.2b (right). In this type of problem, the cost function is supposed fixed and it is not calibrated. The demand function is calibrated considering the observed flows. The method to calibrate the function derives from different theoretical approaches (i.e., generalized least square, maximum likelihood, maximum entropy, Bayesian). The demand function and the cost function, before the calibration, are reported in the figure. After the calibration, in relation to the data and the calibration method, a new demand function is obtained. When the assignment problem with the new demand function is solved, considering that the cost function is not modified, the solution is the intersection point between the new demand function and the initial cost function. If, in the minimization function, a measure of the distance between the initial and the observed flow is considered, the solution belongs to the feasible set reported in the figure with a bold line. Considering a demand update, the demand function could change; given that the cost function is a constraint, the solution could be in a mono-dimensional set.

(IV) Cost function calibration and demand updating can be carried out with an approach of single optimizations, or simultaneously through *reverse assignment* approach. The reverse assignment on observed times and flows gives as output the demand values (d) and/or parameters (β) and the cost parameters (α), in the hypothesis that the functional form of the demand at route choice level (P) and supply functions (ξ) are given.

With the reverse assignment model, the classic assignment problem is reversed in respect to the input and output variables: the input variables are the output and vice versa. In a reverse assignment model the cost and demand functions parameters and values are estimated starting from observed data times and flows in a subset of links. In particular the Reverse Dynamic Assignment (RDA) model reverses the input with the output variables of a dynamic assignment model.

The inputs of the RDA model are the observed data in each time slice in terms of user flows on some link sections (i.e., users on given alighting and boarding links, users in given vehicles) and transit travel time on given links (i.e., dwell time). The application of RDA can be supported by the wide availability of real-time data for the transport system (Chapter 1 of this book).

The general reverse assignment model, which aims to obtain at the same time demand and cost function parameters, was proposed for a steady state road system (Russo and Vitetta, 2011a) with a least square method for the solution procedure and it is applied to transit systems in this chapter.

The reverse assignment problem is presented in geometric terms in Fig. 8.2c. In this type of problem, the demand and the cost functions are calibrated considering the observed times and flows (Fig. 8.3). The method to calibrate the function derives from different theoretical approaches (i.e., generalized least square, maximum likelihood, Bayesian). The demand and the cost functions before the calibration are reported in the figure. After a calibration procedure, in function of the data and the calibration method, new cost and demand functions are obtained. When the assignment problem with the new functions is solved, the solution is not bound to stay on one of the two starting functions. If, in the minimization function, some measures of the distances between the initial and the observed flow and cost are considered, the solution belongs to the feasible bi-dimensional set represented in the figure.

The crucial difference between the single optimization approach presented in literature and the RDA consists in the feasible set and in the solution that is in a wider set containing as edge the solution set of the single optimization problem. In reverse assignment, the solution is obtained with a constrained one-level problem, while the use of an

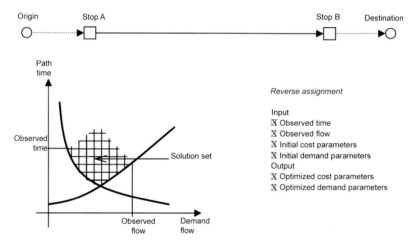

Figure 8.2c Geometric interpretation of the reverse assignment problems.

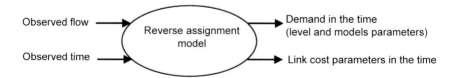

Figure 8.3 Reverse assignment model.

iterative single optimization solution does not guarantee the convergence to the same solution.

The RDA model is defined in terms of minimization of an objective function. It minimizes a measure of distance between the observed value and the modelled values for flows and time, considering also a measure of distance between the initial and optimized values of the demand variables.

In the minimization, the optimization variables could be: the demand values for each O/D pair (split or not in relation to the departure time) and the demand parameters; the route choice parameters; some link cost parameters. The variables could be considered generic for all the users or specific for each user class.

This chapter is divided in the following sections. After this introduction, which describes the RDA problem and state of the art, the assignment model, assumed as topological-behavioral constraint of the optimization problem, is reported in Section 8.2. The RDA in terms of model, variables and solutions procedure is specified in Section 8.3. A numeric application in a test system is presented in Section 8.4 and in Section 8.5, conclusions with further proposal developments are reported.

8.2 Assignment

The route choice and assignment models are recalled in this section, considering only the main elements useful for the RDA model; more general route choice and assignment models are presented in depth in Chapter 6. The type of model chosen for the RDA is dynamic, considering a mesoscopic simulation of network loading.

The dynamic path choice model used in this section is one of the specifications that can be adopted for the dynamic assignment model, but other kinds of models can be adopted and specified in the proposed reverse assignment model.

8.2.1 Supply and Demand: Definition and Notation

The RDA is applied considering a defined supply system because it is observed. The data are relative to the time preceding the instant of simulation. The supply state, in terms of observed services, is modelled and considered for the assignment. The demand is obtained from a simulation topological-behavioral model.

8.2.1.a Supply

The supply is modelled with a stochastic diachronic network. For the user's strategy, the hyperpath is considered (Nguyen and Pallottino, 1988; Spiess and Florian, 1989).

Let:

a) general

- t the period of simulation (i.e., one hour);
- τ the time instant, of the day t, when the system has to be modelled;
- $\Omega^{t,\tau}$ the network graph, represented with a diachronic approach;
- $\kappa_{M'}$ the set of hyperpaths defined at the origin (named origin hyperpaths);
- $\kappa_{s'}$ the set of hyperpaths defined in the decision node (named stop hyperpaths);
- κ, the set of available alternative hyperpaths (in the origin or in the stop);
- k, an alternative (hyperpath) belonging to the set κ of the alternatives;
- s, the stop;
- r, the run;
- j the elementary path, belonging to the hyperpath;
- n the link;

b) hyperpath

- $x^{t,\tau}_k$ the vector of the user's attributes relative to the instant of time τ of t, and to the alternatives k (between O-D);
- $x^{t,\tau}_{k,rs}$ the vector of the user's attributes relative to the instant time τ of t, to the run r, belonging to k, in the stops s;
- x the vector of user's attributes with all the $x^{t,\tau}_k$ and $x^{t,\tau}_{k,rs}$ vectors;
- a the vector of the parameters relative to link cost attributes;
- $\xi(a)$, the vector of function that allow evaluating x from a ($x = \xi(a)$);
- $u^{t,\tau}_{k,exp}$ the utility of the user in τ of t, to the hyperpath k (it depends on the experienced utility and information received); it is a random variable;
- β the vector of the parameters relative to expected user's utility;
- $E(u^{t,\tau}_{k,exp}) = \psi_{exp}(x^{t,\tau}_{k,rs}, \beta)$ the expected value of $u^{t,\tau}_{k,exp}$ specifies as a linear function respect β;

c) elementary path

- $u^{t,\tau}_{k,rs}$ the user utility relative to the instant of time τ of t, to the run r, belonging to k, in the stops s; it is a random variable;
- $E(u^{t,\tau}_{k,rs}) = \varphi(x^{t,\tau}_{k,rs}, \beta)$ the expected value of $u^{t,\tau}_{k,rs}$ specifies as a linear function respect β;
- $q^{t,\tau d}_n$ the flow on the link n at time τ_d of τ of t;
- q, the flow values vector; each entry is $q^{\tau d,t}_n$.

8.2.1.b Demand

The demand is modelled with behavioral models. On the formulation of the reverse assignment problem, the demand in terms of users for the transit system at origin destination level and departure time is assumed known and is assumed to be elastic at path choice level. It will be considered elastic or un-elastic at departure time level, according to the consideration or not of a measure of the distance of the initial and modelled demand values in each time slice in the objective function of the reverse assignment (first term of Eq. 8.8).

Let:

a) (general)

- *O-D* an origin and destination pair for a homogenous group of users;
- *i* a generic time slice;
- τ_{TTi} the characterized desired (arrival or departure) target time of the time slice *i*;

b) (demand)

- $d_{O\text{-}D,\tau TTi}$ the demand on O-D and τ_{TTi} in term of number of homogenous users;
- *d* is the modelled demand vector relative to the day t, with generic entry $d_{O\text{-}D,\tau TTi}$;

c) (probability)

- $p^{t,\tau}(k \mid \kappa)$ the choice probability for the hyperpath *k*, belonging to the choice set κ, in the time τ of *t*;
- $p^{t,\tau}(r \mid s,k,\kappa)$ the choice probability for the run *r*, belonging to the hyperpath *k*, in the stop *s*, in the time τ of *t*;
- $p^{t,\tau}(j \mid k,\kappa)$ the choice probability for the elementary path *j* in the time τ of *t* conditioned to the choice of the hyperpath *k*;
- $p^{t,\tau}(j \mid \kappa)$ the choice probability for the elementary path *j* in the time τ of *t*;
- $p_n(\tau_d \mid j,\tau_{TTi})$ the choice probability of travelling on stay in the link *n*, belonging to the elementary path *j*, at τ_d, conditional upon departing at τ_{TTi};
- *P*, the assignment matrix; each entry (raw-column) is $p_n(\tau_d \mid j,\tau_{TTi}) \, p^{t,\tau}(j \mid \kappa)$; it is a function of β.

8.2.2 Supply/Demand Interaction

The supply/demand interaction model could be specialized for each user class (i.e., informed/not-informed user; high/low journey frequency). The specialization of the model for each class of user could modify the path choice model in terms of pre-trip and en-route strategy (or choice set); choice model.

The user at the origin defines a pre-defined strategy (pre-trip). Starting from the pre-defined strategy at the origin, the choice is sequential (en-route) in the decision nodes (stops). The decisions are influenced by the state of supply, the experienced attributes and predicted attributes received by the information system.

At the origin the elementary alternative is the origin hyperpath belonging to the set κ_M of available hyperpaths; it connects the origin and the destination. At the decision node, the elementary alternative is the stop hyperpath belonging to set κ_s of available hyperpaths; it connects the decision node and the destination.

For simplicity's sake the choice is divided in the case of the origin and in the case of the decision node.

In the origin at the time τ of t, with the network graph $\Omega^{t,\tau}$, the user choices are:

(a) κ_M, the set of origin hyperpaths for O-D pair (with hyperpaths connecting the origin and the destination);
(b) κ, the set of perceived origin hyperpaths considering a subset of κ_M with a selective approach;
(c) k, an origin hyperpath belonging to the set κ;
(d) the first decision node; the probability of choosing the first decision node is defined considering that only one link is considered from the origin to the first decision node.

In a decision node s, different from the origin, at the time τ of t, with the network graph $\Omega^{t,\tau}$, the user choices are:

(a) κ_s, the set of stop hyperpaths of κ_M (with hyperpaths connecting the node s and the destination);
(b) κ, the set of perceived stop hyperpaths of κ_s with a selective approach;
(c) k, a stop hyperpath belonging to the set κ;
(d) r,s, the run r in each stop s (decision nodes), belonging to the hyperpath k (considering that in each stop more than one hyperpath could be considered and in each hyperpath more than one run could start from the same stop).

   (a) At the origin, the set of origin hyperpaths contains all the hyperpaths connecting the origin and the destination; from the origin to the first decision node only one link is considered; the origin hyperpath can be generated with mono-criterion or multi-criteria approach considering the first k-hyperpaths of the hyperpaths with cost lower than a defined upper level of the cost; the origin hyperpaths can be generated considering an approach with master hyperpaths. At the decision node, the set of stop hyperpaths derive from the set of origin hyperpaths considering the hyperpaths starting in the specific decision node.

(b) The set of perceived hyperpaths contains the alternatives generated with behavioral (i.e., avoiding high values of some attributes such as number of transfer, access egress time) and logical (i.e., avoiding loops) constraints extracted from the set generated in the point (a).

(c) The choice probability for hyperpath k is modelled with random utility theory. The choice probability for the hyperpath k, belonging to the choice set κ, in the time τ of t, is:

$$p^{t,\tau}(k \mid \kappa) = prob(u^{t,\tau}_{k,exp} > u^{t,\tau}_{k',exp'} \ \forall k, k' \in \kappa, k \neq k') \tag{8.2a}$$

(d) In each decision node s, the choice probability for the run r is modelled with random utility theory. The choice probability for the run r, belonging to the hyperpath k, in the stop s, in the time τ of t, is:

$$p^{t,\tau}(r \mid s,k,\kappa) = prob(u^{t,\tau}_{k,rs} > u^{t,\tau}_{k,r's'} \ \forall r, r' \in s,\kappa, r \neq r') \tag{8.2b}$$

If the decision node is the origin, instead of the run r, the model gives the probability of choosing the first decision node.

As a result of the choices (a), (b), (c) and (d), the analyst evaluates the elementary path choice probability for the realized path j conditioned to the choice of a specific hyperpath (e.1) or for all the hyperpaths (e.2).

(e.1) Defined a set of hyperpaths κ, in each hyperpath, an elementary path choice probability for the path j, conditioned to the choice of the hyperpath k, can be evaluated. Starting from (8.2a) and (8.2b), the probability associated to each elementary path conditioned to the choice of the hyperpath k can be defined. At the time τ of t, it is the product of the probability of taking the run r (and the first decision node) belonging to the path j ($r \in j$), conditioned to stay in the decision node s belonging to the path j ($s \in j$):

$$p^{t,\tau}(j \mid k,\kappa) = \Pi_{rs \in j} p^{t,\tau}(r \mid s,k,\kappa) \tag{8.3}$$

(e.2) The probability choices for the elementary path j in the time τ of t, is the sum of (8.3) respect to all the hyperpaths:

$$p^{t,\tau}(j \mid \kappa) = \Sigma_{k \in \kappa} p^{t,\tau}(k \mid \kappa) \, p^{t,\tau}(j \mid k,\kappa) \tag{8.4}$$

The alternatives available could be spatially overlapped. If the overlapping effect is considered, the covariance between the alternatives has to be simulated in the models (8.2a), (8.2b) and (8.3). The models could belong to the general class of GEV. The Probit model (Daganzo,

1979) considers the covariance in a general form but it cannot allow the probability evaluation in a closed form. A model that considers the overlapping between the alternatives, specified in a closed form and similar to the logit is the C-Logit (Cascetta et al., 1996). It can be applied at elementary path level without path enumeration considering the D-C Logit update (Russo and Vitetta, 2003). Inside the class of GEV, one type of models for (8.2a), (8.2b) and another type of model for (8.3) could be considered. At route choice a model that considers the interference effect (Vitetta, 2016) can be considered.

The demand flow $d_{O\text{-}D,\tau TTi}$, departing at time instant τ_{TTi}, is distributed with path flows in the elementary paths j connecting the O-D. The path flow is distributed between links n in the time τ_d following τ_{TTi}. Let $p_n(\tau_d | j, \tau_{TTi})$ the probability of travelling on stay in the link n, belonging to the elementary path j at τ_d, conditional upon departing at τ_{TTi}, the link flow $q^{\tau d,t}_n$ can be defined. Note that $p_n(\tau_d | j, \tau_{TTi})$ depends on traffic conditions (i.e., capacity constraint), on the network costs (i.e., scheduled time) with value observed before τ and value forecasted after τ. The sum of probabilities $p_n(\tau_d | j, \tau_{TTi})$ for all links and time τ_d, is equal to one: $\Sigma_n \Sigma_{td} p_n(\tau_d | j, \tau_{TTi}) = 1$, $\forall$ n, $\tau_d | (\tau_d \geq \tau_{TTi})$. The link flow $q^{t,\tau d}_n$ is obtained with a mesoscopic dynamic assignment model:

$$q^{t,\tau d}_n = \Sigma_{O\text{-}D} \Sigma_{\tau TTi < \tau_d} d_{O\text{-}D,\tau_{TTi}} \, p_n(\tau_d | j, \tau_{TTi}) \, p^{t,\tau}(j | \kappa) \tag{8.5}$$

For simplicity's sake, in the notation τ_d and t are not reported. In the RDA the assignment model is defined with the assignment matrix.

Let:

- d, the demand values vector for transit system;
- q, the flow values vector;
- P, the assignment matrix;

the assignment model can be defined as introduced also in the Eq. (8.1)

$$q(\alpha, \beta, d) = P(\xi(\alpha), \beta) \cdot d \tag{8.6}$$

The assignment matrix is obtained with a numerical simulation from the assignment and it is considered un-elastic with respect to the optimization variables.

The assignment matrix depends on a set of attributes and on a set of parameters. If the system is observed at time τ, the attributes can be evaluated (a part of them are observed and a part of them are estimated) and it is not reported as argument of the assignment matrix for simplicity's sake. The attributes of supply (α) and demand (β) are the optimization variables of the model, together with the demand values (d).

8.3 RDA Model

The RDA considered in this section is applied for simulating a rolling horizon at the end of the rolling period. It is applied during the day in successive overlapped rolling horizons (Mahmassani, 2001; Musolino and Vitetta, 2011). Each rolling horizon has a fixed length in terms of time (stage length). The stage length depends on the accuracy of the calibrated optimization variables. The method to adopt derives from a compromise between the accuracy of the parameter to calibrate and the elapsed time available for the optimization. In this context, the availability of fast optimization and simulation procedures is crucial. The optimal variables obtained in a given rolling horizon are adopted in the successive rolling horizon (parameter of the models for system forecast; initial values in the reverse assignment method).

For the system forecast, some devices are available in the market (see Chapter 3). In these devices, generally, the system state at time t is observed and a new strategy for travel is communicated to the users. The users receive the information, make their choices and determine the new state of the system in time interval [t, t + Δt] (Chapter 4). The system is observed again in the new time t + Δt. In some cases, the information is given in relation to the observed values and the system forecast considering regressive models on the historical services. In the general case, the user in time t receives the information for the time interval [t, t + Δt] relative also to other links in the system, considering the other user behaviors. RDA models, based on behavioral models, are suitable for this challenge. The calibrated models estimate the new parameters starting from observed data and considering behavioral models. The logic of these systems is similar to the right part of the Fig. 8.4.

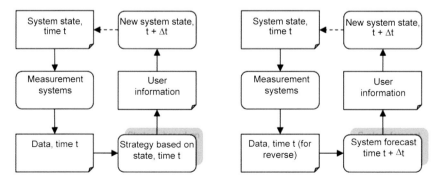

Figure 8.4 System state and forecasted information to the users.

8.3.1 Starting Values

The input variables are (starting values):

- the demand in term of initial values (d_i), functional form and initial parameters (β_i);
- the real time services obtained from real time observations (observed for the instant of time previous τ);
- the cost function initial parameters (α_i);
- the observed flows in term of number of users (q_m) on some links (i.e., alighting, boarding, board flow);
- the observed cost (c_m) on some links (i.e., dwell time, run time).

As reported before, the optimal values obtained in the previous rolling horizon can be adopted as initial values for the demand values, demand and cost parameter.

In relation to the observed cost, in transit system the dwell time can be measured and modelled as a function of the type of vehicle and the number of users in the alighting, boarding and board links. The time on running link depends mainly on the flows of other modes (for transit on shared lanes). If the mode choice is assumed un-elastic, the running time can be calibrated as a function of external variables with respect to the transit system.

8.3.2 Optimization Variables

With RDA the optimal values are relative to:

- demand in terms of value (d) and/or model parameters (β),
- link cost parameters (α).

From the optimization problem, the performances in the optimal point are obtained in term of flow (q) and attributes (x) values.

In the model, several values and parameters can be considered as optimization variables. In relation to the problem studied, a possible set of values and parameters considered are:

- demand value for each O/D pair in each time slice (relative to one or more class of users);
- demand choice parameters (relative to one or more class of users) relative to route choice (at hyperpaths and/or elementary path levels);
- attribute parameters relative to the dwell and/or travel time functions.

8.3.3 Objective Function and Optimization Model

The distance function as a measure of the total distance between observed and modelled variables is reported below:

- initial and modelled demand values (d_i, d) and/or parameters (β_i, β);
- observed and modelled attributes values (x_m, $\xi(\alpha)$) and/or parameters (α_i, α);
- observed and modelled (with Eq. 8.6) flows (q_m, $q(\alpha, \beta, d)$).

Several specifications for the distance function could be assumed. In this chapter, the specification of a square distance is defined as:

$$z(y_a, y_b(\delta)) = (y_a^T \cdot y_b(\delta)) / (c_y \cdot y_{am})^2 \qquad (8.7)$$

with

- y_a a vector of initial variables;
- $y_b(\delta)$ a vector of modelled variables, with the possibility of dependency a vector of other variables δ;
- y_{am} the average values of the elements of the vector;
- c_y the coefficient of variation of the variable y (it is the inverse of the level of credibility given to the initial value of the variable).

Starting from the specification of z, in the general case, considering all the components of the optimization, the objective function can be defined as:

$$z_{of}(d, \beta, \alpha) = z(d_i, d) + z(\beta_i, \beta) + z(q_m, q(\alpha, \beta, d)) + z(x_m, \xi(\alpha)) + z(\alpha_i, \alpha) \qquad (8.8)$$

The optimal values of the demand values (d_o) and parameters (β_o) and attributes parameters (α_o) can be obtained from the model reported below, subject to some constraints:

$$(d_o, \beta_o, \alpha_o) = arg\ min_{do, \beta o, \alpha o}\ z_{of}(d, \beta, \alpha) \qquad (8.9)$$

The generalized least squares method is adopted considering the nature of the problem. This method is robust as it allows to obtain results in real-time calibration.

The constraints of the optimization model are relative to the optimization variables. Each of them belongs to a specific feasible set. The optimal demand, attributes and flow have to belong to specific functions, defined *a priori*. The optimal demand is constrained by the models specified at the departure time and path choice. The cost is constrained by the cost models specified at link level. The flow is constrained by the demand-supply interaction model and user behavior model at route choice level reported in Chapter 5 and shortly recalled in the previous section.

8.3.4 Solution Procedure

The calibration depends on the hypothesis of the probabilistic distribution of the difference between observed and optimal values. Some methods consider hypothesis on the *a posteriori* and/or *a priori* probabilistic distribution.

The solution procedure is reported in Fig. 8.5. Input and output data are reported in the rectangles; models are reported in the ellipses. The variables reported are same as used for the model. It is divided in three sections.

In small size system, the optimization can be developed with a gradient projection algorithm. This algorithm allows to respect the constraints on the design variables (i.e., positive values for the link cost parameters and demand values). This type of algorithm does not have high efficiency, and quasi-Newton algorithm or other optimization methods can be more appropriate with big data in real-size problems.

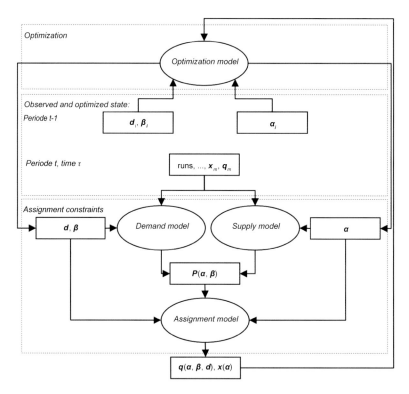

Figure 8.5 Solution procedure.

8.4 Numerical Test

The presented procedure is applied in a test system (Fig. 8.6) in order to verify the applicability.

    The numerical test is developed in the case of one of the user classes presented. The sub-model chosen (all available loop-less alternatives and logit model) and the test network adopted allow implementing the simulation with easy evaluation, without the use of an oriented transport decision support system. It is developed in order to allow readers to test the models. All data needs for the application are reported in this section. More detailed models (i.e., models proposed in Chapters 2, 5 and 6) can be adopted in real application in relation to the available data and elaboration power.

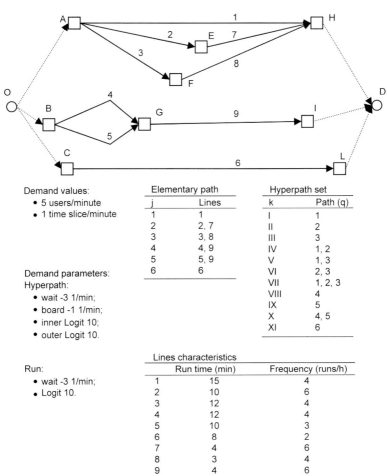

Demand values:
- 5 users/minute
- 1 time slice/minute

Demand parameters:
Hyperpath:
- wait -3 1/min;
- board -1 1/min;
- inner Logit 10;
- outer Logit 10.

Run:
- wait -3 1/min;
- Logit 10.

Elementary path	
j	Lines
1	1
2	2, 7
3	3, 8
4	4, 9
5	5, 9
6	6

Hyperpath set	
k	Path (q)
I	1
II	2
III	3
IV	1, 2
V	1, 3
VI	2, 3
VII	1, 2, 3
VIII	4
IX	5
X	4, 5
XI	6

Lines characteristics		
	Run time (min)	Frequency (runs/h)
1	15	4
2	10	6
3	12	4
4	12	4
5	10	3
6	8	2
7	4	6
8	3	4
9	4	6

Figure 8.6 Test system, true values, elementary paths and hyperpaths.

In this numerical test, it is assumed in a fixed stage length of 60 minutes for the rolling horizon. In the successive rolling horizon, the same method can be applied with new observed data and initial optimization variables (the initial optimization variables could be the optimum values obtained in the current rolling horizon).

The system considered in this numeric application has one origin, one destination and nine lines. To compare the result's accuracy, the flows and times obtained from the assignment are assumed as observed values, with supposed true values for the optimization variables. The supposed true characteristics and optimization variables are reported in Fig. 8.6.

The supposed true optimized variables are changed with a uniform random variation of 30 per cent. These values are considered as initial values of the optimization procedure.

The RDA is applied to obtain the optimal values of the optimized variables and indicators between the supposed true values and the optimized values are evaluated.

The following path choice model is considered in the dynamic mesoscopic assignment:

- in the available set of the alternatives (point b of Section 8.2.2), a selective approach is adopted;
- for the hyperpath choice (point c of Section 8.2.2), a logit model is adopted with waiting and boarding time attributes and inclusive utility, with respect to all the elementary paths contained in the hyperpath;
- for the elementary path choice in the stop (point d of Section 8.2.2), a logit model is adopted with a waiting time attribute; in the assumption that the decision is taken in the instant of arrival of the user at stop.

Some considerations relative to the adopted model are:

- in the available set of alternatives, in relation to the real case studied, all the hyperpaths, or a sub-set obtained by means of other selective approaches can be adopted;
- for the hyperpath choice other models that consider overlap and covariance between alternatives can be adopted (i.e., C-Logit, Neste Logit, Probit);
- for the elementary path choice logit model, the waiting time attribute is congruent with the objective of this numerical application; in real application other attributes (i.e., estimated time to reach the destination) and/or other types of models can be considered.

The following optimization variables are considered:

- for demand, the parameters for waiting and boarding time and the values in one hour of simulation (stage length of the rolling horizon) consider a time slice of one minute;
- for supply, the running time of the line is supposed that the travel speed is the optimization variable.

Some considerations relative to the optimization variables adopted are:

- for demand in relation of the demand specification, other parameters can be adopted; in relation to availability and reliability of the observed data and in relation to the elaboration power, other stage lengths for the rolling horizon and the time slice can be adopted;
- for supply, the parameters of the dwelling time, function of the users flow can be considered (Di Gangi and Nuzzolo, 1991); in this application, the parameters for the running time are considered.

The objective function has four components in terms of square error under variance of the observation between: demand parameters (initial, optimized); demand values (initial, optimized); flow values (observed, optimized); run time values (observed, optimized).

It is assumed that the coefficient of variation is equal to 0.1 for the demand values and parameters, 0.1 for the flow values and 0.01 for the time values.

For testing different scenarios in term of optimization variables and links observed, it is assumed that the observed flows and the run time are in the lines (it is supposed that in the other lines the flow and time are true): all, 1 to 6, 1.

The main results are reported in Table 8.1. In term of optimal demand values, the calibration is stable with respect to the lines considered; the Average SQuare error (ASQ) between true and optimal values is in the range of 0.62–0.73, and the best values are given when 6 lines are observed. In relation to the demand parameters, the best calibration is obtained when only a subset of lines is observed. The travel times in every case are with zero values of ASQ, which is justified considering also the low value of the coefficient of variations assumed. The traffic flows are well reproduced only if a subset of lines is observed.

From a general point of view, but limited to the test case considered, it appears that it is better to observe data only in a subset of lines well defined (i.e., a screen line). In any case, the high reduction in the value of the objective function gives a good indication of the quality of the methodology reported.

Table 8.1 Results of the RDA on the Test System Reported in Fig. 8.6.

	Lines	1..9	1..6	1
ASQ	flow	76.59	53.12	46.18
(Initial-True)	time	2.28	3.02	3.06
	beta	0.46	0.46	0.46
	demand	0.60	0.60	0.60
FO	flow	1720.88	699.73	54.21
(Initial)	time	2726.56	1454.96	136.11
	beta	0.16	0.16	0.16
	demand	0.00	0.00	0.00
ASQ	flow	12.24	0.84	0.13
(Optimal-True)	time	0.00	0.00	0.00
	beta	4.97	0.25	0.34
	demand	0.66	0.62	0.73
FO	flow	295,76	11.07	0.09
(Optimal)	time	2.10	0.15	0.00
	beta	26.44	0.18	0.23
	demand	0.04	0.02	0.55

8.5 Conclusions and Further Developments

The RDA model in the presented version can be directly implemented for different classes of users, also for different classes of transit lines (i.e., high/low frequency, high/low punctuality). Therefore the presented model is defined multiservice.

The RDA multiservice could be extended in future works also for multimodal transit systems. In this case the demand model and the relative assignment (simulation) model have to be defined also for the other modes (i.e., bike, pedestrian, park ride). In this case also, the observed data in terms of flow and time have to be available also to the others modes.

In relation to the available data and to the specified models, other optimization variables could be considered: parameters relative to the choice models for other modes/services (i.e., cicle, walking, park'n ride); link cost parameters relative to the running time for the transit service.

Keywords: assignment; link cost calibration; demand update; reverse assignment; transit system.

References

Cascetta, E. 1984. Estimation of a trip matrices from traffic counts and survey data: A generalized least square estimator. Transportation Research Part B, No. 18: 289–299.

Cascetta, E. and S. Nguyen. 1988. A unified framework for estimating or updating origin/destination matrices from traffic counts. Transportation Research Part B, No. 22: 437–455.

Cascetta, E., D. Inaudi and G. Marquis. 1993. Dynamic estimators of origin-destination matrices using traffic counts. Transportation Science 27(4): 363–373.

Cascetta, E., A. Nuzzolo, F. Russo and A. Vitetta. 1996. A new route choice logit model overcoming IIA problems: Specification and some calibration results for interurban networks. *In*: Jean-Baptiste Lesort (ed.). Proceedings of the 13th International Symposium on Transportation Traffic Theory, Pergamon Press.

Cascetta, E. and F. Russo. 1997. Calibrating aggregate travel demand models with traffic counts: Estimators and statistical performance. Transportation 24: 271–293.

Cathey, F. W. and D. J. Dailey. 2003. A prescription for transit arrival/departure prediction using automatic vehicle location data. Transportation Research Part C 11(3): 241–264.

Daganzo, C. F. 1979. Multinomial Probit: The Theory and its Applications to Travel Demand Forecasting. Academic Press, Nueva York.

Di Gangi, M. and A. Nuzzolo. 1991. Un modello di deflusso per linee di trasporto collettivo. Metodi e modelli per la pianificazione e la gestione dei sistemi di trasporto collettivo, a cura di A. Nuzzolo e F. Russo, pp. 221–247. Franco Angeli. ISBN 88-204-8378-5.

Fisk, C. S. 1988. On combining maximum entropy trip matrix estimation with user optimal assignment. Transportation Research, Part B 22: 69–79.

Horbury, A. 1999. Using non-real-time automatic vehicle location data to improve bus services. Transportation Research Part B 33(8): 559–579.

Mahmassani, H. S. 2001. Dynamic network traffic assignment and simulation methodology for advanced system management applications. Networks and Spatial Economics 1(3): 267–292.

Maher, M. I. 1983. Interferences on trip matrices from observations on link volumes: a Bayesian statistical approach. Transportation Research Part B 17: 435–447.

Moreira-Matias, L., J. Mendes-Moreira, J. Freire de Sousa and J. Gama. 2015. Improving mass transit operations by using AVL-based systems: A survey. IEEE Transactions on Intelligent Transportation Systems 16(4): 1636–1653.

Musolino, G. and A. Vitetta. 2011. Short-term forecasting in road evacuation: calibration of a travel time function. WIT Transactions on the Built Environment 116: 615–626.

Nguyen, S. and S. Pallottino. 1988. Equilibrium traffic assignment for large-scale transit networks. European Journal of Operational Research 37: 176–186.

Nguyen, S., E. Morello and S. Pallottino. 1988. Discrete time dynamic estimation model for passenger origin/destination matrices on transit networks. Transportation Research Part B, No. 22: 251–260.

Parveen, M., A. Shalaby and M. Wahba. 2007. G-EMME/2: Automatic calibration tool of the EMME/2 transit assignment using genetic algorithms. J. Transp. Eng. 133(10): 549–555.

Russo, F. and A. Vitetta. 2003. An assignment model with modified logit, which obviates enumeration and overlapping problems, in Transportation 30: 177–201.

Russo, F. and A. Vitetta. 2011a. Reverse assignment: Calibrating link cost functions and updating demand from traffic counts and time measurements. Inverse Problems in Science and Engineering 19(7): 921–950.

Russo, F. and A. Vitetta. 2011b. Reverse assignment problem a geometric interpretation. Transportation Dynamics. pp. 307–314. *In*: W. Y. Szeto, S. C. Wong and N. N. Sze (eds.). Proceedings of the 16th International Conference of Hong Kong Society for Transportation Studies (HKSTS), Hong Kong Society for Transportation Studies Limited.

Spiess, H. and M. Florian. 1989. Optimal strategies: a new assignment model for transit networks. Transportation Research 23B: 83–102.

Van Zuylen, J. H. and L. G. Willumsen. 1980. The most likely trip matrix estimated from traffic counts. Transportation Research Part B 14: 281–293.

Vitetta, A. 2016. A quantum utility model for route choice in transport systems. Travel Behavior and Society 3: 29–37.

Wall, Z. and D. J. Dailey. 1999. An Algorithm for Predicting the Arrival Time of Mass Transit Vehicles Using Automatic Vehicle Location Data. Paper No. 990870, Preprint Transportation Research Board 78th Annual Meeting, Washington, D.C.

9

Optimal Schedules for Multimodal Transit Services

An Activity-based Approach

William H. K. Lam[1],* and *Zhi-Chun Li*[2]

ABSTRACT

In this chapter, an activity-based Stochastic User Equilibrium (SUE) transit assignment model is developed to investigate transit service scheduling problems in multi-modal transit networks. The proposed model can be used to generate short-term and long-term timetables for multimodal transit lines to aid transit operation and service planning. The model also captures the interaction between transit timetables and passengers' activity- and travel-scheduling behavior, as passengers' activity choices and travel choices (departure time, activity/trip chain, activity duration, transit line and mode) are considered simultaneously under the SUE condition. A heuristic solution algorithm combining

---

[1] Department of Civil and Environmental Engineering, The Hong Kong Polytechnic University, Hung Hom, Kowloon, Hong Kong, China; and School of Traffic and Transportation, Beijing Jiaotong University, Beijing 100044, China.
E-mail: william.lam@polyu.edu.hk

[2] School of Management, Huazhong University of Science and Technology, Wuhan 430074, China.
E-mail: smzcli@hust.edu.cn

* Corresponding author

the Hooke-Jeeves method with an iterative supply-demand equilibrium approach is developed to solve the proposed model. Two numerical examples are presented to illustrate the differences between the proposed activity-based approach and traditional trip-based methods, while comparing the effects of optimal timetables with even and uneven headways. The effects of transit-service disruption on the operation of the transit system are also explored. The SUE passenger travel scheduling pattern derived using the activity-based approach is shown to be significantly different from that obtained by using the trip-based method and a demand-sensitive timetable (with uneven headway) is found to be more efficient than an even-headway timetable. Together, the proposed activity-based approach and schedule-based SUE transit assignment model can be used for short-term and long-term transit planning without incurring the common-line problem endemic to frequency-based transit assignment models.

9.1 Introduction

A number of studies have been conducted on transit network assignment problems. The methods adopted in previous studies can be classified into two main categories: frequency-based methods (see, e.g., De Cea and Fernandez, 1993; Wu et al., 1994; Lam et al., 1999, 2002; Kurauchi et al., 2003; Uchida et al., 2005; Schmöcker et al., 2008; Li et al., 2009) and schedule-based methods (see, e.g., Wong and Tong, 1998; Tong and Wong, 1999; Tong et al., 2001; Nuzzolo et al., 2001; Poon et al., 2004). In general, the assumptions of frequency-based methods are as follows—first, each transit line in a network operates with a fixed even headway; and second, the passenger waiting time before boarding a transit vehicle at a stop is a probabilistic function of the vehicle headway on all of the transit lines in the set of common lines or attractive lines on the transit link (i.e., all of the feasible lines passing through the particular transit link). In contrast, the schedule-based method is used to precisely model the movements of vehicles in the transit network and create passenger-demand profiles based on the time of the day. Passengers' waiting time at a transit stop is then dependent on the line schedule or timetable and the time of their arrival at the stop. Although the schedule-based method is computationally demanding, recent rapid improvements in electronic and computational techniques have made this method applicable to the operational planning of large-scale transit networks (e.g., see Crisalli and Rosati, 2004) or urban railway networks, such as metropolitan (metro) subway networks (see, e.g., Wong and Tong, 1998; Tong and Wong, 1999; Tong et al., 2001).

In practice, frequency-based and schedule-based transit assignment methods are most commonly used to model transit networks for long-term and short-term transit planning, respectively. Both can be classified as traditional trip-based modelling frameworks, in which the standard travel unit is a discrete trip. These trip-based models do not reflect (1) the connections between trips and activities, (2) the temporal constraints and dependencies of activity scheduling, or (3) the underlying activity behavior that generates travel demand, measured as the number of people taking trips. In the last two decades, the activity-based modelling approach has received widespread attention and its use has contributed substantially to understanding of travel-choice behavior (Kitamura, 1988; Jones et al., 1990; Axhausen and Gärling, 1992; Ettema and Timmermans, 1997; Yamamoto and Kitamura, 1999; Yamamoto et al., 2000; Miller and Roorda, 2003; Timmermans, 2005; Hensher, 2007; Chow and Recker, 2012; Zhang and Timmermans, 2012). In an activity-based model, travel demand is derived from participation in activities, which are sequenced by the time of the day. Travel patterns are organized as sets of related trips known as 'tours'. Tours are made up of multiple trips anchored at important beginning and end points, such as the home or the workplace. Activity-based models reflect the interdependence of trips, household influence and the scheduling of activities in time and space (Pendyala et al., 2002).

Recently, analytical activity-based user equilibrium approaches were incorporated into dynamic traffic assignment (DTA) models (Lam and Yin, 2001; Lam and Huang, 2002; Huang et al., 2005; Ramadurai and Ukkusuri, 2010; Ouyang et al., 2011). The resulting integrated framework can be used to tackle more decision types, such as choice of departure time, travel route, activity location and activity duration and thus surpasses traditional DTA models in terms of modelling ability and accuracy. However, scholars working in this area have focused mainly on modelling the activity-choice and travel-choice behavior of automobile users in road networks. Little attention has been paid to the activity-travel scheduling behavior of passengers travelling in multimodal transit networks (e.g., bus or metro passengers). It is thus necessary to explore the interaction between the scheduling of transit services and passengers' activity and travel behavior (Li et al., 2010, 2016; Fu and Lam, 2014).

Scheduling or timetabling transit services (e.g., buses or metro lines) as one of the major components of transit operation and service planning, has been shown to be an effective means of coordinating transit passenger demand and alleviating road traffic congestion (De Palma and Lindsey, 2001; Ceder, 2003; Zhao and Zeng, 2008; Chen et al., 2010), due to the significant influence of timetables on passengers' activity and travel choices and vice versa. An efficient transit-service timetable not only enhances the efficiency of transit provision, but helps to increase passengers' flexibility in scheduling activities and making travel choices.

As a result, the average cost of travelling in the transit network is reduced and the quality of transit services is increased.

Over a typical day, passengers' temporal and spatial distribution in a transit network is clearly affected by their choice of activities. Passengers' activity schedules have some effect on transit-service timetables and vice versa. For instance, an individual may, in practice, have to give up or change his or her activity plan as a result of missing a desired transit run or experiencing an unexpected delay in the arrival of a transit vehicle. In contrast, an efficient transit line schedule or timetable that ensures high levels of service and accessibility provides travellers with timely connections, enabling them to reach their desired activity locations as planned. Therefore, it is necessary to investigate scheduling/ timetabling problems in a multimodal transit network with reference to passengers' activity-travel scheduling behavior, particularly in Asia's transit-oriented cities.

In reality, travel demand derives from people's desire to participate in various economic and social activities, such as working, eating and shopping. Therefore, the interaction between passengers' activities and travel-choice behavior plays an important role in short-term and long-term transit service planning. In this chapter, an activity-based stochastic user equilibrium (SUE) model used to schedule transit services in a multimodal transit network is proposed. The proposed model enables explicit consideration of the interaction between transit-service timetables and passenger-activity schedules. It has the potential to help authorities and transit operators to generate optimal transit-service timetables for both short-term transit operations and long-term service planning. In particular, once timetables have been produced for future transit lines, the schedule-based transit assignment method can be used to plan transit services in the short term and the long term without incurring the common-line problem endemic to the frequency-based transit assignment method (see below for details). Therefore, the proposed model serves as a robust tool for assessing the effects of new transit routes (such as high-speed rail routes) and the performance of alternative transit networks for urban or regional travel.

In light of the above, an activity-based approach is used to investigate scheduling/timetabling problems in a multimodal transit network. The three major developments reported in this chapter are as follows. (1) Using an activity-based approach to SUE modelling, we first formulate the SUE problem of activity-travel scheduling in a multimodal transit network in terms of transit passengers' choice of departure time, activity/trip chain, activity duration and transit line and mode, organized by the time of the day. (2) Next, we develop a heuristic algorithm to solve the transit services timetabling problem. (3) The proposed modelling approach is then used to illustrate the merits of the activity-based model in comparison to the traditional trip-based method. Several test scenarios are constructed to assess the effects of optimal timetables with even and uneven headways and the effects of transit-service disruption due to factors, such as traffic incidents, vehicle breakdown and signaling malfunction.

The remainder of the chapter is organized as follows. In Section 2, the basic components of the proposed model are described, such as the activity-time-space (ATS) network representation and the assumptions of the model. In Section 3, the formulation of the model is described with reference to the transit network supply-demand equilibrium problem and the timetable-optimization problem. In Section 4, two numerical examples are provided to illustrate the applications of the proposed activity-based model and solution algorithm. Concluding remarks are made in Section 5, accompanied by suggestions for further study in the light of the recent development of multimodal transit systems in Asia.

9.2 Basic Considerations

9.2.1 Activity-time-space Network Representation

Before explaining the ATS network representation, the following concepts must be defined. A *trip chain* consists of an ordered set of *sojourn* nodes and *trip legs* (i.e., segments between consecutive sojourns). Individuals pursue activities in the sojourn nodes. Each sojourn or activity node in a trip chain is both the destination of the previous trip leg and the origin of the next trip leg. Let c denote a trip chain that traverses N sojourn nodes, as follows: $c = \{n_1, n_2,..., n_i,... n_N\}$, where n_i is the ith activity node of the chain. A trip chain that begins and ends at the same node (i.e., $n_1 = n_N$) is called a *tour*. Obviously, different orders of activity nodes result in different trip chains. The set of all chains starting at zone r is given as Ω_r. As an illustration, Fig. 9.1 shows typical home-based trip chains defined by the following series of activity nodes: {home – work – home} (H-W-H), {home – work – entertainment – home} (H-W-E-H), {home – school – work – home} (H-S-W-H) and {home – school – work – entertainment – home} (H-S-W-E-H). The H-S-W-E-H chain represents the activities of a commuter taking his or her children to school on the way to work in the morning and pursuing leisure activities after returning home from work in the evening.

To describe potential passenger movement in a transit network within a given period, an ATS network approach is used to construct passenger-flow networks associated with specific activity/trip chains. Figure 9.2 shows the passenger-flow ATS network associated with the H-W-H chain. The horizontal axis represents the spatial dimension, which comprises activity locations and transit stations while the vertical axis represents the temporal dimension; that is, the duration of an activity or waiting time at a

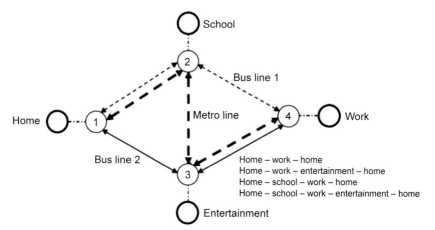

Figure 9.1 Home-based trip chains.

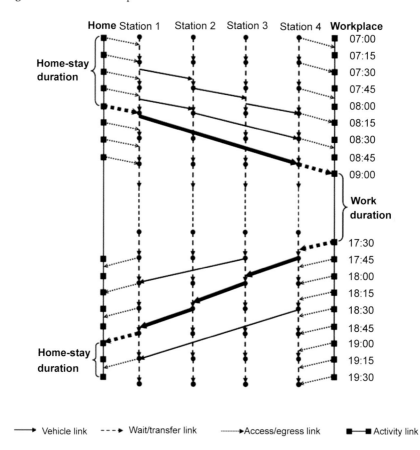

Figure 9.2 Passenger flow in an ATS network for trip chain H-W-H.

station. *Nodes* and *links* are the two main components of the ATS network. A link designates an activity or a movement using a specific transit mode; a square node represents an activity location; and a circular node represents a transit station. Stations may be linked by different transit modes (e.g., buses or metro lines). In the time-space supernetwork representation, a trip chain is depicted as a set of *time-space paths* (or simply paths). In essence, modelling the passenger activity-travel scheduling problem in a multimodal transit network is equivalent to modelling the path choice of a passenger in the ATS network. Therefore, passengers' departure-time and travel-mode decisions are implicit in their path decisions in the ATS network representation. In the next section, the utility of a path in a trip chain is addressed.

9.2.2 Assumptions

The key assumptions underlying the work reported in this chapter are as follows:

A1 Transit systems are the focus of the study; non-transit modes (e.g., automobiles and trucks) are not considered. Transit vehicles in the network are assumed to fully follow a scheduled timetable (Tong and Wong, 1999; Tong et al., 2001) and stochastic disturbances in in-vehicle travel time due to supply or demand uncertainty are not considered, although they will be explored in a future study.

A2 The set of feasible activity/trip chains is assumed to be pre-specified. This assumption has been adopted in related studies, such as that of Maruyama and Sumalee (2007), but can be relaxed by integrating a dynamic activity/trip chain generator with the proposed model (Arentze and Timmermans, 2009; Habib and Miller, 2009).

A3 Trip-makers' daily activity-travel schedules are assumed to involve the following SUE decisions—choice of trip/activity chain, choice of time of departure from origin (e.g., home), choice of transit line/path to reach an activity location and choice of activity duration. Trip-makers base their activity and travel scheduling on a tradeoff between the perceived utility or benefits of participating in activities in particular locations and the perceived disutility of travelling between these locations. We assume that all individuals in the transit system are expected utility maximizing decision makers; that is, they schedule their activity-travel patterns to maximize their perceived expected trip utility.

A4 The utility of an activity depends on the activity's start time and duration. The expected utility of each activity is determined, using a modified form of the bell-shaped marginal timing utility function recently proposed by Joh et al. (2002). This function has been adopted in related studies, such as those of Ettema and Timmermans (2003), Ashiru et al. (2004) and Zhang et al. (2005).

A5 The expected disutility of travel between activity locations is determined by the following factors—expected in-vehicle time (with in-vehicle congestion effects), expected waiting/transfer time, expected access/egress time (i.e., walking time), fare, line-change penalty and expected schedule delay costs of early or late arrival (Tong and Wong, 1999; Poon et al., 2004; Sumalee et al., 2009).

A6 The number of transit vehicles on each line (i.e., fleet size) is taken as given. Transit-system operators then choose a timetable for each line that maximizes the total expected user net utility in the system (De Palma and Lindsey, 2001; Zhao and Zeng, 2008).

9.3 Model Formulation

9.3.1 Transit Network Supply-Demand Equilibrium

Supply side

Consider the multimodal transit network $G = (N, A, L)$, where N is the set of all nodes, including transit stations and activity locations, A is the set of all links and L is the set of all lines. In the expanded ATS network shown in Fig. 9.2, there are four types of links, which are defined below.

Vehicle link. A vehicle link represents a run of a transit vehicle between two consecutive stations. The tail/head nodes of each vehicle link are associated with the station (time) of departure/arrival for the vehicle run on the horizontal (vertical) axis, that is, each vehicle link contains information on the departure time, the station of departure, arrival time, the station of arrival and the vehicle service time on that link. The service time of a transit vehicle on a vehicle link is equal to the difference between the time of departure of the transit vehicle from the link and the time of its arrival at the link. Let A_1 be the set of vehicle links in the time-space supernetwork and t_a the travel time of a transit vehicle on vehicle link a. This gives the following:

$$t_a = \tau_a^D - \tau_a^A, \quad \forall a \in A_1, \tag{9.1}$$

where τ_a^D and τ_a^A denote the time at which the transit vehicle departs from the head node of link a and the time at which the vehicle arrives at the tail node of link a, respectively.

Wait/transfer link. A wait or transfer link indicates that the passenger stays at a station to wait for a service. The passenger can decide to board the first-arriving vehicle or wait for subsequent vehicles if he or she is unable or unwilling to board the first-arriving vehicle due to overcrowding. If the passenger boards the first-arriving vehicle, he or she will traverse the vehicle link to the next station. Otherwise, the passenger will continue to wait for the next vehicle to arrive at the station in the next time interval. The waiting time at wait/transfer links, similar to that at vehicle links, can be determined by passengers' time of entry to the link (τ_a^A) and time of exit from the link (τ_a^D), i.e., $t_a = \tau_a^D - \tau_a^A$, $\forall a \in A_2$, where A_2 is the set of wait/transfer links in the time-space supernetwork.

Access link/egress link. An access link connects an origin node with a station of departure. An egress link connects a station of arrival with the final destination. The walking time at an access or egress link can be calculated from the length of the link and passengers' walking speed; that is,

$$t_a = \frac{L_a}{\overline{S}}, \quad \forall a \in A_3, \tag{9.2}$$

where A_3 is the set of walk links, L_a is the length of walk link a and $\overline{S}$ is passengers' average walking speed (km/h).

Activity link. An activity link represents a place in which an individual performs a certain activity. Activities may occur at home (e.g., sleeping, childcare, watching television) or outside the home (e.g., work, education, shopping, entertainment). The perceived utility or benefit of participating in an activity is dependent on the activity's start time and duration. The duration of an activity at an activity link can be calculated from the difference between the passenger's time of arrival at and time of departure from the link, i.e., $t_a = \tau_a^D - \tau_{a'}^A$, $\forall a \in A_4$, where A_4 is the set of activity links, τ_a^A is the time of the passenger's arrival at the tail node of link a (or the activity start time) and τ_a^D is the time of the passenger's departure from the head node of link a (or the activity end time).

Let $MU_i(x)$ denote the marginal utility of activity i, which expresses the utility of participating in activity i at the single time point x. Accordingly, the utility gained from a passenger's performing activity i at activity link a with a start time of τ_a^A and a duration of t_a (or the activity end time $\tau_a^A + t_a$) can be calculated as follows:

$$U_a(\tau_a^A, t_a) = \int_{\tau_a^A}^{\tau_a^A + t_a} MU_i(x)dx, \quad \forall a \in A_4, \tag{9.3}$$

where $MU_i(x)$ is determined using the modified bell-shaped marginal utility function recently proposed by Joh et al. (2002), with the additional term U_i^o, as follows:

$$MU_i(x) = U_i^0 + \frac{\sigma_i \lambda_i U_i^{max}}{\exp\left[\sigma_i\left(x - \xi_i\right)\right]\left\{1 + \exp\left[-\sigma_i\left(x - \xi_i\right)\right]\right\}^{\lambda_i + 1}}, \quad \forall i \in I, \qquad (9.4)$$

where x is the time of day, I is the set of all activities and U_i^0 is the baseline utility level of activity i. U_i^{max} is the maximum utility of activity i, and σ_i, λ_i and ξ_i are activity-specific parameters. The parameter ξ_i determines the time of day at which the marginal utility reaches its maximum value (i.e., the inflection point), σ_i determines the slope or steepness of the curve and λ_i determines the relative position of the inflection point. These parameters can be estimated from survey data (see for, e.g., Ettema and Timmermans, 2003; Ashiru et al., 2004). Figure 9.3 provides illustrative examples of the marginal-utility distributions obtained for four activities with the input parameters given in Table 9.1. The four activity types are

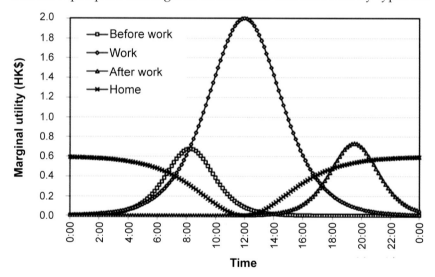

Figure 9.3 Examples of marginal-utility functions for various activities.

Table 9.1 Input Parameters for Marginal-Utility Functions in Fig. 9.3.

Parameter	Activity Type			
	Staying at Home	Working	Before-Work Activity	After-Work Activity
σ	−0.008	0.01	0.015	0.015
λ	1.0	1.0	1.0	1.0
ξ (min)	720	720	495	1,170
U^{max} (HK\$/h)	300	800	180	195
U^0 (HK\$/h)	0.6	0.0	0.0	0.0

Note: Before-work activity: taking children to school; after-work activity: entertainment.

as follows: staying at home, before-work activities (e.g., taking children to school, visiting the dentist), working, and after-work activities (e.g., shopping, entertainment).

Thus far, we have formulated the (dis)utility functions for various links. Subsequently we define the total expected utility of a path in a trip chain. This is the difference between the total expected utility derived from all instances of activity participation along the path minus the total expected disutility of travel along the path, and is expressed as follows:

$$U_{pc} = U_{pc}^A - U_{pc}^T, \quad \forall p \in P_c, c \in \Omega_r, r \in R, \tag{9.5}$$

where U_{pc} is the total expected utility of path p in trip chain c, R is the set of all origin zones in the network and P_c is the set of all possible paths related to trip chain c. U_{pc}^A is the expected activity utility component, which is given by the following:

$$U_{pc}^A = \sum_{a \in A_4} U_a \delta_{apc}, \quad \forall p \in P_c, c \in \Omega_r, r \in R \tag{9.6}$$

where δ_{apc} equals 1 if link a is on path p in chain c, and 0 otherwise. U_a can be computed using Eqs. 9.3 and 9.4.

According to assumption **A5**, the expected disutility U_{pc}^T of travel along path p in chain c is determined by the generalized expected in-vehicle time (including congestion effects), φ_{pc}^1; the expected waiting time, φ_{pc}^2; the expected walking time, φ_{pc}^3; the number of line changes, φ_{pc}^4; the expected schedule delay costs of early or late arrival, φ_{pc}^5; and the fare, φ_{pc}^6. This can be expressed as follows:

$$U_{pc}^T = \alpha_1 \varphi_{pc}^1 + \alpha_2 \varphi_{pc}^2 + \alpha_3 \varphi_{pc}^3 + \alpha_4 \varphi_{pc}^4 + \varphi_{pc}^5 + \varphi_{pc}^6, \quad \forall p \in P_c, c \in \Omega_r, r \in R, \tag{9.7}$$

where the coefficients denoted as (α) are the reciprocal substitution factors between the expected disutility components, which are used to convert different quantities into the same units.

Note that passenger discomfort is affected by the degree of crowding in transit vehicles (Wu et al., 1994; Lo et al., 2003; Li et al., 2009). To model the effects of crowding discomfort on passengers' activity-travel decisions, we use the following definition of the generalized expected in-vehicle time φ_{pc}^1, measured in equivalent time units, for path p in chain c:

$$\varphi_{pc}^1 = \sum_{a \in A_1} t_a \left(1.0 + \rho_0 \left(\frac{v_a}{C_a}\right)^{\rho_1}\right) \delta_{apc}, \quad \forall p \in P_c, c \in \Omega_r, r \in R, \tag{9.8}$$

where ρ_0 and ρ_1 are positive calibrated parameters, t_a can be calculated using Eq. 9.1 and v_a and C_a are the passenger flow and vehicle capacity at link a, respectively. v_a is obtained as follows:

$$v_a = \sum_{r \in R} \sum_{c \in \Omega_r} \sum_{p \in P_c} f_{pc} \delta_{apc}, \quad \forall a \in A_1, \tag{9.9}$$

where f_{pc} is the passenger flow on path p in chain c (defined later in the chapter).

Following the previous definitions, φ_{pc}^2 and φ_{pc}^3 can be expressed as follows:

$$\varphi_{pc}^j = \sum_{a \in A_j} t_a \delta_{apc}, \quad \forall j = \{2,3\}, p \in P_c, c \in \Omega_r, r \in R. \tag{9.10}$$

In general, for certain activities, particularly compulsory activities (e.g., working), there is a preferred arrival time and early or late arrival will incur a schedule-delay penalty. The total expected schedule delay cost along path p in chain c is the sum of the expected schedule delay costs incurred by all activity participation along that path; that is,

$$\varphi_{pc}^5 = \sum_i SD_{pc}^i, \quad \forall p \in P_c, c \in \Omega_r, r \in R, \tag{9.11}$$

where the expected schedule delay cost incurred by participating in activity I, SD_{pc}^i is defined as follows:

$$SD_{pc}^i = \begin{cases} \beta_i \left(t_i^* - \Delta_i - \mu_{pc}^i \right), & \text{if } t_i^* - \Delta_i > \mu_{pc}^i \\ \gamma_i \left(\mu_{pc}^i - t_i^* - \Delta_i \right), & \text{if } t_i^* + \Delta_i < \mu_{pc}^i, \quad \forall p \in P_c, c \in \Omega_r, r \in R, i \in I, \\ 0, \text{otherwise} \end{cases} \tag{9.12}$$

where $[t_i^* - \Delta_i, t_i^* + \Delta_i]$ is the desired time window for starting activity I, which does not incur a schedule delay penalty; t_i^* is the mid-point of the time window and β_i and γ_i are the unit costs of early and late arrival, respectively; μ_{pc}^i is the arrival time at i, the activity location of trip-makers departing from origin zone r; μ_{pc}^i is the sum of the departure times and journey times for the links from origin r to location i along path p and is expressed as follows:

$$\mu_{pc}^i = t_r^0 + \sum_{a \in A_p^{ri}} t_a, \quad \forall p \in P_c, c \in \Omega_r, r \in R, i \in I, \tag{9.13}$$

where t^0_r is the time of trip-makers' departure from zone r and A^{ri}_p is the set of all links from origin zone r to activity location i along path p.

Demand side

According to assumption **A3**, individuals schedule their activity and travel choices to maximize their perceived expected trip utility. Note that individuals' decisions on departure time, transit mode and activity duration are implicit in their path choices in the ATS network. Therefore, an individual's choice of departure time, activity/trip chain, activity duration, transit line and mode can be reduced to a two-dimensional choice of activity/trip chain and path, which can be modelled as a nested logit formulation under random-utility theory (Ben-Akiva and Lerman, 1985), as follows:

$$f_{pc} = q_r \, \Pr_{pc|r} = q_r \frac{\exp(\theta_2 U_c)}{\displaystyle\sum_{c \in \Omega_r} \exp(\theta_2 U_c)} \frac{\exp(\theta_1 U_{pc})}{\displaystyle\sum_{p \in P_c} \exp(\theta_1 U_{pc})}, \quad \forall p \in P_c, c \in \Omega_r, r \in R,$$

$$(9.14)$$

where q_r is the total number of trip-makers starting at origin zone r and $\Pr_{pc|r}$ is the probability that trip-makers starting at zone r will choose path p in chain c; θ_1 and θ_2 are scaling parameters that measure variation in trip-makers' choice behavior. To fulfill the criteria set by random-utility theory and ensure reasonable signs of cross-elasticity, $\theta_2 \leq \theta_1$ must hold. U_{pc} is determined by Eq. 9.5. U_c is the expected utility of chain c, which is given as follows:

$$U_c = \frac{1}{\theta_1} \ln \sum_{p \in P_c} \exp(\theta_1 U_{pc}), \quad \forall c \in \Omega_r, r \in R.$$

$$(9.15)$$

According to random-utility theory, the expected utility U^{out}_r of trip-makers leaving zone r to perform various activity chains can be expressed as follows:

$$U^{\text{out}}_r = \frac{1}{\theta_2} \ln \sum_{c \in \Omega_r} \exp(\theta_2 U_c), \quad \forall r \in R.$$

$$(9.16)$$

Let Q_r be the total number of seed individuals or potential trip-makers at zone r. These individuals make decisions on whether to leave zone r to participate in other activities or to stay in zone r. Let U^{in}_r be the daily expected utility obtained from staying in zone r, which is calculated by integrating the marginal-utility function of staying in zone r for different times of day (see Eqs. 9.3 and 9.4). q_r, the number of trip-makers originating from r can then be estimated as follows:

$$q_r = Q_r \frac{\exp(\theta_3 U_r^{\text{out}})}{\exp(\theta_3 U_r^{\text{out}}) + \exp(\theta_3 U_r^{\text{in}})}, \quad \forall r \in R, \tag{9.17}$$

where θ_3 measures the variation in the perceived utility of making or not making a trip. The relationship $\theta_3 \leq \theta_2 \leq \theta_1$ must hold. Once the value of Q_r has been obtained, the total number of individuals who stay at r can be calculated by $(Q_r - q_r)$.

Supply-demand equilibrium

According to Eqs. 9.14–9.17, the transit passenger path flow f_{pc} is a function of the expected utility U_{pc} of using path p in chain c, which is a function of the link passenger flow v_a and thus the path flow f_{pc} itself, according to the definition of generalized expected in-vehicle time φ_{pc}^1 in Eqs. 9.8–9.9. Therefore, supply-demand interaction leads to the following fixed-point network equilibrium problem:

$$\mathbf{f}^* = \mathbf{q} \cdot \mathbf{Pr}\left(\mathbf{U}\left(\boldsymbol{\varphi}\left(\mathbf{v}(\mathbf{f}^*)\right)\right)\right), \tag{9.18}$$

where the symbols in bold represent the vectors of the corresponding variables.

Proposition 1. For a given transit timetable, a simultaneous activity/trip chain and path choice equilibrium for transit passengers in the ATS network (which implicitly include passengers' choice of departure time, activity duration and transit mode) is reached if and only if the fixed-point formulation shown in 9.18 holds.

Note that the vector functions $\mathbf{Pr}(\cdot)$, $\mathbf{U}(\cdot)$ and $\boldsymbol{\varphi}(\cdot)$ in Eq. 9.18 are continuous with regard to the path flow vector $\mathbf{f}$. According to Brouwer's fixed-point theorem (Granas and Dugundji, 2003), there is at least one feasible solution to the fixed-point problem shown in 9.18. However, the expected path utility function U_{pc} defined in Eq. 9.5 is not a strictly monotonic function, so the fixed-point model in 9.18 is non-convex. This implies that the proposed model may offer multiple local solutions to the fixed-point problem.

Fixed-point problem 9.18 is the result of interaction between the supply and demand sides. Its stationary solution can be obtained by iteratively solving the supply model in 9.1–9.13 and the demand model in 9.14–9.17. In this study, the method of successive averages (MSA) is used to solve fixed-point problem 9.18. This iterative approach is conducted as follows:

Step 1. Initialize. Choose an initial trip demand pattern $\{q_r^{(0)}\}$ and a feasible passenger path flow pattern $\{f_{pc}^{(0)}\}$, and set n as 1.

Step 2. Determine expected path utility. Following the supply model shown in 9.1–9.13, calculate the expected utility component for each activity and the expected disutility component for each trip between activity locations, then obtain the expected utility $U_{pc}^{(n)}$ for each path of each chain in the ATS network.

Step 3. Assign nested logit based flow. Calculate the expected utility U_r^{out} of performing activities outside zone r, and use Eq. 9.17 to determine the auxiliary demand pattern $\{g_r^{(n)}\}$. Next, assign $\{g_r^{(n)}\}$ to each path of each chain in the ATS network using Eq. 9.14, and obtain the auxiliary path flow pattern $\{h_{pc}^{(n)}\}$.

Step 4. Update. Use the MSA to update the trip-demand pattern and the path-flow pattern, i.e.,

$$q_r^{(n+1)} = q_r^{(n)} + (g_r^{(n)} - q_r^{(n)})/n, \quad f_{pc}^{(n+1)} = f_{pc}^{(n)} + (h_{pc}^{(n)} - f_{pc}^{(n)})/n.$$

Step 5. Check convergence. If the gap function $G = \| \mathbf{f}^{(n)} - \mathbf{q} \cdot \mathbf{Pr}(\mathbf{f}^{(n)}) \| / \| \mathbf{f}^{(n)} \|$ falls below a pre-specified tolerance threshold, stop; otherwise, set n as $n + 1$ and return to Step 2.

It should be noted that the MSA-based solution algorithm above may converge to a stationary point of the fixed-point problem in 9.18 (for more details, see Magnanti and Perakis, 2002). Recently, the MSA has been widely used to find solutions to fixed-point problems (e.g., see Zhang et al., 2005; Sumalee et al., 2009).

9.3.2 Transit-timetabling Problem

The goal of solving the transit-timetabling problem is to determine a periodic timetable for a set of transit vehicles (comprising the times of arrival at and departure from each station) that satisfies a certain objective and several operational constraints (Caprara et al., 2002; Zhao and Zeng, 2008). Let $(t_{arr})_{l,m}^k$ represent the arrival time of the kth transit vehicle at the mth station on line l. A rounded arrival timetable for the kth transit vehicle on line l can then be expressed as the following vector:

$$\left((t_{arr})_{l,1}^k, (t_{arr})_{l,2}^k, ..., (t_{arr})_{l,M}^k, (t_{arr})_{l,(M+1)}^k, ..., (t_{arr})_{l,(2M-1)}^k\right), \tag{9.19}$$

where M is the number of transit stations along line l. Vector 9.19 includes two one-way trips—an uptrend trip (number of stations from 1 to M) and a downtrend trip (number of stations from $(M + 1)$ to $(2M – 1)$). The number of stations in the downtrend trip, m (i.e., $M + 1 \leq m \leq 2M – 1$), actually corresponds to the number of stations in the uptrend trip, $(2M – m)$.

Let $(t_{\text{hold}})^k_{l,m}$ and $(t_{\text{dep}})^k_{l,m}$ be the waiting time and departure time of the kth transit vehicle at and from the mth station on line l, respectively. As the vehicle holding time $(t_{\text{hold}})^k_{l,m}$ is assumed to be constant, the departure time $(t_{\text{dep}})^k_{l,m}$ can be calculated as follows:

$$(t_{\text{dep}})^k_{l,m} = (t_{\text{arr}})^k_{l,m} + (t_{\text{hold}})^k_{l,m}. \tag{9.20}$$

Using Eqs. 9.19 and 9.20, the round-trip timetable for a given line can be calculated by determining the arrival timetable (using Eq. 9.19).

Let K be the fleet size for line l. The round-trip timetable for K transit vehicles on line l is determined by identifying the $K \times (2M-1)$-dimensional matrix, as follows.

$$\mathbf{T}_l = \left[(t_{\text{arr}})^k_{l,m} \right]_{K \times (2M-1)}. \tag{9.21}$$

The timetable for the whole transit network can then be represented as

$$\mathbf{T} = (\mathbf{T}_1, \mathbf{T}_2, ..., \mathbf{T}_L), \tag{9.22}$$

where L is the total number of lines in the network.

According to assumption **A6**, the transit-system operator aims to choose a timetable **T** that maximizes the total expected user net utility in the system governed by timetable **T**. For a given timetable **T**, the total expected user net utility can be expressed as follows:

$$Z(\mathbf{T}) = \sum_r \sum_c \sum_p f_{pc}(\mathbf{T}) U_{pc}(\mathbf{T}) + \sum_r (Q_r - q_r(\mathbf{T})) U_r^{\text{in}}. \tag{9.23}$$

In Eq. 9.23, f_{pc}, U_{pc} and q_r depend on timetable **T** and are thus functions of **T**. The terms on the right-hand side of the equation denote the total utility gained by trip-makers and non-trip-makers, respectively.

Given the foregoing statements, the transit-timetabling problem can be formulated as follows:

$$\max_{\mathbf{T}} \ Z(\mathbf{T}, \mathbf{f}(\mathbf{T}), \mathbf{U}(\mathbf{T}), \mathbf{q}(\mathbf{T})). \tag{9.24}$$

$\mathbf{f}(\mathbf{T})$, $\mathbf{U}(\mathbf{T})$ and $\mathbf{q}(\mathbf{T})$ can be obtained by solving the following:

$$\mathbf{f}(\mathbf{T}) = \mathbf{q}(\mathbf{T}) \cdot \Pr\left(\mathbf{U}\left(\varphi\left(\mathbf{v}(\mathbf{f}(\mathbf{T})) \right) \right) \right). \tag{9.25}$$

In addition, the following headway constraint is considered:

$$(t_{\text{arr}})_{l,m}^{k} - (t_{\text{arr}})_{l,m}^{k-1} \geq h_{\min}^{l}, \quad \forall m, k, \tag{9.26}$$

where $h_{\min}^{l}$ is the allowable minimum headway on line l. The left-hand side of Eq. 9.26 represents the difference in the arrival times of two consecutive transit vehicles (i.e., the headway) at station m on line l. It should be pointed out that in the above model, the fleet size of each transit line is fixed and exogenously given. The timetable is optimized to maximize the total expected user net utility, using the existing fleet to match passenger demand.

The timetabling model shown in 9.24–9.26 is a bi-level mathematical programming problem with supply-demand equilibrium constraints. As the problem is intrinsically non-convex, it may be difficult to obtain a globally optimal solution. Many heuristic algorithms for solving bi-level programming problems have been proposed (Luo et al., 1996). In this study, the Hooke-Jeeves approach, a multidimensional search procedure, is adapted to solve the timetabling model shown in 9.24–9.26. This approach does not require explicit knowledge of the derivative characteristics of the objective function. In essence, the proposed procedure involves performing two types of search consecutively—an exploratory search and a pattern search. The proposed solution algorithm is outlined below.

Step 1. Initialize.

1.1. Choose an initial solution $\mathbf{T}^{(0)}$. Let $\mathbf{T} = \mathbf{T}^{(0)}$. Solve the passenger activity-travel choice network equilibrium problem in (9.18), then compute $Z(\mathbf{T})$ using Eq. 9.23.

1.2. Choose an initial step size $\eta > 0$ and acceleration factor ω, with $\zeta = 1$.

1.3. Set index j as 1 and index k as 0.

Step 2. Conduct exploratory search.

2.1. If $k > K$, go to Step 3.

2.2. Let $\hat{\mathbf{T}} = \mathbf{T} + \zeta \eta e_j$ (e_j is a vector with 1 in the jth position and 0 elsewhere). Solve fixed-point problem 9.18 with and compute $Z(\hat{\mathbf{T}})$.

2.3. If $Z(\hat{\mathbf{T}}) > Z(\mathbf{T})$, let $\mathbf{T} = \hat{\mathbf{T}}$, $Z(\mathbf{T}) = Z(\hat{\mathbf{T}})$, $j = j + 1$ and $\zeta = 1$, and go to Step 2.1; otherwise, go to Step 2.4.

2.4. If $\zeta = 1$, let $\zeta = -1$ and go to Step 2.2; otherwise (i.e., $\zeta = -1$), let $j = j + 1$ and $\zeta = 1$, and go to Step 2.1.

Step 3. Conduct pattern search.

3.1. If $Z(\hat{\mathbf{T}}) > Z(\mathbf{T}^{(k)})$, let $\mathbf{T}^{(k+1)} = \hat{\mathbf{T}}$, $Z(\mathbf{T}^{(k+1)}) = Z(\hat{\mathbf{T}})$, $\mathbf{T} = \mathbf{T}^{(k)} + \omega(\mathbf{T}^{(k+1)} - \mathbf{T}^{(k)})$, $j = 1$ and $k = k + 1$ and go to Step 2; otherwise, go to Step 3.2.

3.2. If η is 'sufficiently' small, stop and output the optimal solution $\mathbf{T}^{(k)}$; otherwise, let $\eta = 0.5 \eta$, $j = 1$ and $\mathbf{T} = \mathbf{T}^{(k)}$ and go to Step 2.

The Hooke-Jeeves approach above can converge to a (local) optimal solution of the timetabling problem in 9.24–9.26 (for proof of the convergence, please refer to Bazaraa et al., 2006, p. 370). Note that the original Hooke-Jeeves approach is designed to solve unconstrained optimization problems. The problem examined here is subject to simple bound constraints such as minimum headway constraint 9.26. We modify the Hooke-Jeeves approach to accommodate these constraints by restricting the departure (and thus arrival) times of two consecutive transit vehicles to give no less than the minimum headway. Specifically, when choosing an initial feasible solution in Step 1.1 and performing the exploratory search (Step 2.2) and pattern search (Step 3.1), the departure/arrival times of all transit vehicles from/at each station on a line are updated in turn while the minimum headway requirement is satisfied (i.e., once this constraint has been violated, the current headway is set at the minimum headway). Most of the central processing unit (CPU) time required by the proposed solution algorithm occurs in Step 2.2, because the objective-function evaluation and thus the process of solving the passenger activity-travel choice network equilibrium problem in 9.18 must be carried out repeatedly during this step.

9.4 Numerical Studies

In this section, two test scenarios are used to illustrate the applications of the proposed model and the solution algorithm. The first scenario is designed to clarify the differences between the activity-based approach and the traditional trip-based method. The second scenario is used to illustrate the use of the model for timetable optimization. We also compare the outcomes generated by optimized timetables subject to even and uneven headway and investigate the effects of transit-service disruption on total user utility. The study horizon is a typical weekday divided into 1,440 one-minute intervals. The proposed solution algorithm is coded in the programming language C and run on a personal computer with an Intel Pentium 1.4 GHz CPU and 1 GB of random-access memory. In both test scenarios, the convergence precision of the solution algorithm is set at 10^{-5}.

9.4.1 Scenario 1

The transit network in Scenario 1 is shown in Fig. 9.4 and consists of a typical activity chain (H-W-H), a metro line and a bus line on a typical working day (i.e., Monday to Friday). The one-way in-vehicle time for the metro line is 20 minutes, with a fare of HK$20 (note that US$1.00 = HK$7.80). The corresponding values for the bus line are 40 minutes and HK$5. The metro has a 1,500-passenger capacity and the bus has a

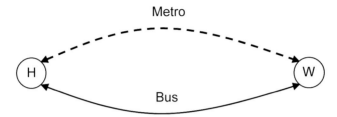

Figure 9.4 Transit network in Scenario 1.

120-passenger capacity. There are 20,000 potential (or latent) trip-makers in the residential area, Q. The majority of these individuals (commuters) travel to the city center (the area marked as 'W' in Fig. 9.4) to work during the daytime and return home after work, whereas others stay at home to carry out telework (telecommuters) or other activities. The parameters for the marginal-utility functions of staying at home and working are shown in Table 9.1. In the transit network in Scenario 1, the (even) headway of both the metro line and the bus line is assumed to be fixed at 10 minutes and the departure times of the first bus and train runs are assumed to be the same: 06:30. The average walking distance from home to bus stops and railway stations is assumed to be 0.4 km and 0.6 km, respectively. Passengers' average walking speed is 4.8 km/h. The parameters of crowding-discomfort function 9.8, ρ_0 and ρ_1, are 0.15 and 4.0, respectively. The other model parameters are as follows: α_1 = 60 (HK\$/h); α_2 = 120 (HK\$/h); α_3 = 120 (HK\$/h); α_4 = 5.0 (HK\$) (i.e., the transfer penalty due to a line change); β = 30 (HK\$/h); γ = 90 (HK\$/h); t^* = 09: 00; Δ = 0.25 h; θ_1 = 0.3; θ_2 = 0.2; and θ_3 = 0.005. The CPU time required to numerically obtain the converged solution is approximately 109 seconds.

Figure 9.5a shows the distribution of commuter work duration as endogenously derived from the activity-based model. The mean and standard deviation (SD) of the commuter work duration distribution are 8.30 hours and 0.78 hours, respectively. The total number of trip-makers is 17,777 (see also Table 9.3).

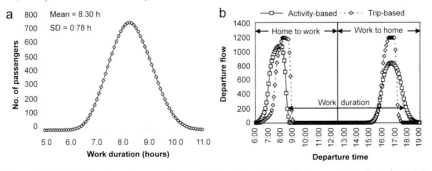

Figure 9.5 (a) Distribution of commuter work duration obtained using activity-based model; (b) departure-flow patterns for activity-based model and trip-based model.

To distinguish between the activity-based and trip-based models, Fig. 9.5b shows the temporal departure flow patterns obtained using the two modelling approaches. As the trip-based approach cannot be used to endogenously model commuters' work duration distribution, we adopt Lam and Huang's (2002) assumption that work duration is fixed at 8.30 hours (identical to the expected work duration in the activity-based model). To aid comparison, the number of trip-makers in the trip-based model is fixed at 17,777. As shown in Fig. 9.5b, there are significant differences in the departure-flow patterns obtained by using the two approaches. The departure-flow curve from home to work derived from the activity-based model (represented by the solid line) is farther to the left than that derived from the trip-based model (represented by the dotted line). This means that commuters modelled using the activity-based approach leave home earlier than those modelled using the trip-based approach. Based on the tradeoff between in-home utility and working utility, commuters prefer to leave home earlier in the morning; however, the trip-based approach does not capture commuters' perceptions of the tradeoff in utility between activities.

In Table 9.2, the optimal (even) train headway generated by the activity-based and trip-based models, respectively, is compared in terms of total user net utility (bus headway is fixed at 10 minutes). Again, the work duration of all commuters in the trip-based model is pre-specified and equivalent to the expected work duration in the activity-based model. Table 9.2 shows that the optimal train headway for the activity-based and trip-based models is 0.10 hours and 0.15 hours, respectively. With these headway values, the total user net utility yielded by the activity-based model is HK$3.1 $\times$ 10^5 greater than that yielded by the trip-based model. Using the activity-based model, commuters' average time of departure from home to work is 25 minutes earlier than that obtained using the trip-based model, which supports the results shown in Fig. 9.5b.

Table 9.2 Comparison of Results for Activity-based and Trip-based Models.

Modeling Approach	Optimal Even Headway (hour)	Total User Net Utility (10^6 HK$)	Average Departure time from home	Work Duration (hours)
Activity-based	0.10	18.97	07:47	8.2
Trip-based	0.15	18.66	08:12	8.2

Note: To aid comparison, the value of work duration used in the trip-based model is pre-specified as the expected value of work duration in the activity-based model.

We now examine the effects of train-timetable changes on the number of trip-makers and their allocation of time to activities by increasing train headway from 10 minutes (base case (I)) to 20 minutes (alternative case (II)), as shown in Table 9.3. When the train headway is doubled, the total number of trip-makers decreases by 439; the average journey time (comprising in-vehicle travel time, walking time and waiting time) increases by 0.17 hours; the average work duration increases by 0.18 hours; and the average time spent at home decreases by 0.35 hours. An increase in train headway leads to a long average waiting time and thus a poor level of train service. Therefore, some commuters decide to use buses rather than trains. As a result, the average in-vehicle travel time increases by 0.06 hours and the average walking time decreases slightly, by 0.02 hours (because the average distance from home to the nearest bus stop is smaller than that from home to the nearest train station). In addition, commuters have to leave home earlier to avoid traffic congestion in the morning peak hours and the large penalty incurred by late arrival, as well as leaving their offices later to avoid traffic congestion in the afternoon peak hour. As a result, commuters spend more time at work and less time at home on an average.

Table 9.3 Effects of Train-timetable Changes on Number of Trip-makers and Time Allocated to At-home and Outside-home Activities.

	Number of Trip-makers	Number of Teleworkers	Average Time (hours)				
			Travel	Walking	Waiting	Work	Home
Base case (I)	17,777	2,223	0.74	0.48	0.17	8.30	14.31
Alternative case (II)	17,338	2,662	0.80	0.46	0.30	8.48	13.96
Difference (II-I)	−439	439	0.06	−0.02	0.13	0.18	−0.35

Note: Base case (I) is associated with a 10-minute train headway; alternative case (II) is associated with a 20-minute train headway.

As illustrated above, the activity-based modelling approach can be used to model the distribution of trip-makers' activity duration and their daily allocation of time to activities at home and outside the home. The trip-based approach cannot be used to model these important variables. However, the activity-based approach demands more detailed input data than the trip-based approach. For instance, the activity-based approach requires the parameters of the marginal-utility function for each activity (see Eq. 9.4) to be calibrated, necessitating an extensive survey of the demographic, socio-economic and trip characteristics of the areas under study.

9.4.2 Scenario 2

The transit network in Scenario 2 is shown in Fig. 9.1. It consists of one metro line, two bus lines (line 1 and line 2) and four activities conducted on a regular working day (Monday to Friday): staying at home, taking children to school (a before-work activity), working, and entertainment (an after-work activity). School and work trips are treated as compulsory, because the failure to meet their rigid arrival time requirements incurs a penalty. Returning home and leisure activities (entertainment) rarely have strict arrival time restrictions and are thus considered non-compulsory (or discretionary) activities without a schedule-delay penalty. In this example, the desired arrival time window is 09:00–09:30 for work trips and 08:00–08:30 for school trips. The one-way in-vehicle times and fares for the bus and metro lines are given in Tables 9.4 and 9.5, respectively. There are 30,000 potential trip-makers, Q. For simplicity, the headway on bus line 1 is assumed to be fixed at 10 minutes. As the metro line and bus line 2 have the same fleet size, five vehicles, it is only necessary to optimize the timetables for the metro line and bus line 2. The minimum headway on the metro line and bus line 2 is 5 minutes and 10 minutes, respectively. The other parameters are the same as those used in Scenario 1. The CPU time required to numerically obtain the converged solution is approximately 1,316 seconds.

Table 9.4 Fares for Bus and Metro Lines (HK$) (one direction).

From Node		To Node			
		1	2	3	4
Bus	1		5	5	8
	2				5
	3				5
Metro	1		10	15	25
	2			10	15
	3				10

Table 9.5 Travel Times for Bus and Metro Links.

Bus Link	Travel Time (min)	Metro Link	Travel Time (min)
(1, 2)	20	(1, 2)	5
(2, 4)	20	(2, 3)	5
(1, 3)	15	(3, 4)	5
(3, 4)	15		

Table 9.6 provides the optimal timetables with even and uneven headways, respectively, for the metro and bus services during morning peak hours. To save space, the corresponding timetables for the evening peak hours are not given here, but can be provided on request. Table 9.6 shows that the optimal even headway for the metro line is 6 minutes and that for bus line 2 is 12 minutes. In the optimal uneven headway timetable, the vehicle headway of the metro line and bus line 2 during

Table 9.6 Optimal Timetables with Even and Uneven Headways for Metro Line and Bus Line 2 During Morning Peak Hours.

	Station	Time of Arrival of Transit Vehicles at Station																			
Even headway	Metro line																				
	1	**07:11**	**07:17**	**07:23**	**07:29**	**07:35**	**07:41**	**07:47**	**07:53**	**07:59**	**08:05**	**08:11**	**08:17**	**08:23**	**08:29**	**08:35**	**08:41**	**08:47**	**08:53**	**08:59**	**09:05**
	2	07:16	07:22	07:28	07:34	07:40	07:46	07:52	07:58	08:04	08:10	08:16	08:22	08:28	08:34	08:40	08:46	08:52	08:58	09:04	09:10
	3	07:21	07:27	07:33	07:39	07:45	07:51	07:57	08:03	08:09	08:15	08:21	08:27	08:33	08:39	08:45	08:51	08:57	09:03	09:09	09:15
	4	07:26	07:32	07:38	07:44	07:50	07:56	08:02	08:08	08:14	08:20	08:26	08:32	08:38	08:44	08:50	08:56	09:02	09:08	09:14	09:20
	Bus line 2																				
	1	07:01	07:13	07:25	07:37	07:49	08:01	08:13	08:25	08:37	08:49	09:01	09:13	09:25	09:37	09:49					
	3	07:16	07:28	07:40	07:52	08:04	08:16	08:28	08:40	08:52	09:04	09:16	09:28	09:40	09:52	10:04					
	4	07:31	07:43	07:55	08:07	08:19	08:31	08:43	08:55	09:07	09:19	09:31	09:43	09:55	10:07	10:19					
Uneven headway	Metro line																				
	1	**07:15**	**07:21**	**07:27**	**07:33**	**07:39**	**07:45**	**07:52**	**07:58**	**08:04**	**08:09**	**08:15**	**08:23**	**08:28**	**08:35**	**08:40**	**08:45**	**08:52**	**08:58**	**09:05**	**09:10**
	2	07:20	07:26	07:32	07:38	07:44	07:50	07:57	08:03	08:09	08:14	08:20	08:28	08:33	08:40	08:45	08:50	08:57	09:03	09:10	09:15
	3	07:25	07:31	07:37	07:43	07:49	07:55	08:02	08:08	08:14	08:19	08:25	08:33	08:38	08:45	08:50	08:55	09:02	09:08	09:15	09:20
	4	07:30	07:36	07:42	07:48	07:54	08:00	08:07	08:13	08:19	08:24	08:30	08:38	08:43	08:50	08:55	09:00	09:07	09:13	09:20	09:25
	Bus line 2																				
	1	07:07	07:19	07:31	07:43	07:55	08:14	08:25	08:45	08:55	09:14	09:26	09:38	09:50	10:02						
	3	07:22	07:34	07:46	07:58	08:10	08:29	08:40	09:00	09:10	09:29	09:41	09:53	10:05	10:17						
	4	07:37	07:49	08:01	08:13	08:25	08:44	08:55	09:15	09:25	09:44	09:56	10:08	10:20	10:32						

Note: All numbers in the table rounded up to nearest minute.

peak hours reaches the minimum allowable headway in response to high passenger demand. Table 9.6 also shows that there are two peak intervals in passenger demand for the metro service, but one peak interval for bus line 2, because some commuters using the metro undertake an additional before-work activity (i.e., taking children to school). The passengers departing from home during these intervals arrive at their destinations (school or workplace) punctually. Specifically, the two peak intervals for the even headway metro service are 07:47–08:11 and 08:35–09:05, and those for the uneven headway metro service are 07:45–08:15 and 08:35–09:05. The peak intervals for bus line 2 with even and uneven headway are 08:25–08:49 and 08:25–08:55, respectively. In contrast with the even-headway timetable, the flexible-headway timetable permits more bus and metro runs to be scheduled during peak intervals, allowing more passengers to reach punctually their final destinations.

To trace the differences resulting from the two timetables, the distributions of individuals at different activity locations are depicted in Fig. 9.6 while Figs. 9.6a and 9.6b show that only a small proportion of commuters choose to perform before-work activities (e.g., taking children to school) during the morning peak hours, whereas a large proportion of commuters choose to participate in after-work activities (e.g., entertainment) during evening peak hours. Working is a compulsory activity with a strict start time when late arrival incurs a large schedule-delay penalty. Commuters may have to give up before-work activities to avoid schedule delay. However, a freer time budget at the end of the working day allows them to perform additional after-work activities. Compared with the even-headway timetable, the uneven-headway timetable attracts more commuters who undertake non-work activities both before and after work. As a result, the uneven-headway timetable allows far fewer individuals to return home between 17:00 and 20:00.

Table 9.7 indicates further differences between the even- and uneven-headway timetables. Although headway type has a trivial effect on the

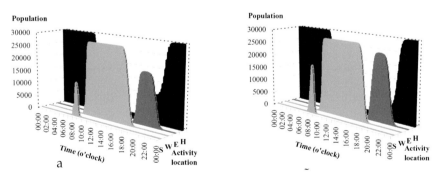

Figure 9.6 Population distributions at different locations: (a) even headway, (b) uneven headway.

total number of trip-makers, its influence on individual trip-makers' choice of activity/trip chain is significant. With uneven and demand-sensitive headway, the number of trip-makers using the longest chain, H-S-W-E-H, increases by 5,899 (from 12,668 to 18,567), whereas the number of trip-makers using the other three shorter chains, H-W-H, H-S-W-H and H-W-E-H decreases by 4,775, 574 and 407, respectively. This implies that trip-makers using a demand-sensitive timetable choose to visit more activity locations. Here what happens is that the flexible-headway timetable offers more benefits than the fixed-headway timetable for both system users and the community at large. In particular, the average passenger waiting time decreases by about 2 minutes and the average passenger schedule delay cost decreases by about HK$7 (from HK$11.61 to HK$4.53). As the total operator revenue and the total user net utility increase by HK$1.1 × 10^5 and HK$5.7 × 10^5, respectively, the total system utility increases by HK$6.8 × 10^5.

Table 9.7 Results for Even- and Uneven-Headway Timetables.

		Even Headway	Uneven Headway
Number of trip-makers		28,012	28,155
Passenger demand in different activity chains	H-W-H	6,777	2,002
	H-S-W-H	1,693	1,119
	H-W-E-H	6,874	6,467
	H-S-W-E-H	**12,668**	**18,567**
Average duration of activity (hours)	S	0.75	0.84
	W	8.33	8.46
	E	3.12	3.27
	H	**11.80**	**11.43**
Average waiting time (minutes)		4.78	2.64
Average schedule delay cost (HK$)		11.61	4.53
Total operating revenue (10^6 HK$)		1.07	1.18
Total user net utility (10^6 HK$)		**31.02**	**31.59**
Total system utility (10^6 HK$)		32.09	32.77

Table 9.7 also indicates that the headway type has an important effect on the use and allocation of trip-makers' activity time. With a flexible timetable, trip-makers spend more time on out-of-home activities. Specifically, the average time spent on before-work activities increases by 0.09 hours; that spent at work increases by 0.13 hours; and that spent on after-work activities increases by 0.15 hours. As a result, the average time spent at home decreases by 0.37 hours. Again, the timetable with demand-sensitive headway is shown to increase access to transit services, relaxing the restrictions on trip-makers' time allocation and giving them greater freedom to perform activities with time constraints, such as working or taking children to school.

It should be noted, however, that although uneven scheduling (i.e., demand-adjusted timetables) may be more satisfying for passengers, extra effort is required due to the irregular operation of the signal system, differences in the time spent at the terminal (for train cleaning and drivers' rest) and more intensive training provided for drivers. This creates a tradeoff between the additional benefits and the complexity of uneven transit vehicle operational schedules—an interesting and important topic for future research.

Finally, we investigate the effects of transit-service disruption on total system user utility. Transit-service disruption can be caused by various random factors, such as vehicle breakdown, signal-system malfunction or traffic accidents. As a result of transit-service disruption, some transit runs may be cancelled. The greater the disruption, the more transit runs are cancelled, and vice versa. Figure 9.7 displays the change in total user utility with increasing transit-service disruption in the case of even headway (see Table 9.6), which is measured by the number of transit-service runs cancelled. On the horizontal axis, '0' represents zero transit-service disruption, i.e., the base case, and '1' to '5' represent the number of transit service runs cancelled. Specifically, for bus line 2, '1', '2' and '3' indicate that the bus run(s) at 08:25, 08:25 and 08:37, and 08:25, 08:37 and 08:49 are cancelled, respectively. For the metro line, the numbers '1' to '5' represent the cancellation of one to five train runs from 08:29; that is, 08:29; 08:29 and 08:35; 08:29, 08:35 and 08:41; 08:29, 08:35, 08:41 and 08:47; and 08:29, 08:35, 08:41, 08:47 and 08:53, respectively. Figure 9.7 shows that as the number of the disrupted peak-period transit runs increases, the total user utility

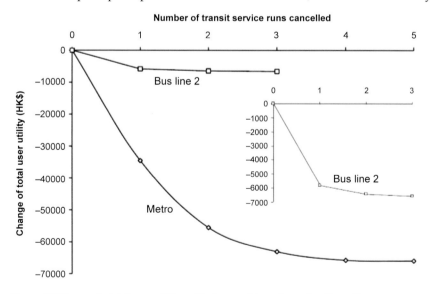

Figure 9.7 Change in total user utility with degree of transit-service disruption.

decreases. For example, total user utility decreases by HK$34,498 when the train run at 08:29 is cancelled, and by HK$5,828 when the bus run at 08:25 is cancelled. However, the impacts of the transit service disruptions on the total user utility are marginally decreasing. In addition, Fig. 9.7 shows that the adverse effect of metro-service disruptions is much larger than that of bus-service disruptions. Therefore, it is important for the transit authority to perform regular inspections and maintenance checks of transit vehicles to guarantee the efficient operation of transit services, particularly mass public transportation modes such as the metro.

9.5 Conclusions and Further Studies

In this chapter, we described the use of an activity-based approach integrated with a schedule-based dynamic transit assignment model to determine timetables for future transit services, simultaneously considering the stochastic activity and travel-choice behavior of transit passengers under the SUE condition. A new activity-based SUE transit assignment model capturing the interaction between transit timetables and passenger activity-travel scheduling behavior over a typical day is proposed. The model can be used to explicitly consider passengers' choices of departure time, activity/trip chain, activity duration, transit line and transport mode. A heuristic algorithm is developed to solve the proposed model and thereby determine timetables for future transit lines (such as metro lines for urban travel and high-speed rail for regional travel). Integrated with this novel approach, the schedule-based SUE transit assignment model can be used for short-term and long-term transit planning without incurring the common-line problem endemic to frequency-based transit assignment models.

The results of the two numerical tests offer important insights into the applications of the proposed model. First, the traditional trip-based model may lead to significant bias in the estimation of passengers' travel-scheduling patterns, as it does not take into account the interaction between passengers' activity choices and travel choices. The optimal headway generated by the activity-based approach is more productive and efficient in terms of the total user net utility of the transit system. Second, the demand-sensitive (uneven headway) timetable has a greater influence than the even-headway timetable on trip-makers' activity-travel schedules, comprising their activity-/trip-chain choices, their activity patterns before and after work and their daily allocation of time to activities. Third, given a fixed fleet size for each transit line, the demand-sensitive timetable is more efficient than the even- or fixed-headway timetable, as the former offers greater benefits for both users and society at large.

The second numerical example shows that the proposed activity-based SUE transit assignment model offers a useful tool for determining timetables for multimodal transit lines on short-term transit operations. The new activity-based approach allows the interaction between passenger activity schedules and transit timetables to be modelled explicitly. Therefore, given the resulting future transit timetables, the proposed model can be used for the long-term strategic planning of transit services and the evaluation of various transit policies. In addition, once sufficient data are available to calibrate passenger-activity patterns, the proposed approach will offer a new method of integrating passenger activity choice models with software used to plan future transit services, such as the Dynamic Regional Transit tool (see Crisalli and Rosati, 2004).

To facilitate the presentation of the essential ideas, the model proposed in this chapter does not accommodate the following aspects of real transit networks, which should be considered in future studies:

1. A heuristic solution algorithm is presented and small networks are used as illustrative examples. Although the numerical results for small networks are sensible, a more efficient solution algorithm should be developed and tested with realistic large-scale transit networks.

2. Only transit modes are considered. In reality, travellers' activity schedules are affected by the accessibility and level of service of various competitive modes of transport (e.g., bicycle, car, taxi, bus and metro). Therefore, it is necessary to extend the proposed model to multi-modal transport networks, including road networks. Road congestion and travel-time variations should be considered in the advanced models to investigate the activity-travel scheduling problems in multi-modal transportation networks.

3. In the proposed model, all individuals are assumed to be homogenous in the value they place on time and their perception of utility. This assumption should be relaxed to accommodate the heterogeneity of different user groups, with particular consideration to the interaction between household activities and travel schedules (Zhang et al., 2009).

4. In the proposed model, an individual's activities and travel-choice behavior are assumed to be independent (at the so-called 'one individual' level). The results of travel surveys indicate that joint participation in activities and travel makes up a substantial proportion of individuals' activity-travel scheduling patterns. Less attention has been paid to the interdependence between individuals' activities and travel behavior. It is clearly necessary to investigate the effects of joint activity-travel scheduling choices on transit-service planning.

5. To capture the effects of activity location capacity constraints on transit-service planning, the proposed model should be extended to incorporate crowding discomfort at a given activity location.

6. It is assumed that transit vehicles adhere strictly to timetables and the potential demand (i.e., the total number of seed individuals) is deterministic. In reality, the line-haul travel time or vehicle dwelling time at stops are random due to various uncertain factors, such as variations in road traffic congestion, fluctuations in the number of passengers boarding and alighting at stops, vehicle breakdowns, roadworks, traffic-signal failure and adverse weather (Fu et al., 2014). The use of stochastic (or robust) optimization to incorporate uncertainty in supply and demand is an interesting and important topic for further research.

7. The fleet size (or number of transit vehicles) is treated as exogenous in this chapter. To optimize the fleet size on transit lines for long-term strategic planning, future researchers should incorporate the capital and operational costs of inducting additional transit vehicles (buses or trains) into the proposed model.

8. In the model, the maximization of total user net utility is considered from the user's perspective. However, other objective functions, such as the maximization of profits by operators and social-welfare maximization, should also be considered. This extension will lead to more comprehensive policy analysis (Ceder, 1986).

9. To apply the proposed novel activity-based transit assignment model to real networks, it will be necessary to empirically calibrate the utility functions of various activities for different groups of people.

10. The proposed activity-based SUE network assignment model can be extended to the scheduling of activity-travel patterns in multi-modal transit networks under adverse weather conditions with varying rainfall intensity. This will aid the long-term planning of multi-modal transit services in cities with frequent rainy periods (e.g., Bangkok and Hong Kong). Weather-forecast information should be incorporated into the existing model to solve the passenger activity-travel scheduling problem. The extended model should enable explicit consideration of the effects of adverse weather on the performance of different transit modes, particularly in terms of service reliability and on the utility of various activities.

11. The proposed model can be extended to consider not only the effects of rain but the influence of other adverse weather conditions, such as typhoons or heavy snow.

12. Individuals' activity travel choice behavior is significantly affected by activity location capacity constraints and transportation-network congestion. In conditions of severe traffic congestion due to adverse weather, individuals tend to choose the more reliable transit mode, e.g., the subway, which has an exclusive right of way, preventing congestion due to interaction with other modes. It is thus necessary

to extend the proposed model and solution algorithm to consider demand and supply uncertainties in multi-modal transportation networks.

Given the diversity of research and development (R&D) in the field of analytical activity-based approaches to short-term and long-term transit service planning, the model and solution algorithm presented in this chapter are by no means exhaustive. However, they do provide a basic coverage of various important directions of R&D in this field, with particular attention given to the solution of ATS problems to aid transit operation and service planning. The authors hope that this chapter will provide a state-of-the-art methodology for the development of new activity-based transit assignment models and solution algorithms by both practitioners and researchers, as well as indicating new R&D opportunities and inspiring further efforts in this field. The new activity-based approach is expected to improve the planning, design and operation of multi-modal transit systems and thereby increase the efficiency and reliability of transportation networks in our cities.

Acknowledgments

The work described in this chapter was jointly supported by grants from the Research Grants Council of the Hong Kong Special Administrative Region (PolyU 5181/13E and PolyU 152057/15E) and the National Natural Science Foundation of China (71471013).

Keywords: activity-based model; scheduling/timetabling problem; transit assignment; transit-service disruption.

References

Ashiru, O., J. W. Polak and R. B. Noland. 2004. Utility of schedules: Theoretical model of departure-time choice and activity-time allocation with application to individual activity schedules. Transp. Res. Rec. 1894: 84–98.

Arentze, T. A. and H. J. P. Timmermans. 2009. A need-based model of multi-day, multi-person activity generation. Transp. Res. Part B 43: 251–265.

Axhausen, K. W. and T. Gärling. 1992. Activity-based approaches to travel analysis: Conceptual frameworks, models, and research problems. Transport Rev. 12(4): 323–341.

Bazaraa, M. S., H. D. Sherali and C. M. Shetty. 2006. Nonlinear Programming: Theory and Algorithms. Wiley, Hoboken, New Jersey.

Ben-Akiva, M. and S. R. Lerman. 1985. Discrete Choice Analysis: Theory and Application to Travel Demand. MIT Press, Cambridge, MA.

Caprara, A., M. Fischetti and P. Toth. 2002. Modelling and solving the train timetabling problem. Oper. Res. 50(5): 851–861.

Ceder, A. 1986. Methods for creating bus timetables. Transp. Res. Part A 21(1): 59–83.

Ceder, A. 2003. Public transport timetabling and vehicle scheduling. pp. 31–57. *In*: W. H. K. Lam and M. G. H. Bell (eds.). Advanced Modeling for Transit Operations and Service Planning, Pergamon.

Chen, C. H., S. Yan and C. H. Tseng. 2010. Inter-city bus scheduling for allied carriers. Transportmetrica 6(3): 161–185.

Chow, J. Y. and W. W. Recker. 2012. Inverse optimization with endogenous arrival time constraints to calibrate the household activity pattern problem. Transp. Res. Part B 46(3): 463–479.

Crisalli, U. and L. Rosati. 2004. DY-RT: A tool for schedule-based planning of regional transit networks. pp. 135–158. *In*: N. H. M. Wilson and A. Nuzzolo (eds.). Schedule-based Dynamic Transit Modeling: Theory and Applications, Kluwer Academic Publishers.

De Cea, J. and E. Fernandez. 1993. Transit assignment for congested public transport system: An equilibrium model. Transp. Sci. 27(2): 133–147.

De Palma, A. and R. Lindsey. 2001. Optimal timetables for public transportation. Transp. Res. Part B 35(8): 789–813.

Ettema, D. and H. J. P. Timmermans. 1997. Activity-based Approaches to Travel Analysis. Oxford, Pergamon.

Ettema, D. and H. J. P. Timmermans. 2003. Modeling departure time choice in the context of activity scheduling behavior. Transp. Res. Rec. 1831: 39–46.

Fu, X. and W. H. K. Lam. 2014. A network equilibrium approach for modelling activity-travel pattern scheduling problems in multi-modal transit networks with uncertainty. Transportation 41(1): 37–55.

Fu, X., W. H. K. Lam and Q. Meng. 2014. Modelling impacts of adverse weather conditions on activity–travel pattern scheduling in multi-modal transit networks. Transportmetrica B: Transport Dynamics 2: 151–167.

Granas, A. and J. Dugundji. 2003. Fixed Point Theory. New York, Springer.

Habib, K. M. N. and E. J. Miller. 2009. Modelling activity generation: A utility-based model for activity-agenda formulation. Transportmetrica 5(1): 3–23.

Hensher, D. A. 2007. Some insights into the key influences on trip-chaining activity and public transport use of seniors and the elderly. Int. J. Sustain. Transp. 1(1): 53–68.

Huang, H. J., Z. C. Li, W. H. K. Lam and S. C. Wong. 2005. A time-dependent activity and travel choice model with multiple parking options. pp. 717–739. *In*: H. Mahmassani (ed.). Transportation and Traffic Theory. Elsevier, Oxford.

Joh, C. H., T. A. Arentze and H. J. P. Timmermans. 2002. Modeling individuals' activity-travel rescheduling heuristics: Theory and numerical experiments. Transp. Res. Rec. 18007: 16–25.

Jones, P. M., F. S. Koppelman and J. P. Orfeuil. 1990. Activity analysis: State of the art and future directions. pp. 34–55. *In*: P. Jones (ed.). Developments in Dynamic and Activity-based Approaches to Travel Analysis. Avebury, Aldershot, England.

Kitamura, R. 1988. An evaluation of activity-based travel analysis. Transportation 15(1): 9–34.

Kurauchi, F., M. G. H. Bell and J. -D. Schmöcker. 2003. Capacity constrained transit assignment with common lines. J. Math. Model. Algor. 2(4): 309–327.

Lam, W. H. K., Z. Y. Gao, K. S. Chan and H. Yang. 1999. A stochastic user equilibrium assignment model for congested transit networks. Transp. Res. Part B 33(5): 351–368.

Lam, W. H. K., J. Zhou and Z. Sheng. 2002. A capacity restraint transit assignment with elastic line frequency. Transp. Res. Part B 36(10): 919–938.

Lam, W. H. K. and Y. Yin. 2001. An activity-based time-dependent traffic assignment model. Transp. Res. Part B 35(6): 549–574.

Lam, W. H. K. and H. J. Huang. 2002. A combined activity/travel choice model for congested road networks with queues. Transportation 29(1): 5–29.

Li, Z. C., W. H. K. Lam and S. C. Wong. 2009. The optimal transit fare structure under different market regimes with uncertainty in the network. Netw. Spat. Econ. 9(2): 191–216.

Li, Z. C., W. H. K. Lam, S. C. Wong and A. Sumalee. 2010. An activity-based approach for scheduling multimodal transit services. Transportation 37(5): 751–774.

Li, Z. C., Y. Yin, W. H. K. Lam and A. Sumalee. 2016. Simultaneous optimization of fuel surcharges and transit service runs in a multi-modal transport network: A time-dependent activity-based approach. Transportation Letters 8(1): 35–46.

Lo, H. K., C. W. Yip and K. H. Wan. 2003. Modeling transfer and non-linear fare structure in multi-modal network. Transp. Res. Part B 37(2): 149–170.

Luo, Z. Q., J. S. Pang and D. Ralph. 1996. Mathematical Programs with Equilibrium Constraints. Cambridge University Press.

Magnanti, T. L. and G. Perakis. 2002. Computing fixed points by averaging. pp. 181–198. *In*: M. Gendreau and P. Marcotte (eds.). Transportation and Network Analysis: Current Trends. Kluwer Academic Publishers, Netherlands.

Maruyama, T. and A. Sumalee. 2007. Efficiency and equity comparison of cordon- and area-based road pricing schemes using a trip-chain equilibrium model. Transp. Res. Part A 41(7): 655–671.

Miller, E. J. and M. J. Roorda. 2003. Prototype model of household activity-travel scheduling. Transportation Research Record 1831: 114–121.

Nuzzolo, A., F. Russo and U. Crisalli. 2001. A doubly dynamic schedule-based assignment model for transit networks. Transp. Sci. 35(3): 268–285.

Ouyang, L., W. H. K. Lam, Z. C. Li and D. Huang. 2011. Network user equilibrium model for scheduling daily activity travel patterns in congested networks. Transp. Res. Rec. 2254: 131–139.

Pendyala, R. M., T. Yamamoto and R. Kitamura. 2002. On the formulation of time-space prisms to model constraints on personal activity-travel engagement. Transportation 29(1): 73–94.

Poon, M. H., S. C. Wong and C. O. Tong. 2004. A dynamic schedule-based model for congested transit networks. Transp. Res. Part B 38(4): 343–368.

Ramadurai, G. and S. Ukkusuri. 2010. Dynamic user equilibrium model for combined activity-travel choices using activity-travel supernetwork representation. Netw. Spat. Econ. 10(2): 273–292.

Schmöcker, J. -D., M. G. H. Bell and F. Kurauchi. 2008. A quasi-dynamic capacity constrained frequency-based transit assignment model. Transp. Res. Part B 42(10): 925–945.

Sumalee, A., Z. J. Tan and W. H. K. Lam. 2009. Dynamic stochastic transit assignment with explicit seat allocation model. Transp. Res. Part B 43(8-9): 895–912.

Timmermans, H. J. P. 2005. Progress in Activity-based Analysis. Elsevier, Amsterdam.

Tong, C. O. and S. C. Wong. 1999. A stochastic transit assignment model using a dynamic schedule-based network. Transp. Res. Part B 33(2): 107–121.

Tong, C. O., S. C. Wong, M. H. Poon and M. C. Tan. 2001. A schedule-based dynamic transit network model–recent advances and prospective future research. J. Adv. Transp. 35(2): 175–195.

Uchida, K., A. Sumalee, D. P. Watling and R. Connors. 2005. A study on optimal frequency design problem for multi-modal network using probit-based user equilibrium assignment. Transp. Res. Rec. 1923: 236–245.

Wong, S. C. and C. O. Tong. 1998. Estimation of time-dependent origin–destination matrices and transit networks. Transp. Res. Part B 32(1): 35–48.

Wu, J. H., M. Florian and P. Marcotte. 1994. Transit equilibrium assignment: A model and solution algorithm. Transp. Sci. 28(3): 193–203.

Yamamoto, T. and R. Kitamura. 1999. An analysis of time allocation to in-home and out-of-home discretionary activities across working days and non-working days. Transportation 26(2): 211–230.

Yamamoto, T., S. Fujii, R. Kitamura and H. Yoshida. 2000. Analysis of time allocation, departure time, and route choice behavior under congestion pricing. Transp. Res. Rec. 1725: 95–101.

Zhang, X., H. Yang, H. J. Huang and H. M. Zhang. 2005. Integrated scheduling of daily work activities and morning-evening commutes with bottleneck congestion. Transp. Res. Part A 39(1): 41–60.

Zhang, J. Y., M. Kuwano, B. Lee and A. Fujiwara. 2009. Modeling household discrete choice behavior incorporating heterogeneous group decision-making mechanisms. Transp. Res. Part B 43(2): 230–250.

Zhang, J. and H. Timmermans. 2012. Activity-travel behavior analysis for universal mobility design. Transportmetrica 8(3): 149–156.

Zhao, F. and X. G. Zeng. 2008. Optimization of transit route network, vehicle headways and timetables for large-scale transit networks. Eur. J. Oper. Res. 186(2): 841–855.

10

Transit Network Design with Stochastic Demand

K. An[1] *and H. K. Lo*[2],*

ABSTRACT

This chapter studies transit network design with stochastic demand by considering two types of services—rapid transit services, such as rail, and flexible services, such as dial-a-ride shuttles. Rapid transit services operate on fixed routes and dedicated lanes, and with fixed schedules, whereas dial-a-ride services can make use of the existing road network; hence are much more economical to implement. We integrate these two service networks into one multi-modal network and then determine the optimal combination of these two service types under user equilibrium (UE) flows. Two approaches are used to address the issue of stochastic demand: one is robust optimization; the other is stochastic programming. The robust optimization approach assumes that the stochastic demand is captured in a polyhedral uncertainty set. The UE principle is represented by a set of variational inequality (VI) constraints.

---

[1] Department of Civil Engineering, Monash University, Clayton, VIC 3800, Australia.
  E-mail: kun.an@monash.edu
[2] Department of Civil and Environmental Engineering, The Hong Kong University of Science and Technology, Clear Water Bay, Hong Kong.
* Corresponding author: cehklo@ust.hk

Eventually, the whole problem is linearized and formulated as a mixed-integer linear program. A cutting constraint algorithm is adopted to address the computational difficulty arising from the VI constraints. The stochastic programming approach assumes that the distribution of the stochastic demand is given. A service reliability (SR) based two-phase stochastic program is formulated. The transit line alignment and frequencies are determined in Phase-1 for a specified SR; the flexible services are determined in Phase-2 depending on the demand realization to capture the cost of demand overflow beyond the specified SR. The objective is to optimize SR so as to strike a balance between the ability of the transit services to carry the demand and the resultant system cost. The proposed model is applied in a small illustrative network to demonstrate their applicability and compare the solution performances of the two approaches.

10.1 Introduction

The transit network design problem (TNDP) is to decide the locations of stations, route alignment as well as frequency to serve the travel demands between specific origin-destination (OD) pairs. Due to high construction and operating costs of rapid transit line (RTL), some lines may face low passenger load and some may even require government subsidy for their operations. What's more, the intrinsic uncertainty in travel demand makes it uneconomical to rely on RTL alone to serve the demand. Dial-a-ride (DAR) services, in contrast, are able to utilize the existing road network, thus having relatively lower capital costs, which involve the procurement, operations and maintenance of the vehicle fleet. Meanwhile, they have great flexibility to cater for demand fluctuation. However, the congestion effect of dial-a-ride services cannot be neglected. Thus, it may not be economical and environmentally efficient to rely heavily on dial-a-ride services for areas with large and relatively stable travel demands. Therefore, how to determine the optimal combination of RTL, i.e., those on fixed routes and fixed schedules, and DAR services, i.e., those making use of existing roadways without fixed routes nor schedule commitment, to serve a metropolitan area is an important topic to study. This optimal combination of two different types of services, if designed to be cost effective and of high quality, would encourage modal shift to public transport, perhaps eventually break the vicious cycle of auto dependence in many metropolitan areas.

The Network Design Problem (NDP) can be classified into discrete, continuous and mixed, as discussed in Yang and Bell (1998). The discrete NDP is concerned with the network topology itself (Wang et al., 2013; Gao et al., 2005; Lai and Lo, 2004). Examples include scheduling or routing of a service network. The continuous NDP takes the network topology as given and is concerned with optimizing the network parameters. Examples include enhancing the link capacity or setting the toll charges (Gao et al., 2004; Ekström et al., 2012). Bellei et al. (2002) investigated the pricing problem in a multi-user and multi-modal network with elastic demand, where user equilibrium is formulated as a fixed point problem considering travellers' hierarchical route choices and road network congestion. The mixed NDP combines the two types to simultaneously determine new links to be added and capacity increases of existing roads (Luathep et al., 2011). The transit network design problem falls in the category of mixed NDP which involves determining discrete and continuous variables, namely, transit line alignment and frequency. Most existing studies focus on the deterministic TNDP, assuming that the OD demand is fixed and known. It is typically formulated as a mixed integer linear program (MILP) where the station selection, line alignment and frequencies are determined simultaneously to achieve a certain objective, such as cost minimization or coverage maximization (Wan and Lo, 2009; Bruno et al., 1998).

The literature on NDP concerning uncertain demand can be classified into two categories. The first approach is stochastic programming, which assumes known demand distributions and utilizes Monte-Carlo simulation to decompose the random demands into a finite number of scenarios for approximating the cost expectation. It is typically formulated as a MILP to minimize the total expected cost (Ruszczynski, 2008; Birge and Louveaux, 1988; Benders, 1962) and solved by a commercial software, such as CPLEX, or the L-shaped method. An and Lo (2014a, 2014b, 2016) and Lo et al. (2013) proposed an alternative method, namely, the service reliability (SR)-based approach, which separates the large-size MILP into two phases for solution efficiency. The second approach is robust optimization, which focuses on the min-max problem, namely, optimization of the worst case scenario. It requires that the network design solutions, determined before the demand realization, are feasible for all demand realizations. The side effect is that the solutions may be overly conservative. Some studies thus turned to refining the uncertainty set such that all the realized demands within the set are satisfied while those outside are ignored. It is important to trade off the size of the uncertainty set and the robustness level (Bertsimas and Sim, 2004; Ben-Tal and Nemirovski, 1999; An and Lo, 2015). Robust and stochastic optimizations are two complementary ways to deal with uncertainty, each having its own advantages and disadvantages. In spite of their differences on the underlying distribution requirement and solution algorithm, researchers also made renewed efforts to bridge the robust and the stochastic optimization paradigms. Bertsimas and Goyal (2010) quantified the performance of robust optimization solution in the stochastic world for different distribution patterns. Chen et al. (2007)

investigated stochastic programming from the perspective of robust optimization and proposed forward and backward deviation measures for the involved random parameters.

In addition to the challenge of stochastic demand, this chapter also incorporates user equilibrium (UE) passenger flows in a multi-modal transit network. The NDP with UE flows is typically formulated as a bi-level problem, in which the upper level problem focuses on generating the optimal network design; whereas the lower level represents travellers' travel choices. This bi-level problem is typically non-linear and non-convex. Some studies formulate the UE principle as variational inequality (VI) constraints, which reduce the bi-level problem into a single level problem with VI constraints. By applying a cutting constraint algorithm, the single level problem can be solved iteratively, alleviating the onerous task on feasible paths enumeration (Ekström et al., 2012; Luathep et al., 2011). Various solution approaches have been developed to deal with bi-level problems, such as heuristic approach, global optimization approach (e.g., Wang et al., 2013; Wang and Lo, 2010). Nguyen and Pallottino (1988) first introduced the hyper-path concept in transit assignment problem in a multi-modal network. Marcotte and Nguyen (1998) utilized the hyper path concept to formulate a user equilibrium model considering passenger strategies on route selection. Although the TNDP with UE flows has been studied intensively, few studies have investigated the influence of demand uncertainty.

This chapter studies TNDP with stochastic demand under UE flows by considering two types of services—rapid transit line (RTL) and dial-a-ride (DAR) services. RTL services operate on fixed routes and fixed schedules, which may include multiple lines, whereas DAR services are demand responsive to carry the demand realized on a particular day that exceeds the capacity of RTL services. We integrate RTL and DAR services into one multi-modal network, allowing for intra-modal transfers and then determine the optimal combination of these two service types under UE passenger flows. The UE principle is represented by a set of variational inequality (VI) constraints. Two methods are utilized to handle demand uncertainty—robust optimization and stochastic programming, and evaluate their individual advantages and disadvantages. In the robust optimization approach, the demand uncertainty is captured as a polyhedral uncertainty set. It is reformulated as an MILP by replacing the uncertainty set by its dual problem. The RTL configuration and frequencies, or the here-and-now variables, are determined in the MILP and fixed for the planning horizon, while DAR services deployment for the worst case scenario is determined, with their congestion effects accounted for in the road network. Eventually, the whole problem is linearized and formulated as a mixed-integer linear program. A cutting constraint algorithm is adopted to address the computational difficulty arising from the VI constraints. In the stochastic programming approach, we utilize the notion of service reliability to address the issue of stochastic demand and formulate the problem as a two-phase stochastic program. The network and frequencies of RTL services are determined in Phase-1 while DAR services are determined in Phase-2, depending on the demand realization. The objective is to decide the optimal combination of these two service types to minimize the total expected operating cost while serving all realized OD demands. A gradient solution approach is developed to solve the problem. Finally, we compare the performances of robust optimization solution with various robustness levels and two-stage stochastic solutions, assuming normally distributed stochastic demands.

The reminder of this chapter is as follows: Section 10.2 presents the robust model formulation and solution algorithm. Section 10.3 presents the service reliability-based two stage stochastic formulation and solution algorithm. Section 10.4 illustrates the model and solution algorithm with a case study. Section 10.5 concludes our work and proposes future extensions.

10.2 Robust Model

10.2.1 Problem Setting

Let $G(N, A)$ be the candidate transportation network with a node set N and an arc set A. A is the set of feasible arcs $(i, j) \in A$ linking stations i and j for $i, j \in N, i \neq j$. Each feasible arc (i, j) is associated with a link distance l_{ij}. DAR services use the same road network as that of RTL. Let q_{ij} denote the nominal value of travel demand from station i to j. To simplify the notation, we use d to represent a specific OD pair instead of ij in the following.

To integrate the two services into one multi-modal network, we segregate each node into three sub-nodes, namely, station sub-node, RTL sub-node and DAR sub-node, with their corresponding set denoted as N_s, N_{RTL}, N_{DAR}. These three types of sub nodes are interconnected with each other, allowing for passenger boarding, alighting and transferring between the two modes. OD demands are loaded or unloaded from the station sub-nodes. The RTL (DAR) sub nodes are connected by RTL (DAR) arcs. The feasible arc set A can thus be separated into two subsets accordingly, RTL arc set A_{RTL} and DAR arc set A_{DAR}. Figure 10.1 illustrates the multi-modal network with a three station network—A_0, B_0, C_0 represent station sub-nodes; A_1, B_1, C_1 represent RTL sub-nodes; and A_2, B_2, C_2 represent DAR sub-nodes. The arcs connecting A_1, B_1, C_1 (A_2, B_2, C_2) are RTL (DAR) arcs and the flows on them represent the amounts of passengers taking RTL (DAR) services. The problem is to determine a set of RTL routes $r \in R$ and their corresponding frequencies $f_r \in \mathbf{f}$ as well as the deployment of DAR services to meet the stochastic OD demands so that the total expected cost is minimized. The maximum number of lines R_{max} is predefined. To select the origin and destination station for each transit route flexibly, we introduce a dummy starting node set $S = \{S_r, r \in R\}$ and ending node set $T = \{T_r, r \in R\}$ so that every RTL route r has a fixed dummy origin S_r and destination T_r. The exposition of this network structure formulation can be found in An and Lo (2014a). For brevity here, we skip the details and only mention the assumptions for this multi-modal problem:

(a) Each route serves both directions with the same frequency.
(b) DAR services operate on the existing road network, with congestion modelled by the Bureau of Public Roads (BPR) function.

Assumption (a) is a common practice; (b) shows how DAR services operate, serving as shuttles to carry the demands not covered by RTL. Meanwhile, they can also serve as another mode for certain congested RTL sections to mitigate crowdedness or as additional segment capacity. The road network for DAR is the same as the candidate network for RTL. The impact of private vehicles on the road network is considered directly through deducting road link capacity by their background traffic flow. The UE principle is upheld while including the congestion of DAR services. Although there is no explicit limit on the number of transfers, it can be partially remedied by imposing a sufficient large penalty for passenger transfers between RTL and DAR as in Lo et al. (2003, 2004).

The decision variables are as follows. A binary variable set $\mathbf{Y} = \{Y_{ij}^r\}$, $ij \in A_{RTL}$ denotes whether a link is on a RTL. Y_{ij}^r is 1 if link (i, j) is on line

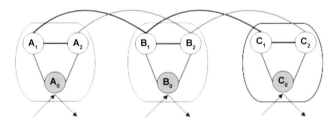

Figure 10.1 Multi-modal network representation.

r; 0 otherwise. The binary variable $\tilde{Y}_{ij}$ equals 1 if link ij is covered by any line $r \in R$, zero otherwise, with its set denoted by $\tilde{\mathbf{Y}} = \{\tilde{Y}_{ij}\}$. $\mathbf{f} = \{f_r\}$ stands for the RTL route frequency. The binary variable set $\mathbf{W} = \{W_i\}, i \in N_{RTL}$ indicates whether station i is on a RTL. Note that $\tilde{\mathbf{Y}}$ and $\mathbf{W}$ can be uniquely determined by $\mathbf{Y}$. $X_{ij}^d, ij \in A_{RTL}$, represents the passenger flow on RTL from station i to j for OD pair d whose set is denoted as $\mathbf{X}$. $Z_{ij}^d, ij \in A_{DAR}$, represents the passenger flow on DAR from station i to j for OD pair d, whose set is denoted as $\mathbf{Z}$. V_{i+} stands for the amount of passengers transferring from

RTL to DAR at station i, whereas V_{i-} stands for the amount of passengers transferring from DAR to RTL. The transfer passenger flow set is denoted by **V**. Note that the first two decision variables have directions. The problem is to determine each transit line for the forward direction, with services for the backward direction included automatically.

10.2.2 A Robust Formulation with Equilibrium Constraints

The stochastic OD demand, Q^d, fluctuates from day to day with the mean q^d or q_{ij}: $Q^d \in U^d$ and $E(Q^d) = q^d$. Let D^d be the destination node of OD pair d. U is known as the uncertainty set in robust optimization. Let θ be the uncertainty level. In particular, a polyhedral uncertainty set is defined as:

$$\left\{Q^d, \forall d\right\} \in U \equiv \left\{ \mathbf{Q}^d : q^d(1-\theta) \le Q^d \le q^d(1+\theta), \forall d \sum_{\forall d \in \{d:D^d=j\}} Q^d \le B_j \right\} \quad (10.1)$$

The polyhedral uncertainty set is less conservative than the box uncertainty set as it includes the joint constraint $\sum_{\forall d \in \{d:D^d=j\}} Q^d \le B_j$ (Ben-Tal et al., 2011), which is more realistic as it limits the total travellers heading for the same destination j by an upper bound B_j. For instance, the total demand arriving at a city center is limited by the amount of jobs, amount of retail shops or parking, etc. The essence of robust optimization is to find a sub-optimal solution for the RTL alignment as well as DAR services deployment such that the solutions are feasible even for the worst case scenario.

The goal of this section is to minimize the system cost under the worst case scenario by locating the RTL stations, deciding the line frequencies as well as introducing DAR services as needed. To the company, these two services have different unit costs. To passengers, crowding discomfort on RTL and road congestion of DAR services are considered simultaneously to achieve a UE passenger distribution between these two modes. The network design problem with stochastic demand under UE can be formulated as a robust bi-level program, where the network design variables {**Y,f,W,Z**} are determined in the upper level, and the UE passenger flows {**X,V**} in the lower level. The UE principle can be represented by variational inequality constraints to reduce the bi-level problem into a single-level problem with equilibrium constraints. We formulate the robust TNDP with equilibrium constraints in **P1**. Let $\bar{N} = N_{RTL} \cup S \cup T$ be the set of all RTL nodes including the dummy origin and destination nodes and $\bar{A}$ be the set of all RTL links including the dummy arcs. Since passenger flows on RTL, DAR or transfer links are represented by different variables X_{ij}^d, Z_{ij}^d, V_{i+}^d, V_{i-}^d it is redundant to specify the sub-node set that the subscripts of X_{ij}^d, Z_{ij}^d, V_{i+}^d, V_{i-}^d belong to.

Namely, for X_{ij}^d we must have $i, j \in N_{RTL}$ and for Z_{ij}^d, we must have $i, j \in N_{DAR}$. To simplify the notation, we simply use N to represent the station index without specifying the specific mode the flow variables X_{ij}^d, Z_{ij}^d, V_{i+}^d, V_{i-}^d belong to.

$$\textbf{(P1)} \quad \min_{\mathbf{f}, \mathbf{Y}, \tilde{\mathbf{Y}}, \mathbf{W}, \mathbf{X}, \mathbf{Z}, \mathbf{V}} \sum_{ij \in A_{RTL}} \left(c^1 l_{ij} \sum_{r \in R} f_r Y_{ij}^r + c^2 l_{ij} \tilde{Y}_{ij} \right) + c^3 \sum_{i \in N_{RTL}} W_i + c^4 \sum_{d \in D} \sum_{ij \in A_{DAR}} l_{ij} Z_{ij}^d$$
$$+ c^5 \sum_{d \in D} \sum_{ij \in A_{RTL}} t_{ij} X_{ij}^d + c^5 \sum_{d \in D} \sum_{ij \in A_{DAR}} t_{ij} Z_{ij}^d + c^5 \sum_{d \in D} \sum_{i \in N} t^0 \left(V_{i+}^d + V_{i-}^d \right) \tag{10.2}$$

$$\text{s.t.} \sum_{j \in N} Y_{ij}^r \leq 1, \quad \forall i \in N_{RTL} \cup S, r \in R \tag{10.3}$$

$$\sum_{j \in N} Y_{ji}^r \leq 1, \quad \forall i \in N_{RTL} \cup T, r \in R \tag{10.4}$$

$$\sum_{i \in N} Y_{ik}^r = \sum_{j \in N} Y_{kj}^r, \quad \forall k \in N_{RTL}, r \in R \tag{10.5}$$

$$Y_{ij}^r + Y_{ji}^r \leq 1, \quad \forall ij \in A_{RTL}, r \in R \tag{10.6}$$

$$\tilde{Y}_{ij} \leq \sum_{r \in R} \left(Y_{ij}^r + Y_{ji}^r \right), \quad \forall ij \in A_{RTL} \text{ and } i < j \tag{10.7}$$

$$\tilde{Y}_{ij} \geq Y_{ij}^r + Y_{ji}^r, \quad \forall r \in R, \ ij \in A_{RTL} \text{ and } i < j \tag{10.8}$$

$$W_i \geq \tilde{Y}_{ij}, W_i \geq \tilde{Y}_{ji}, \ \forall i \in N_{RTL}, j \in N_{RTL} \tag{10.9}$$

$$Y_{ij}^r = 0 \text{ or } 1, \quad \forall r \in R, ij \in A_{RTL} \tag{10.10}$$

$$\tilde{Y}_{ij} = 0 \text{ or } 1 \quad \forall ij \in A_{RTL} \tag{10.11}$$

$$W_i = 0 \text{ or } 1, \quad \forall i \in N_{RTL} \tag{10.12}$$

$$f_{\min} \leq \sum_{r \in R} \left(Y_{ij}^r + Y_{ji}^r \right) f_r \leq f_{\max}, \quad \forall ij \in A_{RTL} \tag{10.13}$$

$$\sum_{j \in N} X_{kj}^d + \sum_{j \in N} Z_{kj}^d \geq Q^d , \text{if } k = O^d \tag{10.14}$$

$$\sum_{i \in N} X_{ik}^d + \sum_{i \in N} Z_{ik}^d \geq Q^d , \text{if } k = D^d \tag{10.15}$$

$$\sum_{j \in N} X_{kj}^d - \sum_{i \in N} X_{ik}^d + V_{k+}^d - V_{k-}^d = 0,\, k \neq O^d \text{ or } D^d , \quad \forall k \in N, d \in D, Q^d \in U \tag{10.16}$$

$$\sum_{j \in N} Z_{kj}^d - \sum_{i \in N} Z_{ik}^d + V_{k-}^d - V_{k+}^d = 0,\, k \neq O^d \text{ or } D^d \tag{10.17}$$

$$t_{ij} = t_{ij}^0 \left(1 + 0.1 \left(\frac{\sum_d X_{ij}^d}{\sum_r f_r Y_{ij}^r \xi + \varepsilon} \right)^2 \right), ij \in A_{RTL} \tag{10.18}$$

$$t_{ij} = t_{ij}^0 \left(1 + 0.15 \left(\frac{\sum_d Z_{ij}^d}{C_{ij}} \right)^4 \right), ij \in A_{DAR} \tag{10.19}$$

$$\sum_{ij \in A_{RTL}} t_{ij} * (x_{ij} - \tilde{x}_{ij}) + \sum_{ij \in A_{DAR}} t_{ij} * (z_{ij} - \tilde{z}_{ij}) + \sum_{i \in N} t^0 * (v_i - \tilde{v}_i) \leq 0, \quad \forall \left(\tilde{x}_{ij}, \tilde{z}_{ij}, \tilde{v}_i \right) \in \Omega \tag{10.20}$$

$$x_{ij} = \sum_d X_{ij}^d ,\; z_{ij} = \sum_d Z_{ij}^d ,\; v_i = \sum_d \left(V_{i+}^d + V_{i-}^d \right) \tag{10.21}$$

c^1, c^2, c^3, c^4, c^5 are, respectively, the coefficients for RTL operating cost, RTL construction cost, station construction cost, DAR operating cost and passenger value of time. Also t^0 denotes the transfer penalty which is a constant. The RTL link operating and construction costs are proportional to the link distance l_{ij}. The objective function is to minimize the combined RTL operating cost, RTL construction cost, RTL station construction cost, DAR operating cost and passenger cost in order to serve the random demand $Q^d \in U^d$. The RTL connectivity is represented by constraints (10.3)–(10.12). Constraints (10.3) and (10.4) indicate that only one RTL sub-node is directly connected to RTL sub-node i from upstream and downstream, respectively, if i is on route r. Constraints (10.3) and (10.4) also ensure that at most one RTL link can be generated from the dummy origin S_r and ended at destination T_r. Constraint (10.5) states that there are exactly two RTL links connecting each RTL node on route r. Constraint (10.6) represents that at most one direction of a link can be occupied by one transit line. Constraints (10.7) and (10.8) ensure that a rail track is constructed on link ij that is if any line goes through it. Constraint (10.9) ensures that a station is constructed if any line traverses through it. Constraints (10.10)–(10.12) are integrality constraints. Constraint (10.13) restricts that the total frequencies of the lines running on shared track segments are bounded. Constraints (10.14)–(10.17) represent the passenger flow balancing conditions. The demand Q^d in (10.14)–(10.17) is stochastic and bounded by the uncertainty set U^d. $Q^d \in U^d$ requires that the optimal solution of **P1** must be feasible for any demand realization in U^d, which leads to a min-max problem, i.e., the worst case or maximum demand scenario within the uncertainty set. O^d and D^d, respectively, represent the origin or destination node index of OD pair d. To accommodate the approach of robust formulation, the equality constraints for origin and destination nodes are replaced by inequalities. It is easy to see it will be pushed to equality as a consequence of the optimization. The inequality (10.14) states that the total amount of passengers flowing out from station k either by RTL or DAR is greater than the demand Q^d if k is the origin of OD pair d. The second inequality (10.15) follows the same logic for destination nodes. Constraints (10.16) and (10.17) are the flow conservation constraint for RTL and DAR sub nodes, respectively. (10.18)–(10.21) are standard VI constraints to achieve UE. The passenger cost on RTL is modeled as a non-linear function (10.18), where $t_{ij}^0, t_{ij}, \xi, \varepsilon$, respectively, are the free flow travel time, actual in vehicle time on RTL, vehicle capacity and a sufficient small positive number to avoid the case of infeasibility when $Y_{ij}^r = 0$. The passenger cost on DAR services is represented by the BPR function (10.19), where t_{ij}^0, t_{ij}, C_{ij} are, respectively, free flow travel time, actual in vehicle time on DAR and road link capacity. In the VI constraints, t_{ij} the link travel times on RTL and DAR are, respectively, calculated by (10.18) and (10.19). Link flow is calculated by adding up flows from all OD pairs d as shown in (10.21). $\tilde{x}_{ij}$, $\tilde{z}_{ij}$, $\tilde{v}_i$ are feasible link flows on RTL, DAR, and transfer links, respectively, with their feasible region denoted by Ω. The VI constraints state that for any feasible link flow $(\tilde{x}_{ij}, \tilde{z}_{ij}, \tilde{v}_i) \in \Omega$, the optimal link flow solution (x_{ij}, z_{ij}, v_i) must satisfy (10.20). We note that Ω is shaped by linear constraints (10.14)–(10.17) with stochastic demand Q^d, which renders Ω a polyhedron without simple, explicit boundaries. Namely, it is difficult to determine the boundaries and extreme points of Ω. To deal with this challenge, we reformulate constraints (10.14) and (10.15) associated with the demand uncertainty set into a set of linear constraints with deterministic parameters as described in the next section.

(1) Linearization of stochastic demand constraints

We note that **P1** is a non-linear model with stochastic demand. In this section, we first turn to robust optimization which focuses on the worst case scenario. Note that the paradox could occur under UE in the sense that a smaller demand could incur a larger system cost. In order words, the maximum demand combination may not constitute the worst case scenario, which imposes great computational difficulty in searching for the actual worst case demand scenario under UE. One possible way is to extend the cutting plane method (Lou et al., 2009; Chung et al., 2012) designed for robust discrete network design problem to our model. Nevertheless, the cutting plane method involves a tedious process of cutting and resolving the formulation for different demand combinations. As such, we simply ignore the possible existence of paradox in the sense that lower demand could yield a larger cost. However, this assumption can be relaxed and addressed by the cutting plan method in the further. For stochastic demand described by a polyhedral uncertainty set as in (10.1), **P1** can be linearized by reformulating the constraints related to the stochastic demand as a LP via its dual problem. The worst case scenario (highest demand combination ignoring paradox) associated with constraints (10.14) and (10.15) are equivalent to find the maximum Q^d in U^d, which still satisfies (10.14) and (10.15):

$$\max_{Q^d} Q^d,\ s.t.\ Q^d \le q^d (1+\theta),\ -Q^d \le -q^d (1-\theta),\ \sum_{\forall d \in \{d:D^d = j\}} Q^d \le B_j \qquad (10.22)$$

We rewrite the polyhedral uncertainty set U^d, i.e., the constraints in (10.22), as $\mathbf{AQ^d} \le \mathbf{b_1}$ for simplicity, where $\mathbf{Q^d} = \{Q^d, \forall d \in \{d : D^d = j\}\}$ is the vector of Q^d involved in U^d; $\mathbf{A}$ and $\mathbf{b}$ are, respectively, the parameter matrix and the RHS vector of the linear constraints. An equivalent constraint can be obtained by its dual problem (Ben-Tal et al., 2011; Bertsimas and Sim, 2004).

$$\max_{Q^d} Q^d\ s.t.\ \mathbf{AQ^d} \le \mathbf{b} \Leftrightarrow \min_{\lambda^d} \mathbf{b}^T \lambda^d\ s.t.\ \mathbf{A}^T \lambda^d = 1,\ \lambda^d \ge 0 \qquad (10.23)$$

where $\lambda^d = \{\lambda_1^d, \lambda_2^d, \lambda_3^{Dd}\}$ is the vector of dual variables corresponding to the three constraints in (10.22), and D^d is the destination node index of OD pair d. The constraint objective of (10.14) changes from finding the maximum Q^d that is less than $\sum_{j \in N} X_{kj}^d + \sum_{j \in N} Z_{kj}^d$, if $k = O^d$ to finding the minimum $b^T \lambda^d$ that is less than $\sum_{j \in N} X_{kj}^d + \sum_{j \in N} Z_{kj}^d$, if $k = O^d$. Constraint (10.15) follows the same logic. It enables us to directly add the dual problem to **P1** as constraints. By applying this method, constraints (10.14) and (10.15) are replaced by:

$$\lambda_1^d q^d (1+\theta) - \lambda_2^d q^d (1-\theta) + \lambda_3^{D^d} B_{D^d} \le \sum_{j \in N} X_{kj}^d + \sum_{j \in N} Z_{kj}^d,\ k \text{ is the origin of OD } d, \forall d$$

$$(10.24)$$

$$\lambda_1^d q^d (1+\theta) - \lambda_2^d q^d (1-\theta) + \lambda_3^{D^d} B_{D^d} \le \sum_{i \in N} X_{ik}^d + \sum_{i \in N} Z_{ik}^d,\ k \text{ is the destination of OD d}, \forall d$$

$$(10.25)$$

$$\lambda_1^d - \lambda_2^d + \lambda_3^{D^d} B_{D^d} = 1, \forall d \tag{10.26}$$

$$\lambda_1^d, \lambda_2^d, \lambda_3^{D^d} \geq 0, \forall d \tag{10.27}$$

After the linearization, the feasible passenger flow set Ω is shaped by a set of linear constraints (10.16), (10.17) and (10.24)–(10.27). Now Ω is a bounded polyhedron with finite vertexes. This attribute of Ω will assist us in developing efficient solution algorithm.

(2) Linearization of the VI constraint

In the VI constraint, we seek to find passenger link flow x_{ij}, z_{ij}, v_i so that for any feasible flow $(\tilde{x}_{ij}, \tilde{z}_{ij}, \tilde{v}_i) \in \Omega$, (10.20) is satisfied. However, it is computationally formidable to enumerate all the feasible flows in Ω. Since Ω is a bounded polyhedron, the feasible flows can be calculated by the convex combination of the vertices or extreme points of Ω. Take $\tilde{x}_{ij}$ for example; then $\tilde{x}_{ij} = \sum_e \eta_e \tilde{x}_{ij,e}$, $\sum_e \eta_e = 1$, $0 \leq \eta_e \leq 1$, where $\tilde{x}_{ij,e}$ represents the eth extreme point or vertex of Ω. Let $\tilde{z}_{ij,e}$ and $\tilde{v}_e$ be defined in the same way and the number of extreme points be denoted by E. For RTL passenger flow, we have:

$$t_{ij} * (x_{ij} - \tilde{x}_{ij}) \Leftrightarrow t_{ij} * (x_{ij} - \sum_e \eta_e \tilde{x}_{ij,e}) \Leftrightarrow t_{ij} * (\sum_e \eta_e x_{ij} - \sum_e \eta_e \tilde{x}_{ij,e}) \Leftrightarrow \sum_e \eta_e \left(t_{ij} * (x_{ij} - \tilde{x}_{ij,e}) \right) \tag{10.28}$$

The VI constraint (10.20) is reformulated as:

$$\sum_e \eta_e \left(\sum_{ij \in A_{RTL}} t_{ij} * (x_{ij} - \tilde{x}_{ij,e}) + \sum_{ij \in A_{DAR}} t_{ij} * (z_{ij} - \tilde{z}_{ij,e}) + \sum_{i \in N} t^0 * (v_i - \tilde{v}_{i,e}) \right) \leq 0 \tag{10.29}$$

It is obvious that if the main bracket on the LHS of (10.29) is less than or equal to zero for any $e = 1...E$, then (10.29) must hold. (10.20) is equivalent to the following constraints:

$$\sum_{ij \in A_{RTL}} t_{ij} * (x_{ij} - \tilde{x}_{ij,e}) + \sum_{ij \in A_{DAR}} t_{ij} * (z_{ij} - \tilde{z}_{ij,e}) + \sum_{i \in N} t^0 * (v_i - \tilde{v}_{i,e}) \leq 0, \ \forall e = 1,...,E \tag{10.30}$$

After the reformulation, we only need to ensure the feasibility of (10.30) for all extreme points of Ω instead of all the points in the whole region of Ω, which substantially reduce the actual number of constraints to be added to problem **P1**. Note that **P1** is a mixed integer non-linear program, with all its nonlinear terms involved in the objective function and the VI constraints. The following section describes the procedure to linearize the nonlinear terms:

(3) Linearization of $f_r Y_{ij}^r$

A real variable y_{ij}^r is introduced to replace the product of frequency f_r and RTL construction variable Y_{ij}^r, i.e., $y_{ij}^r = f_r Y_{ij}^r$. y_{ij}^r can be interpreted as the RTL link frequency which is 0 when the link is not covered by RTL and equal to f_r when a line r traverses it. A set of mixed integer linear constraints are employed to realize the transformation.

$$y_{ij}^r - f_r \leq 0, \ y_{ij}^r - \varpi Y_{ij}^r \leq 0, \ \varpi\left(Y_{ij}^r - 1\right) - y_{ij}^r + f_r \leq 0, \ y_{ij}^r \geq 0, \tag{10.31}$$

where ϖ is an extremely large positive number.

(4) Linearization of t_{ij} and $t_{ij} * z_{ij}$ on dial-a-ride services, $ij \in A_{DAR}$

A continuous real variable $\bar{t}_{ij}$ is introduced to represent the total passenger travel time on link ij using DAR services, i.e., $\bar{t}_{ij} = t_{ij} * z_{ij}$. The link travel time t_{ij} and total travel time $\bar{t}_{ij}$ only depend on link flow z_{ij}. We adopt a piecewise linear function to approximate the nonlinear functions of t_{ij} and $\bar{t}_{ij}$. The idea is to partition the passenger flow z_{ij} into $\mathbb{M}$ segments first. The passenger flows at the breaking points are denoted as z_{ij}^m, $m = 0,...,\mathbb{M}$. The link travel time t_{ij}^m and total link travel time $\bar{t}_{ij}^m$ at breaking points are obtained through plugging z_{ij}^m into the corresponding travel time functions. The arc between two break points is approximated by a straight line connecting the two adjacent breaking points, with the slope of $\dfrac{t_{ij}^m - t_{ij}^{m-1}}{z_{ij}^m - z_{ij}^{m-1}}$ for t_{ij} function and $\dfrac{\bar{t}_{ij}^m - \bar{t}_{ij}^{m-1}}{z_{ij}^m - z_{ij}^{m-1}}$ for $\bar{t}_{ij}$ function. Now we are ready to formulate the piecewise linear functions for each link $ij \in A_{DAR}$:

$$t_{ij} = t_{ij}^0 + \sum_{m=1}^{\mathbb{M}} \frac{t_{ij}^m - t_{ij}^{m-1}}{z_{ij}^m - z_{ij}^{m-1}} \mu_{ij}^m, \ \ \bar{t}_{ij} = \bar{t}_{ij}^0 + \sum_{m=1}^{\mathbb{M}} \frac{\bar{t}_{ij}^m - \bar{t}_{ij}^{m-1}}{z_{ij}^m - z_{ij}^{m-1}} \mu_{ij}^m \tag{10.32}$$

$$z_{ij} = \sum_{m=1}^{\mathbb{M}} \mu_{ij}^m \tag{10.33}$$

$$\left(z_{ij}^m - z_{ij}^{m-1}\right) * \kappa_{ij}^m \leq \mu_{ij}^m \leq \left(z_{ij}^m - z_{ij}^{m-1}\right) * \kappa_{ij}^{m-1}, \ \forall m = 1,...,\mathbb{M} \tag{10.34}$$

$$\mu_{ij}^m \geq 0, \forall m = 1,...,\mathbb{M}; \ \kappa_{ij}^0 = 1, \kappa_{ij}^{\mathbb{M}} = 0, \kappa_{ij}^m \in \{0,1\}, \forall m = 1,...,\mathbb{M}-1 \tag{10.35}$$

μ_{ij}^m is the length of the segment m covered by z_{ij}. For instance, if z_{ij} falls in the kth segment, μ_{ij}^m is equal to the length of the segment for $1 \leq m \leq k-1$, i.e., $\mu_{ij}^m = z_{ij}^m - z_{ij}^{m-1}$ and is less than the length of the last segment, i.e., $\mu_{ij}^m \leq z_{ij}^m - z_{ij}^{m-1}$ for $m = k$. For $k + 1 \leq m \leq \mathbb{M}$, $\mu_{ij}^m = 0$. This segment scheme is ensured by constraints (10.33)–(10.35). The piecewise linear functions for t_{ij} and $\bar{t}_{ij}$ are represented by (10.32).

(5) Linearization of t_{ij} and $t_{ij} * x_{ij}$ on rapid transit services, $ij \in A_{RTL}$

The RTL link travel time function is more complicated than that of DAR services since it involves two variables, link capacity $\sum f_r * Y_{ij}^r * \xi$ and link flow x_{ij}. This two-dimensional function requires a different approximation method. We make use of the same piecewise linear approximation method for multi-dimensional functions as described in Luathep et al. (2011). Similarly, the total travel time on link ij through rapid transit services is denoted by $\bar{t}_{ij} = t_{ij} * x_{ij}$. The link capacity is denoted by $\bar{y}_{ij} = \sum f_r * Y_{ij}^r * \xi + \varepsilon$. x_{ij} is segmented into $\mathbb{M}$ intervals while $\bar{y}_{ij}$ is segmented into intervals. They partition the domain into $\mathbb{M} * \mathbb{N}$ rectangles. Each rectangle can be further separated into two triangles by the upward diagonal line. The feasible domain of t_{ij} and $\bar{t}_{ij}$ is thus partitioned into a set of triangles as illustrated in Fig. 10.2.

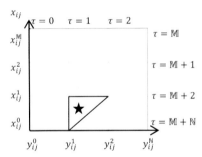

Figure 10.2 Piecewise linear approximation for two dimensional functions.

The key issue is to determine the active triangle that $(x_{ij}, \bar{y}_{ij})$ falls into. Let $(x_{ij}^m, \bar{y}_{ij}^n, t_{ij}^{m,n}, \bar{t}_{ij}^{m,n})$ be the coordinate of an arbitrary corner point (m, n), $\forall 0 \leq m \leq \mathbb{M}, 0 \leq n \leq \mathbb{N}$. t_{ij} and $\bar{t}_{ij}$ are represented by the convex combination of the corner point coordinates of the active triangle. Special ordered sets (SOS) variables are introduced in constraints (10.38)–(10.44) to identify the active triangle that $(x_{ij}, \bar{y}_{ij})$ belongs to. $\{\alpha_{ij}^m\}$, $\forall m = 0,....,\mathbb{M}$ and $\{\beta_{ij}^n\}$, $\forall n = 0,....,\mathbb{N}$ are Special Ordered Set of type one (SOS1) variables, which require that at most one member from the set may be non-zero. SOS1 variable is to represent a set of mutually exclusive alternatives. $\{\gamma_{ij}^\tau\}$, $\forall \tau = 0,....,\mathbb{M}, \mathbb{M} + 1,...,\mathbb{M} + \mathbb{N}$ is Special Ordered Set of type two (SOS2) variables, which requires that at most two adjacent members from the set are non-zeroes.

$$x_{ij} = \sum_{m=0}^{\mathbb{M}} \sum_{n=0}^{\mathbb{N}} \delta_{ij}^{m,n} x_{ij}^m, \quad \bar{y}_{ij} = \sum_{m=0}^{\mathbb{M}} \sum_{n=0}^{\mathbb{N}} \delta_{ij}^{m,n} \bar{y}_{ij}^n \qquad (10.36)$$

$$t_{ij} = \sum_{m=0}^{\mathbb{M}} \sum_{n=0}^{\mathbb{N}} \delta_{ij}^{m,n} t_{ij}^{m,n}, \quad \bar{t}_{ij} = \sum_{m=0}^{\mathbb{M}} \sum_{n=0}^{\mathbb{N}} \delta_{ij}^{m,n} \bar{t}_{ij}^{m,n} \tag{10.37}$$

$$\sum_{m=0}^{\mathbb{M}} \sum_{n=0}^{\mathbb{N}} \delta_{ij}^{m,n} = 1, \; \delta_{ij}^{m,n} \in [0,1], \; \forall m = 0,\dots,\mathbb{M}, \; \forall n = 0,\dots,\mathbb{N} \tag{10.38}$$

$$\sum_{n=0}^{\mathbb{N}} \delta_{ij}^{m,n} \le \alpha_{ij}^{m} + \alpha_{ij}^{m+1}, \quad \forall m = 0,\dots,\mathbb{M} \tag{10.39}$$

$$\alpha_{ij}^{0} = \alpha_{ij}^{\mathbb{M}+1} = 0, \quad \alpha_{ij}^{m} \in [0,1], \; \alpha_{ij}^{m} \in SOS1, \quad \forall m = 1,\dots,\mathbb{M} \tag{10.40}$$

$$\sum_{m=0}^{\mathbb{M}} \delta_{ij}^{m,n} \le \beta_{ij}^{n} + \beta_{ij}^{n+1}, \quad \forall n = 0,\dots,\mathbb{N} \tag{10.41}$$

$$\beta_{ij}^{0} = \beta_{ij}^{\mathbb{N}+1} = 0, \quad \beta_{ij}^{n} \in [0,1], \; \beta_{ij}^{n} \in SOS1, \quad \forall n = 1,\dots,\mathbb{N} \tag{10.42}$$

$$\gamma_{ij}^{\tau} = \sum_{m=0}^{\mathbb{M}} \delta_{ij}^{m,m-\mathbb{M}+\tau}, \quad \gamma_{ij}^{\tau} \in SOS2, \; \forall \tau = 0,\dots,\mathbb{M}, \mathbb{M}+1,\dots,\mathbb{M}+\mathbb{N}, \tag{10.43}$$

$$\max\{0, \mathbb{M}-\tau\} \le m \le \min\{\mathbb{N}+\mathbb{M}-\tau, \mathbb{M}\}$$

$$\sum_{\tau=0}^{\mathbb{M}+\mathbb{N}} \gamma_{ij}^{\tau} = 1 \tag{10.44}$$

Proposition 1: A feasible solution for the SOS1 and SOS2 variables identifies a unique triangle in the domain of $(x_{ij}, \bar{y}_{ij})$.

Proof: Without loss of generality, $\alpha_{ij}^{1}, \beta_{ij}^{2}$ are assumed to be the positive elements in SOS1 set of variables. Substitute the values of SOS1 variables into constraints (10.39) and (10.41), we obtain

$$\sum_{n=0}^{\mathbb{N}} \delta_{ij}^{m,n} = 0, \quad \forall m = 2,3,\dots,\mathbb{M}, \quad \sum_{n=0}^{\mathbb{N}} \delta_{ij}^{0,n} \le \alpha_{ij}^{1}, \quad \sum_{n=0}^{\mathbb{N}} \delta_{ij}^{1,n} \le \alpha_{ij}^{1} \tag{10.45}$$

$$\sum_{m=0}^{\mathbb{M}} \delta_{ij}^{m,n} = 0, \quad \forall n = 0,3,4,\dots,\mathbb{N}, \quad \sum_{m=0}^{\mathbb{M}} \delta_{ij}^{m,1} \le \beta_{ij}^{2}, \quad \sum_{m=0}^{\mathbb{M}} \delta_{ij}^{m,2} \le \beta_{ij}^{2} \tag{10.46}$$

Combining (10.45) and (10.46), we can get

$$\delta_{ij}^{0,1} + \delta_{ij}^{0,2} \le \alpha_{ij}^{1}, \quad \delta_{ij}^{1,1} + \delta_{ij}^{1,2} \le \alpha_{ij}^{1}, \quad \delta_{ij}^{0,1} + \delta_{ij}^{1,1} \le \beta_{ij}^{2}, \quad \delta_{ij}^{0,2} + \delta_{ij}^{1,2} \le \beta_{ij}^{2} \tag{10.47}$$

All the other $\delta_{ij}^{m,n}$ which are not stated in (10.47) are zero.

$\delta_{ij}^{m,n}$ takes on a positive value only at the corner point of the rectangle $\begin{pmatrix} \delta_{ij}^{1,1} & \delta_{ij}^{1,2} \\ \delta_{ij}^{0,1} & \delta_{ij}^{0,2} \end{pmatrix}$ as shown in Fig. 10.2. The active rectangle is thus determined by the two sets of SOS1 variables. The next question is whether the SOS2 variable γ_{ij}^{τ} can identify the active triangle. In constraint (10.43), γ_{ij}^{τ} is defined as the sum of $\delta_{ij}^{m,n}$ along each diagonal line. Hence the nonzero γ_{ij}^{τ} can only occur at the three possible diagonal lines, $\tau = \mathbb{M}$, $\tau = \mathbb{M} + 1$, $\tau = \mathbb{M} + 2$, passing through points (1,1), (0,1) & (1,2), and (0,2), respectively, as shown in Fig. 10.2. It is easy to show that the two feasible solutions $\gamma_{ij}^{\mathbb{M}}$, $\gamma_{ij}^{\mathbb{M}+1} > 0$, or $\gamma_{ij}^{\mathbb{M}+1}, \gamma_{ij}^{\mathbb{M}+2} > 0$ identify the upper and lower triangle, respectively. This finishes the proof. Reversely speaking, to determine the target point in this figure, the positive elements in SOS1 and SOS2 variables can only be $\alpha_{ij}^{1}, \beta_{ij}^{2}, \gamma_{ij}^{\mathbb{M}}, \gamma_{ij}^{\mathbb{M}+1}$ while all the other elements are zero.

After the reformulation, the original robust optimization problem with equilibrium constraints is reduced into a mixed integer linear problem (MILP) which can be readily solved. The equivalent MILP is summarized as follows:

$$\textbf{(P2)} \quad \min_{\mathbf{f,Y,W,X,Z,V}} \sum_{ij \in A_{RTL}} \left(c^1 l_{ij} \sum_{r \in R} f_r Y_{ij}^r + c^2 l_{ij} \tilde{Y}_{ij} \right) + c^3 \sum_{i \in N_{RTL}} W_i + c^4 \tag{10.48}$$

$$\sum_{ij \in A_{DAR}} l_{ij} z_{ij} + c^5 \sum_{ij \in A_{RTL}} \bar{t}_{ij} + c^5 \sum_{ij \in A_{DAR}} \bar{t}_{ij} + c^5 \sum_{i \in N} t^0 v_i$$

s.t.

RTL connectivity constraints: (10.3)–(10.13) and (10.31)	(10.49)
Passenger flow balance constraints: (10.16), (10.17), (10.21), (10.24)–(10.27)	(10.50)
Linear constraints for link $ij \in A_{DAR}$: (10.32)–(10.35)	(10.51)
Linear constraints for link $ij \in A_{RTL}$: (10.36)–(10.44)	(10.52)
The VI constraint: (10.30)	(10.53)

Note that as a MILP, the solution directly obtained in this reformulation actually achieves global optimality, subject to the accuracy of the discretization scheme adopted here.

10.2.3 Solution Algorithm

P2 is a MILP and can be readily solved by a solver such as CPLEX. Nevertheless, the number of extreme points constituting constraint (10.53) could be huge, which makes **P2** often too large to be handled by a commercial solver. Instead, we adopt the cutting constraint algorithm (CCA) to add the extremely points iteratively. The VI constraint for one extreme point of the feasible region Ω is added into **P2** one at a time. This approach substantially decreases the initial scale of **P2** and will shorten the searching time for optimal solution. When a new VI constraint for another extreme point is included, the solver makes use of the current solution as a starting point which greatly expedites the searching process. The relaxed MILP with a reduced number of extreme points $\tilde{E}$, $\tilde{E} \leq E$ is formulated as:

$$\textbf{(P3)} \quad \min_{\textbf{f,Y,W,X,Z,v}} \sum_{ij \in A_{RTL}} \left(c^1 l_{ij} \sum_{r \in R} f_r Y_{ij}^r + c^2 l_{ij} \tilde{Y}_{ij} \right) + c^3 \sum_{i \in N_{RTL}} W_i + c^4 \tag{10.54}$$

$$\sum_{ij \in A_{DAR}} l_{ij} z_{ij} + c^5 \sum_{ij \in A_{RTL}} \bar{t}_{ij} + c^5 \sum_{ij \in A_{DAR}} \bar{t}_{ij} + c^5 \sum_{i \in N} t^0 v_i$$

s.t. (10.49)–(10.52) and

$$\sum_{ij \in A_{RTL}} (\bar{t}_{ij} - t_{ij} * \tilde{x}_{ij,e}) + \sum_{ij \in A_{DAR}} (\bar{t}_{ij} - t_{ij} * \tilde{z}_{ij\,e}) + \sum_{i \in N} t^* (v_i - \tilde{v}_{i,e}) \leq 0, \forall\, e = 1,...,\tilde{E} \tag{10.55}$$

Denote **f***, **Y***, **W***, **t***, **t̄***, **x***, **z***, **v*** as the optimal solution of the reduced problem **P3** with a smaller number of extreme points. Additional extreme points can be found by identifying any feasible solution $(x_{ij}, z_{ij}, v_i) \in \Omega$ that satisfies $\sum_{ij \in A_{RTL}} (\bar{t}_{ij}^* - t_{ij}^* * x_{ij}) + \sum_{ij \in A_{DAR}} (\bar{t}_{ij}^* - t_{ij}^* * z_{ij}) + \sum_{i \in N} t^0 * (v_i^* - v_i) > 0$ (Luathep et al., 2011). Adding this extreme point into **P3** will make the current optimal solution infeasible, which thus leads to a new solution. An equivalent optimization problem **P4** is formulated to find the extreme points:

$$\textbf{(P4)} \quad \max_{\textbf{X,Z,v}} F = \sum_{ij \in A_{RTL}} (\bar{t}_{ij}^* - t_{ij}^* * x_{ij}) + \sum_{ij \in A_{DAR}} (\bar{t}_{ij}^* - t_{ij}^* * z_{ij}) + \sum_{i \in N} t^0 * (v_i^* - v_i) \tag{10.56}$$

s.t. (10.50)

If $F > 0$, its optimal solution $(\tilde{x}_{ij}, \tilde{z}_{ij}, \tilde{v}_i)$ will formulate a new VI constraint (10.53) which is then added into **P3**. Otherwise we can claim that the global optimal solution for **P2** has been found.

After solving **P2**, the line alignment and frequency of rapid transit lines **f***, **Y***, **W*** are fixed for the whole studying horizon (say, a year). Moreover, **P2** also calculates the DAR services needed under the worst-case (or highest) demand scenario. The exact deployment of dial-a-ride services for a particular day will depend on the demand realization. Meanwhile, the passenger cost under UE will change with demand as well. We calculate the average DAR operating and passenger costs by drawing samples of the uncertain demand. With the RTL capacity fixed by **f***, **Y***, **W***, for a specific demand realization, **P2** is reduced to a traditional UE traffic assignment problem. A variety of efficient solution approaches, such as Frank-Wolfe algorithm, Gradient Projection algorithm, etc.

(Chen et al., 2002) can be applied. We adopt the Frank-Wolfe algorithm in this chapter. The procedure is summarized as follows:

Step R1: Determine the uncertainty level θ of stochastic demand Q^d and formulate the MILP problem **P3** without the VI constraint (10.55), namely, the system optimal solution is adopted as the initial solution.

Step R2: Solve the relaxed MILP **P3** with a reduced set of extreme points. The optimal solution is denoted as **f***, **Y***, **W***, **t***, **t̄***, **x***, **z***, **v***.

Step R3: Solve the linear program (LP) problem **P4** for a new extreme point $(\tilde{x}_{ij}, \tilde{z}_{ij}, \tilde{v}_i) \in \Omega$.

Step R4: Convergence check. If $F \leq \varepsilon$, terminate the procedure and the optimal solution under UE flows are maintained as **f***, **Y***, **W***, **t***, **t̄***, **x***, **z***, **v***. Otherwise, add $(\tilde{x}_{ij}, \tilde{z}_{ij}, \tilde{v}_i)$ into **P3** through the VI constraint (10.55) and repeat Step R4 until the convergence criteria is satisfied.

Step R5: Calculate the average passenger cost under UE by sampling the stochastic demand given the RTL network **f***, **Y***, **W***.

10.3 Two-stage Stochastic Model

In the above model, RTL and DAR services jointly handle the stochastic demand and provide a certainty level of robustness to the whole system. The stochastic demand is captured by an uncertainty set with ambiguous demand distribution. In this section, we consider the situation where the stochastic demand distributions are known beforehand and can be obtained by historical data. With known demand distributions, the objective of the problem changes from minimizing the system cost under the worst case scenario to minimizing the expected system cost. Moreover, in the robust model **P2,** the shares of the two modes in carrying passengers are optimized endogenously. In reality, the government may take the equity problem into consideration and provides RTL services in remote regions although their demands may not be large enough to support the RTL construction. High passenger patronage to RTL services is also desired in the central business district (CBD) owing to the large amount of pedestrians on the ground road network in CBD. Shifting more passengers to RTL could significantly reduce utilization of DAR services, mitigate road congestion and ensure pedestrian safety. These additional requirements call for a pre-specified protection of some OD pairs such that their demands are assured to be carried by RTL services. Subsequently we draw upon the convenience of two-stage stochastic programming to address these two issues.

We employ the notion of service reliability ρ to represent the probability that a specific OD stochastic can be covered by RTL services, i.e., the schedule of RTL is designed to cover the stochastic demand up to a certain specified reliability (SR). One can interpret this service reliability as a minimum patronage built in the fixed transit services to carry the stochastic demand. The two types of services, RTL and DAR are combined to handle the demand fluctuation. The RTL services operate on dedicated right-of-ways or tracks with fixed schedules, whereas the DAR services are to carry the demand overflow on a particular day. The transit operator jointly plans for both the line configuration and schedule of the regular services and the extent of deploying third party flexible services. To service operators, flexible services have higher per unit costs; hence, their objective is to determine the optimal combination of these two service types to minimize the overall construction and expected operating cost while serving all demand realizations. The flexible services cost can also be viewed as the penalty of demand loss if only regular transit services are provided. The service reliability chosen in designing the regular services implies a certain level of expected cost in deploying the flexible services. A high (low) service reliability implies a high (low) regular services cost but a low (high) flexible services cost. In this way, the service reliability is not arbitrarily set but is linked to explicit cost terms in deploying the flexible services. And by optimizing the overall combined cost expectation of the two service types, the associated optimal service reliability is also internally determined. In this section, a two-phase solution algorithm is proposed to improve the algorithm efficiency. This formulation also allows us to protect certain OD pairs flexibly by setting their service reliabilities at a predetermined level and excludes these SRs from the optimization process.

10.3.1 Model Formulation

The problem can be formulated as the following: in Phase-1, we determine the rapid line alignment that can accommodate the demand up to a certain specified reliability $\rho = \{\rho^d\}$. Let Ψ_d be the cumulative distribution function of the random demand. The demand guaranteed to be covered by rapid transit line then is: $\bar{q}^d = \Psi_d^{-1}\{\rho^d\}$. The RTL is optimized to serve for the demand up to the level $\bar{q}^d$, which is referred to as the Phase-1 problem.
Phase-1:

$$\textbf{(P5-1)} \min_{f,Y,y,W} \sum_{ij \in A_{RTL}} \left(c^1 l_{ij} \sum_{r \in R} f_r Y_{ij}^r + c^2 l_{ij} \tilde{Y}_{ij} \right) + c^3 \sum_{i \in N_{RTL}} W_i + c^5 \sum_{j \in A_{RTL}} \sum_d t_{ij}^0 X_{ij}^d \tag{10.57}$$

s.t. RTL connectivity constraints: (10.3)–(10.13) and (10.31)

$$\sum_{j \in N} X_{kj}^d = \bar{q}^d = \Psi_d^{-1}\left(\rho^d\right), \quad if\ k = O^d, \forall d \in D \tag{10.58}$$

$$\sum_{i \in N} X_{ik}^d = \bar{q}^d = \Psi_d^{-1}\left(\rho^d\right), \quad \text{if } k = D^d, \forall d \in D \tag{10.59}$$

$$\sum_{j \in N} X_{kj}^d - \sum_{i \in N} X_{ik}^d = 0, \quad \forall k \in N \setminus O^d \cup D^d, d \in D \tag{10.60}$$

$$\sum_d X_{ij}^d \leq \sum_r f_r \xi\left(Y_{ij}^r + Y_{ji}^r\right), \quad \forall ij \in A_{RTL} \tag{10.61}$$

Constraints (10.58)–(10.61) represents the passenger flow balancing conditions. Even though the right hand side of (10.61) sums the forward and backward directions of rapid transit link ij, i.e., $Y_{ij}^r + Y_{ji}^r$, it can only contribute to the constraint once. The reason is that if Y_{ij}^r is one, then Y_{ji}^r must be zero for acyclic lines. This summation means that once the forward direction of a link is assigned to the route, then the backward direction must also be.

After the RTL alignment **Y**, frequency **f** and station locations **W** are fixed by solving the Phase-1 problem, the Phase-2 problem aims to determine the deployment of dial-a-ride services with the RTL predetermined in Phase-1. In Phase-2, we minimize the expected cost of dial-a-ride services and the passenger cost. The approach of scenario simulation is adopted to estimate the expected cost terms for stochastic demands as generated by their continuous probability distributions. Assume that the number of scenarios to be generated is |I|. Each scenario $\iota \in I$ has a probability p_ι, then $\sum_{\iota \in I} p_\iota = 1$. The subscript ι for all variables in the Phase-2 problem below denotes the scenario ι. If the realized demand is greater than the capacity of the RTL determined in Phase 1, then dial-a-ride services are called upon. We solve the following program for each demand scenario $\iota \in I$:
Phase-2:

$$\textbf{(P5-2)} \min_{\textbf{X,Z,V}} \ c^4 \sum_{ij \in A_{DAR}} l_{ij} z_{ij,\iota} + c^5 \sum_{ij \in A_{RTL}} \bar{t}_{ij,\iota} + c^5 \sum_{ij \in A_{DAR}} \bar{t}_{ij,\iota} + c^5 \sum_{i \in N} t^0 v_{i,\iota} \tag{10.62}$$

s.t. (10.30), (10.32)–(10.44)

$$\sum_{j \in N} X_{kj,\iota}^d + \sum_{j \in N} Z_{kj,\iota}^d = q_\iota^d, \quad \text{if } k = O^d, \forall d \in D \tag{10.63}$$

$$\sum_{i \in N} X_{ik,\iota}^d + \sum_{i \in N} Z_{ik,\iota}^d = q_\iota^d, \quad \text{if } k = D^d, \forall d \in D \tag{10.64}$$

$$\sum_{j \in N} X_{kj,\iota}^d - \sum_{i \in N} X_{ik,\iota}^d + V_{k+,\iota}^d - V_{k-,\iota}^d = 0, \quad \forall k \in N \setminus O^d \cup D^d, d \in D \tag{10.65}$$

$$\sum_{j \in N} Z_{kj,\iota}^d - \sum_{i \in N} Z_{ik,\iota}^d + V_{k-,\iota}^d - V_{k+,\iota}^d = 0, \quad \forall k \in N \setminus O^d \cup D^d, d \in D \tag{10.66}$$

Dial-a-ride services are supplementary to RTL services to accommodate demand overflow. Note that the RTL alignment $\mathbf{Y}$ or Y_{ij} and the frequency f_r are already fixed in Phase-1, which fix $\bar{y}_{ij}$ in (10.36) and reduce the linearization from two dimensions ($\bar{y}_{ij}$ and x_{ij}) into one dimension (x_{ij}). UE passenger flows are achieved by solving **P5-2** for each scenario $\iota \in I$ generated. The expected dial-a-ride services cost is then determined through:

$$\varphi(\boldsymbol{\rho}) = \sum_{\iota \in I} p_\iota \left(c^4 \sum_{ij \in A_{DAR}} l_{ij} z_{ij,\iota} + c^5 \sum_{ij \in A_{RTL}} \bar{t}_{ij,\iota} + c^5 \sum_{ij \in A_{DAR}} \bar{t}_{ij,\iota} + c^5 \sum_{i \in N} t^0 v_{i,\iota} \right) \qquad (10.67)$$

Combining the two phases, the objective of the two-phase MILP can be stated as:

$$\textbf{(P5)} \quad \min_{\mathbf{f,Y,y,W,X,Z,v}} \phi(\boldsymbol{\rho}) = \sum_{ij \in A_{RTL}} \left(c^1 l_{ij} \sum_{r \in R} f_r Y_{ij}^r + c^2 l_{ij} \tilde{Y}_{ij} \right) + c^3 \sum_{i \in N_{RTL}} W_i + \varphi(\boldsymbol{\rho}) \quad (10.68)$$

With a fixed service reliability $\boldsymbol{\rho}$, the first two terms in (10.68), which involves the rapid transit services costs, are solved and hence fixed in Phase-1, whereas the second term, which involves the expected dial-a-ride costs and the passenger cost, are solved in Phase-2. The objective is to determine the optimal service reliability $\boldsymbol{\rho}$ so as to minimize the combined cost as described in (10.68).

10.3.2 Service Reliability-based Gradient Method

The proposed solution procedure, named the SR-based gradient solution procedure is to search for the descent direction through the service reliability variable $\boldsymbol{\rho}$. In this problem, as stated earlier, once the RTL alignment $\mathbf{Y}$ and $\mathbf{f}$ are fixed, the corresponding minimum cost is also determined. Thus the primary concern here is to find $\mathbf{Y}^*$ and $\mathbf{f}^*$. Due to the discrete nature of $\mathbf{Y}$, it is not easy to conduct a line search directly over $\mathbf{Y}$ by considering the effect of all of its elements at the same time. That is why we take advantage of the continuous variable $\boldsymbol{\rho}$ to find the descent direction of the objective function. Given a starting point $\boldsymbol{\rho}$, we search for the descent direction to reach the optimal $\boldsymbol{\rho}^*$. From a practical perspective, the physical meaning of $\boldsymbol{\rho}$ is clear, which may help in identifying an initial $\boldsymbol{\rho}$ for faster convergence. The Phase-1 problem is formulated as a MILP problem. The Phase-2 problem is to determine the dial-a-ride services to minimize the operating and passenger cost for each sample demand realization, with the RTL services predetermined in Phase-1. Optimizing $\boldsymbol{\rho}$ as an internal variable determines the optimal cost trade-off between RTL and dial-a-ride services to obtain the expected minimum overall cost.

Moreover, we note that by solving Phase-1 and Phase-2 sequentially offers an efficient way to solve the problem as the interactions between the two service types are separated.

In the following, we develop a line search procedure to optimize for $\boldsymbol{\rho}$. The subscript k of each variable in the procedure denotes the iteration number.

Step S1: Set $k = 1$, initialize $\boldsymbol{\rho}_k = (\rho_k^d)$, $\forall d$.

Step S2: Based on the fixed $\boldsymbol{\rho}_k$, solve the Phase-1 problem to find the optimal RTL line alignment $\mathbf{Y}_k$ and the line frequencies $\mathbf{f}_k$.

Step S3: Solve the Phase-2 problem repeatedly for the all scenarios $\iota \in I$ generated and calculate the expected Phase-2 cost $\varphi(\boldsymbol{\rho}_k)$ through (10.67).

Step S4: Calculate the total system cost $\phi(\boldsymbol{\rho}_k)$ through (10.68).

Step S5: Update the minimum objective value and save it as ϕ^*.

Step S6: If $|\phi(\boldsymbol{\rho}_k) - \phi(\boldsymbol{\rho}_{k-1})| < \varepsilon$, stop; otherwise, proceed to Step S7.

Step S7: Determine the optimal $\boldsymbol{\rho}$.

Step S7.1: Calculate the partial derivative of $\phi(\boldsymbol{\rho}_k)$ with respect to ρ^d.

Since Y_{ij}, $ij \in A$ are integers, it is not possible to analytically find the partial derivatives of ϕ with respect to ρ^d; instead, we use the perturbation analysis to find the partial derivatives as shown in the following numerical procedure.

Step S7.1.0: Given $\boldsymbol{\rho}_k = (\rho^1, \rho^2,... \rho^{d\max})$ and its corresponding solution $(\mathbf{Y}_k, \mathbf{f}_k, \phi(\boldsymbol{\rho}_k))$, where $d_{\max}$ is the number of elements in $\boldsymbol{\rho}$ or the number of OD pairs.

Step S7.1.1:
a) Set $\boldsymbol{\rho}_k^d = (\rho^1, \rho^2,... \rho^d + \varepsilon^d,..., \rho^{d\max})$, where ε^d is a small positive number, say, $\varepsilon^d = 0.1$.
b) If $\rho^d + \varepsilon^d > 1$, go to e); otherwise proceed to c).
c) Solve phase-1 for $\mathbf{Y}_k^d$, $\mathbf{f}_k^d$ with given $\boldsymbol{\rho}_k^d$. If $\mathbf{Y}_k^d \neq \mathbf{Y}_k$ or $\mathbf{f}_k^d \neq \mathbf{f}_k$, go to Step S7.1.2; otherwise set $\varepsilon^d := \varepsilon^d + 0.1$.
d) Go back to a).
e) Set $\boldsymbol{\rho}_k^d = (\rho^1, \rho^2,..., \rho^d + \varepsilon^d,..., \rho^{d\max})$, solve Phase-1 for $\mathbf{Y}_k^d$, $\mathbf{f}_k^d$ with given $\boldsymbol{\rho}_k^d$. If $\mathbf{Y}_k^d \neq \mathbf{Y}_k$ or $\mathbf{f}_k^d \neq \mathbf{f}_k$, go to Step S7.1.2; otherwise set $\varepsilon^d := \varepsilon^d + 0.1$.
f) If $\rho^d + \varepsilon^d \geq 0$, go to e). Otherwise the sensitivity of ρ^d is set to zero.

Step S7.1.2: Solve Phase-2 to obtain $\phi(\boldsymbol{\rho}_k)^d$. Now we can obtain the sensitivity

of ρ^d: $\dfrac{\Delta\phi(\boldsymbol{\rho}_k)}{\Delta\rho^d} = \dfrac{\phi(\boldsymbol{\rho}_k)^d - \phi(\boldsymbol{\rho}_k)}{\rho^{d'} - \rho^d}$, where $\rho^{d'} = \rho^d + \varepsilon^d$, or $\rho^d - \varepsilon^d$.

Step S7.1.3: If $d \neq d_{\max}$, set $d = d + 1$, go to Step 7.1.1. Otherwise, we have obtained the whole set of sensitivities of $\phi(\boldsymbol{\rho}_k)$ over ρ^d, denoted by the vector $\nabla\phi(\boldsymbol{\rho}_k)$.

Step S7.2: Take the negative sensitivity vector $-\nabla \phi(\mathbf{\rho}_k)$ as the descent direction. The step size π_k is chosen in the same way as in Wang and Lo (2012):

$$\pi_k = v_k \frac{\phi(\mathbf{\rho}_k) - v_0 \phi^*}{||\nabla\phi(\mathbf{\rho}_k)||^2},$$ where $||\nabla\phi(\mathbf{\rho}_k)||^2$ represents the Euclidean norm

and v_k is an iteration dependent variable. If ϕ^* is not improved in the 5th iteration, set $v_{k+1} = (2/3)^* v_k$ which will reduce the step size gradually and the convergence is guaranteed consequently. $v_0 \phi^*$ is an estimate of the minimum of $\phi(\mathbf{\rho})$. The value of parameters v_0, v_k should be specified for different problems for improved efficiency.

Step S7.3: Calculate the reliability $\tilde{\mathbf{\rho}}_{k+1} = \mathbf{\rho}_k - \pi_k \nabla \phi(\mathbf{\rho}_k)$. Project the new reliability on to the feasible space $\mathbf{\rho} \in [0,1]$, we denote this result as $\hat{\mathbf{\rho}}_{k+1}$. Set $\mathbf{\rho}_{k+1} = \hat{\mathbf{\rho}}_{k+1}$ to be the new service reliability. Set $k = k + 1$ and, go to Step 2.

In Step S7.1.1, we change the elements in the service reliability vector one at a time until the RTL alignment variable $\mathbf{Y}_k$ or the line frequencies $\mathbf{f}_k$ changes. Since $\mathbf{f}_k$ is a continuous variable, $\mathbf{Y}_k$ or $\mathbf{f}_k$ typically change in three iterations when $\varepsilon^d = 0.1$.

10.4 Numerical Studies

We follow the same network and settings as in Gallo et al. (2011) to illustrate the formulation and algorithm as shown in Fig. 10.3. There are 8 nodes and 24 directed links. The number on each link is the distance l_{ij} of that link. The demand expectations for the 8*8 OD pairs are given by the following matrix $\mathbf{q}$:

$$\mathbf{q} = \begin{pmatrix} 0 & 45 & 130 & 95 & 65 & 60 & 65 & 40 \\ 55 & 0 & 70 & 130 & 35 & 90 & 15 & 30 \\ 150 & 95 & 0 & 150 & 120 & 40 & 75 & 60 \\ 105 & 45 & 55 & 0 & 110 & 80 & 80 & 105 \\ 70 & 70 & 40 & 45 & 0 & 100 & 125 & 110 \\ 130 & 5 & 110 & 120 & 65 & 0 & 70 & 55 \\ 40 & 30 & 45 & 30 & 115 & 55 & 0 & 60 \\ 10 & 45 & 55 & 70 & 90 & 80 & 55 & 0 \end{pmatrix}$$

For comparison purposes, we set the benchmark case to be that with the robust parameter set to be $\theta = 0.3$, and the trips heading for destination j are no more than 1.2 times its expectation:

$$Q^d, \forall d \in U \equiv \left\{ \mathbf{Q} : q^d(1-0.3) \le Q^d \le q^d(1+0.3), \forall d, \sum_{\forall D^d = j} Q^d \le (1+0.2) \sum_{\forall D^d = j} q^d \right\}$$

The frequency boundaries are $f_{max} = 20/hr$ and $f_{min} = 3/hr$. The transit unit capacity is $\xi = 80$ passengers/h. Each road link (i, j) has a capacity of

$C_{ij} = 200$. At most 1 transit line is allowed in the network, i.e., $R_{max} = 1$. The unit RTL operating cost per transit unit is $c^1 = 1$; the unit line construction cost is $c^2 = 4$; the unit station construction cost is $c^3 = 1$; the unit DAR operating cost is proportional to the unit RTL operating cost per passenger $c^4 = c^1/\xi$; the passenger value of time is $c^5 = 0.01$. The transfer penalty is: $t^0 = 3$. The model was programmed in C++ and the MILP was solved using the IBM CPLEX software Academic version 12.3 and run on Intel i7-3370 computer with 8GB RAM, Windows 7, 64 bit; two threads were used. Intuitively, many factors are influential in the RTL construction and dial-a-ride services deployment, such as the robustness level as determined by the uncertainty set and DAR services cost. Subsequently sensitivity analysis is employed to investigate the system performance under different combinations of the influencing factors for both the robust optimization and stochastic programming approaches.

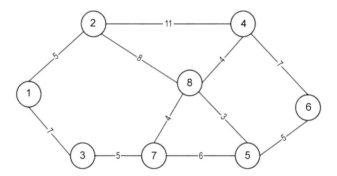

Figure 10.3 Network setting.

In the robust optimization approach, we repeat Steps R2–R4 until the UE condition is satisfied. The algorithm took 1 to 9 iterations to converge to the UE solution. The robust solution for the worst case scenario is obtained in Step R6. With the RTL alignment, frequency and the provided DAR services fixed at the worst case scenario, we calculate the expected passenger cost in Step R5. The computational times for different scenarios vary from 93 to 2023 seconds, with an average of 701 seconds. In this solution instance, the RTL services cover links 1-3-7-8-5-6 with a frequency of 5.6 vehicles per hour. Figure 10.4 shows the optimal solutions for the benchmark case. The resultant passenger flows on RTL and dial-a-ride services are shown separately in Fig. 10.4a and Fig. 10.4b. For both services, we observe the passenger flows on opposite directions for most links do not vary very much, probably due to cost minimization. The average load factor for DAR services is 0.73, varying from 0.43 to 0.95. The lowest load factor occurs at link 1-3. However, demolishing link 1-3 of RTL could increase the passenger flow on DAR services to 312 on link 3-1 and

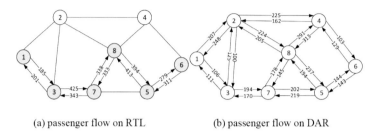

(a) passenger flow on RTL	(b) passenger flow on DAR

Figure 10.4 Optimal solutions for the benchmark case.

to 291 on link 1-3, which could lead to substantial congestion on road links and offset the savings in RTL construction. In Fig. 10.4b, the passengers on DAR services do not exceed the road link capacity—200, owing to the construction of RTL services. This also shows the advantages of combining RTL and dial-a-ride services, especially with road congestion.

To ascertain that the UE flow pattern is achieved by the cutting constraint algorithm, we depict the passenger assignment result for six representative OD pairs under the worst case scenario, which is sufficiently simple to track down the details. Note that the demand for **P2** under the worst case scenario shown in the second column of Table 10.1 is higher than the expected demand in **q**. Column 3 in this table shows the path and transport mode for each OD pair; the number stands for the node and alphabet for the transport mode, i.e., R for RTL, D for DAR. Passengers may choose different paths and the used paths have essentially the same travel time, with a miniscule difference due to the linearization error. For OD pair 1-4, paths 1R-3R-7R-8R-8D-4D and 1D-2D-4D have the same minimum travel time of 25.4. Passengers on OD 1-2 have no choice but to take DAR services since they are the only available services. We observe that most passengers choose direct services; only a few would transfer between the two modes owning to the high transfer penalty, which greatly prohibits their willingness to transfer.

Table 10.1 Path Flow Under UE for Six Representative OD Pairs.

OD	Demand	Path	Path Flow	Travel Time	Transfer
1-2	58	1D-2D	58	5.4	0
1-4	123	1R-3R-7R-8R-8D-4D	20	25.4	20
		1D-2D-4D	103	25.2	
2-6	117	2D-8D-8R-5R-6R	117	19.3	117
3-5	157	3R-7R-8R-5R	126	13.6	0
		3D-7D-5D	30	13.7	0
5-1	91	5R-8R-7R-3R-1R	91	22.7	0
8-7	71	8R-7R	52	4.5	0
		8D-7D	20	4.5	0

Table 10.2 shows the cost comparison between the worst case scenario and the expected cost for the cases with different robustness levels. For each case, the RTL alignment and frequency, hence the RTL costs are maintained the same under the worst case scenario and cost expectation calculation. The expected total cost is 7 per cent lower than the cost estimated by the worst case scenario for the deterministic case with the uncertainty set of (0.0, 0), indicating that the worst case scenario overestimated the cost as expected. We can see that this overestimation is increasing with the robustness level. Among the five cases investigated, the deterministic scenario yields the lowest average system cost, albeit with an insignificant discrepancy with the second lowest system cost. We can expect that the optimal level of robustness such that the expected system cost is minimized may lies in (0.0, 0) to (0.2, 0.1). As for the overall RTL system, the first four cases produce exactly the same network configurations but with increasing frequencies. The DAR services cost also increases to handle the larger robustness level. When the robustness level reaches (0.5, 0.4), the RTL alignment changes, leading to an even larger RTL cost but smaller DAR cost. In particular, the first type of alignment consists of a total transit length of 24 with 5 links, whereas that of the solution without dial-a-ride services has a total length of 27 with 5 links. This result indicates that larger robustness level would increase the company's construction and operating cost (RTL+DAR), but does not help much in decreasing the passenger cost. When the robustness level is very high, i.e., (0.5, 0.4), the passenger cost even increases by 6 per cent.

The difference between the average total cost and the cost under the worst case scenario is shown by the gap between the bars and the black line in Fig. 10.5. The gap can be interpreted as the protection level offered

Table 10.2 Line Alignment and the Cost Components for Solutions with Various Uncertainty Sets.

Uncertainty Set				Comp. Cost		Pass. Cost	Total Cost		
θ	$\sum_{\forall Dd=j} q^d / B_j$	RTL Alignment	Freq.	RTL	DAR	RTL | DAR | Transfer Total	Expectation	Worst case Scenario	Comparison (%)
0.0	0.0	1-3-7-8-5-6	5.1	369	230	144 | 328 | 4 476	1075	1153	7
0.2	0.1	1-3-7-8-5-6	6.0	390	230	149 | 306 | 8 463	1083	1239	15
0.3	0.2*	1-3-7-8-5-6	6.6	404	240	151 | 302 | 8 461	1105	1340	22
0.4	0.3	1-3-7-8-5-6	7.0	414	263	152 | 300 | 9 462	1138	1447	28
0.5	0.4	1-3-7-8-4-6	8.0	492	223	158 | 321 | 10 489	1203	1562	31

* Benchmark case

by robust optimization to hedge against the stochastic demand. We can see that the protection level is much higher for larger robustness levels. The RTL constitutes a major cost component once introduced, whereas the DAR constitutes a small fraction of the cost. The passenger cost first drops a little from the case with robust level (0, 0) to the case with (0.4, 0.3), then increases dramatically for the case with (0.5, 0.4).

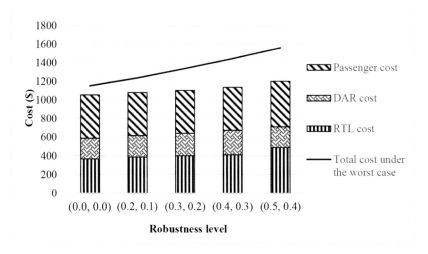

Figure 10.5 Cost components of robust optimization solutions with various robustness levels.

We then examine the effects of DAR unit operating cost on the RTL configurations. Intuitively, higher DAR operating cost will encourage the expansion of RTL services. On the contrary, low DAR operating cost should be able to detain RTL construction. To verify this statement, we did sensitivity analysis under different DAR cost ratios. DAR cost ratio represents the unit operating cost for DAR services, which is proportional to the RTL capacity. For instance, a ratio of 0.2 refers to that the unit DAR cost is $0.2/\xi$, ξ for RTL vehicle capacity. From Table 10.3, we can see that when ratio is less than 1, the RTL alignment as well as the frequency do not change with the cost ratio, which seems to be counter intuitive. However, it can be interpreted in this way—recall that the congestion cost for DAR services is represented by the BPR function. The DAR cost will increase with a power of 4 when demand exceeds the road capacity. When the passenger demand reaches a certain amount, it is not economical to rely on DAR alone even for a low DAR unit cost. When ratio is equal to or greater than 1, increase in cost ratio leads to either higher frequency or more RTL links to be constructed. It coincides with our expectation. In brief, when road congestion is severe, RTL construction is imperative regardless of the unit cost of DAR. However, appropriate pricing on DAR will help in introducing RTL services.

Table 10.3 Line Alignment and the Cost Components for Solutions with Various DAR Cost Ratios.

DAR Cost Ratio	RTL Alignment	Freq.	Company Cost		Passenger Cost	Total Cost	
			RTL	DAR	RTL | DAR | Transfer Total	Expected	Worst Case Scenario
0.2	1-3-7-8-5-6	4.0	342	75	140 | 328 | 0.2 468	885	1100
0.4	1-3-7-8-5-6	4.0	342	111	140 | 328 | 0.2 468	921	1137
0.6	1-3-7-8-5-6	4.0	342	207	140 | 328 | 1.2 468	1018	1243
0.8	1-3-7-8-5-6	4.0	366	269	145 | 315 | 2.4 460	1097	1305
1.0	1-3-7-8-5-6	6.6	404	240	151 | 302 | 8.0 461	1105	1340
2.0	1-2-8-4-6-5-7-3-1	5.0	713	108	185 | 258 | 3.6 443	1267	1536
3.0	1-2-8-4-6-5-7-3-1	5.0	713	162	185 | 258 | 6.6 443	1324	1586
4.0	1-2-8-4-6-5-7-3-1	6.0	758	186	190 | 250 | 4.7 440	1389	1640
5.0	1-2-4-6-5-8-7-3-1	6.0	760	222	188 | 245 | 0.0 433	1415	1697

We apply the two stage stochastic program on the same network in Fig. 10.3 and optimize RTL alignment with various levels of service reliability. The first column in Table 10.4 gives the protected nodes. The OD demands with both origins and destinations belonging to the protected nodes set are assumed to be covered by RTL at a guaranteed service reliability of 0.8, i.e., $\rho^d \geq 0.8$, $d \equiv (i, j)$, $\forall i$ and $j \in$ protected nodes set. The first row refers to the results that no protection is given to any OD; and the service reliabilities are optimized to minimize the expected total system cost, which produces the lowest system cost and lowest company cost. When a certain set of OD pairs is protected, the company cost as well as the total cost increase is as expected. We can see that different protection levels yield markedly different RTL alignments, with their total costs all higher than the one without protection. Specifically, protecting all nodes and all OD pairs produces the highest company cost. However, the passenger cost does not decrease as much as the increase is in company cost.

Table 10.4 Line Alignment and the Corresponding Cost Component for Solutions of the Worst Case Scenario and the Sample Average.

Protected Nodes	RTL Alignment	Freq.	Company Cost		Passenger Cost	Total Cost
			RTL	DAR	RTL I DAR I Transfer Total	
N/A	3-7-8-5-6	4.0	243	297	99 I 396 I 0.2 497	1035
2, 4, 5, 8	3-7-5-8-4-2	4.0	412	215	106 I 380 I 0.3 487	1112
2, 5, 6, 8	1-2-8-5-6	4.0	299	288	117 I 381 I 0.1 498	1086
1, 3, 5, 7, 8	1-3-7-8-5-6	4.0	342	274	139 I 327 I 0.1 466	1082
2, 4, 5, 6, 8	2-4-8-5-6	4.0	327	311	84 I 461 I 2.0 547	1185
1, 2, 3, 4, 5, 6, 7 ,8	2-1-3-7-5-6-4-8-2	4.0	666	224	177 I 268 I 0.0 445	1335

10.5 Conclusion

In this chapter, we formulated the transit network design problem under demand uncertainty through robust optimization and the SR-based stochastic program. Road congestion and transit crowdedness were taken into account to formulate the problem under the UE principle. To solve this highly complex formulation, we developed linearization procedures combined with a cutting constraint algorithm to achieve substantial gains in computation time.

We investigated DAR cost ratios and the robustness level for their effects on rapid transit line alignment. The results showed that higher robustness level led to more RTL and DAR services. However, the passengers obtained marginal additional benefits in reducing their average passenger cost. DAR cost ratio could substantially affect the RTL services deployment. When DAR services are inexpensive, the system tends to rely on DAR services to handle the demand uncertainty. While DAR services become more expensive, more RTL lines with higher frequencies are utilized. In the SR-based stochastic program, protection of certain nodes always led to system cost increases but less savings in passenger costs.

We have successfully applied the two models and solution approaches to an illustrative network with promising results, which may pave the way for larger-scale network applications. The proposed robust and stochastic program approaches are able to address intrinsic uncertainties arising from the demand side and better facilitate rapid transit line alignment and construction sequence decisions.

Although we modeled the TNDP for two types of services—rapid transit services and dial-a-ride services, there are certain limitations that need to be addressed in future studies. The first issue is that passenger waiting time for a transit line should be incorporated, which can be extended by the approach developed in An and Lo (2014a); the second is that the model allows unlimited transfers between transport modes, which is not realistic. We imposed a heavy transfer penalty to mitigate the problem, but we cannot totally avoid it as it is inherent in link-based formulations for the network design problem. This problem can be entirely overcome by a path-based formulation, as developed in Lo et al. (2003). This chapter aimed at minimizing the total system cost. Conceivably, the optimal RTL sequence will differ according to different objectives, such as those of the passengers, RTL companies, DAR companies, or the government. Combining these various objectives in the formation provides another direction for extensions. Finally, the influence of network topologies on the line construction sequence can be further studied by the proposed method, as can be the incorporation of other sources of variations (Watling and Cantarella, 2013).

Acknowledgements

The study is supported by the Public Policy Research grant HKUST6002-PPR-11, General Research Fund #616113 of the Research Grants Council of the HKSAR Government, the Hong Kong PhD Fellowship and Strategic Research Funding Initiative SRFI11EG15.

APPENDIX

Notation Table

Parameters	
d	OD pair from node i to j
Q^d	random demand of OD pair d
q^d	demand expectation of OD pair d
O^d	origin node index of OD pair d
D^d	destination node index of OD pair d
U^d	uncertainty set of OD pair d

Parameters

$\mathbf{Q}^d$	vector of Q^d involved in U^d
θ	uncertainty level
$\underline{q}^d$	demand lower bound of OD pair d with $\underline{q}^d = q^d(1 - \theta)$
$\overline{q}^d$	demand guaranteed to be covered by RTL with $\overline{q}^d = q^d(1 - \theta)$ in robust optimization or $\overline{q}^d = \Psi_d^{-1}(\rho^d)$ in stochastic programming approach
B_j	upper bound of the number of travelers heading for destination j
N	node set
A	arc set
$G(N, A)$	transportation network
$(i, j) \in A$	feasible arcs linking stations i and j
l_{ij}	link distance
N_s	station sub-node
N_{RTL}	rapid transit line (RTL) sub-node
N_{DAR}	dial-a-ride (DAR) sub-node
A_{RTL}	RTL arc set
A_{DAR}	DAR arc set
$r \in R$	rapid transit line set
R_{max}	maximum number of lines allowed
$S = \{S_r, r \in R\}$	dummy starting node set
$T = \{T_r, r \in R\}$	dummy ending node set
$\overline{N} = N_{RTL} \cup S \cup T$	set of all RTL nodes including the dummy origin and destination nodes
$\overline{A}$	set of all RTL links including the dummy arcs
c^1	unit RTL operating cost
c^2	unit RTL construction cost
c^3	unit station construction cost
c^4	unit DAR operating cost
c^5	passenger value of time
t^0	transfer penalty
ξ	vehicle capacity

Parameters	
C_{ij}	capacity of road link ij
$\varepsilon, \varepsilon^d$	small positive number
ϖ	extremely large positive number
Ω	feasible link flow set of **P1**
E	number of extreme points of Ω
$\tilde{E}$	a reduced number of extreme points $\tilde{E} \le E$
$\tilde{x}_{ij}$	feasible link flow on RTL
$\tilde{z}_{ij}$	feasible link flow on DAR
$\tilde{v}_i$	feasible link flow on transfer links
$(\tilde{x}_{ij,e}, \tilde{z}_{ij,e}, \tilde{v}_{i,e})$	e^{th} extreme point or vertex of Ω
Ψ_d	cumulative distribution function (CDF) for demand of OD pair d
$\iota \in I$	demand scenario index
p_ι	probability of scenario ι happens
q_ι^d	demand realization of OD pair d in scenario ι
k	iteration index
v_k, π_k, v_0	step size parameters
Decision Variables	
$\mathbf{Y} = \{Y_{ij}^r\}$	RTL alignment, Y_{ij}^r is 1 if link (i, j) is on line r and 0 otherwise
$\tilde{\mathbf{Y}} = \{\tilde{Y}_{ij}\}$	RTL alignment $\tilde{Y}_{ij} = 1$ if link ij is covered by any line $r \in R$
$\mathbf{f} = \{f_r\}$	frequency of RTL line r
$\mathbf{W} = \{W_i\}$	RTL station selection
$\mathbf{X} = \{X_{ij}^d\}$	passenger flow on RTL from station i to j for OD pair d
$\mathbf{Z} = \{Z_{ij}^d\}$	passenger flow on DAR from station i to j for OD pair d
V_{i+}	amount of passengers transferring from RTL to DAR at station i
V_{i-}	amount of passengers transferring from DAR to RTL at station i
$V = \{V_{i+}, V_{i-}\}$	transfer passenger flow set
x_{ij}	total passenger flow from station i to j on RTL

Parameters	
z_{ij}	total passenger flow from station i to j on DAR
v_i	total passenger transfer flow on station i
y_{ij}^r	RTL link frequency $y_{ij}^r = f_r Y_{ij}^r$
$\bar{y}_{ij}$	RTL link capacity with $\bar{y}_{ij} = \sum_r f_r^* Y_{ij}^r {}^* \xi + \varepsilon$
$\boldsymbol{\rho} = \{\rho^d\}$	service reliability set
Linearization Parameters	
$\mathbb{M}$	number of segments in x_{ij} or z_{ij} partition
$\mathbb{N}$	number of segments in $\bar{y}_{ij}$ partition
t_{ij}^0	free flow link travel time on RTL (DAR) if $ij \in A_{RTL}$ ($ij \in A_{DAR}$)
t_{ij}	actual in vehicle time on RTL (DAR) if $ij \in A_{RTL}$ ($ij \in A_{DAR}$)
$\bar{t}_{ij}$	total passenger travel time on RTL with $\bar{t}_{ij} = t_{ij}^* z_{ij}$ if $ij \in A_{RTL}$ or total passenger travel time on DAR with $\bar{t}_{ij} = t_{ij}^* x_{ij}$ if $ij \in A_{DAR}$
t_{ij}^m	link travel time on DAR at the breaking point m
$\bar{t}_{ij}^m$	total passenger travel time on DAR at the breaking point m
z_{ij}^m	passenger flow on DAR at the breaking point m
μ_{ij}^m	length of the segment m covered by z_{ij}
κ_{ij}^m	active segment identification variable
$t_{ij}^{m,n}$	link travel time on RTL at the breaking point (m, n)
$\bar{t}_{ij}^{m,n}$	total passenger travel time on RTL at the breaking point (m, n)
x_{ij}^m	passenger flow on RTL at the breaking point m
$\bar{y}_{ij}^n$	RTL link capacity at the breaking point n
$\delta_{ij}^{m,n}$	active rectangle identification variable
$\{\alpha_{ij}^m\}, \forall m = 0,..., \mathbb{M}$	special ordered set of type one (SOS1) variables

Parameters

$\{\beta_{ij}^n\},\ \forall n = 0,...,\ \mathbb{N}$	special ordered set of type one (SOS1) variables
$\{y_{ij}^\tau\},\ \forall \iota = 0,...,\ \mathbb{M} + \mathbb{N}$	special ordered set of type two (SOS2) variables
η_e	multiplier of the eth extreme point of Ω
$\boldsymbol{\lambda}^d = \{\lambda_1^d,\ \lambda_2^d,\ \lambda_3^{Dd}\}$	vector of dual variables
$\mathbf{f}^*,\ \mathbf{Y}^*,\ \mathbf{W}^*,\ \mathbf{t}^*,\ \overline{\mathbf{t}}^*,\ \mathbf{x}^*,$ $\mathbf{z}^*,\ \mathbf{v}^*$	optimal solution of the reduced problem **P3** with a smaller number of extreme points
F	objective function value of **P4**
$\boldsymbol{\rho}^*$	optimal service reliability
$\varphi(\boldsymbol{\rho})$	expected phase-2 cost as a function of $\boldsymbol{\rho}$
$\phi(\boldsymbol{\rho})$	overall system cost as a function of $\boldsymbol{\rho}$
ϕ^*	optimal system cost

Keywords: transit network design; stochastic demand; robust; stochastic programming; service reliability; user equilibrium; congestion.

References

An, K. and H. K. Lo. 2014a. Service reliability based transit network design with stochastic demand. Transportation Research Record: Journal of the Transportation Research Board 2467: 101–109.

An, K. and H. K. Lo. 2014b. Ferry service network design with stochastic demand under user equilibrium flows. Transportation Research Part B: Methodological 66: 70–89.

An, K. and H. K. Lo. 2015. Robust transit network design with stochastic demand considering development density. Transportation Research Part B 81(3): 737–754.

An, K. and H. K. Lo. 2016. Two-phase stochastic program for transit network design under demand uncertainty. Transportation Research Part B 84: 157–181.

Bellei, G., G. Gentile and N. Papola. 2002. Network pricing optimization in multi-user and multimodal context with elastic demand. Transportation Research B 36: 779–798.

Benders, J. F. 1962. Partitioning procedures for solving mixed-variables programming problems. Numerische Mathematik 4(1): 238–252.

Ben-Tal, A. and A. Nemirovski. 1999. Robust solutions of uncertain linear programs. Operations Research Letters 25(1): 1–13.

Ben-Tal, A., B. D. Chung, S. R. Mandala and T. Yao. 2011. Robust optimization for emergency logistics planning: Risk mitigation in humanitarian relief supply chains. Transportation Research Part B: Methodological 45(8): 1177–1189.

Bertsimas, D. and V. Goyal. 2010. On the power of robust solutions in two-stage stochastic and adaptive optimization problems. Mathematics of Operations Research 35(2): 284–305.

Bertsimas, D. and M. Sim. 2004. The price of robustness. Operations Research 52(1): 35–53.

Birge, J. R. and F. V. Louveaux. 1988. A multicut algorithm for two-stage stochastic linear programs. European Journal of Operational Research 34(3): 384–392.

Bruno, G., G. Ghiani and G. Improta. 1998. A multi-modal approach to the location of a rapid transit line. European Journal of Operational Research 104(2): 321–332.

Chen, X., M. Sim and P. Sun. 2007. A robust optimization perspective on stochastic programming. Operations Research 55(6): 1058–1071.

Chen, A., D. Lee and R. Jayakrishnan. 2002. Computational study of state-of-the-art path-based traffic assignment algorithms. Mathematics and Computers in Simulation 59(6): 509–518.

Chung, B. D., T. Yao, T. L. Friesz and H. Liu. 2012. Dynamic congestion pricing with demand uncertainty: A robust optimization approach. Transportation Research Part B: Methodological 46(10): 1504–1518.

Ekström, J., A. Sumalee and H. K. Lo. 2012. Optimizing toll locations and levels using a mixed integer linear approximation approach. Transportation Research Part B: Methodological 46(7): 834–854.

Gao, Z., H. Sun and L. L. Shan. 2004. A continuous equilibrium network design model and algorithm for transit systems. Transportation Research Part B: Methodological 38(3): 235–250.

Gao, Z., J. Wu and H. Sun. 2005. Solution algorithm for the bi-level discrete network design problem. Transportation Research Part B: Methodological 39(6): 479–495.

Gallo, M., B. Montella and L. D'Acierno. 2011. The transit network design problem with elastic demand and internalisation of external costs: An application to rail frequency optimization. Transportation Research Part C: Emerging Technologies 19(6): 1276–1305.

Lai, M. F. and H. K. Lo. 2004. Ferry service network design: Optimal fleet size, routing, and scheduling. Transportation Research Part A: Policy and Practice 38(4): 305–328.

Lo, H. K., K. An and W. H. Lin. 2013. Ferry network design under demand uncertainty. Transportation Research Part E: Logistics and Transportation Review 59: 48–70.

Lo, H. K., C. W. Yip and K. H. Wan. 2003. Modeling transfer and nonlinear fare structure in multi-modal network. Transportation Research Part B 37: 149–170.

Lo, H. K., C. W. Yip and Q. K. Wan. 2004. Modeling competitive multi-modal transit services: A nested logit approach. Transportation Research Part C 12: 251–272.

Lou, Y., Y. Yin and S. Lawphongpanich. 2009. Robust approach to discrete network designs with demand uncertainty. Transportation Research Record: Journal of the Transportation Research Board 2090: 86–94.

Luathep, P., A. Sumalee, W. H. K. Lam, Z. Li and H. K. Lo. 2011. Global optimization method for mixed transportation network design problem: A mixed-integer linear programming approach. Transportation Research Part B: Methodological 45(5): 808–827.

Marcotte, P. and S. Nguyen. 1998. Hyperpath formulations of traffic assignment problems. pp. 109–124. *In*: P. Marcotte and S. Nguyen (eds.). Equilibrium and Advanced Transportation Modelling. Kluwer Academic Publisher, Norwell, MA.

Nguyen, S. and S. Pallottino. 1988. Equilibrium traffic assignment for large scale transit network. European Journal of Operational Research 37(2): 176–186.

Ruszczynski, A. 2008. Decomposition methods. pp. 141–211. *In*: Anonymous Handbooks in Operations Research and Management Science Elsevier.

Wan, Q. and H. K. Lo. 2009. Congested multimodal transit network design. Public Transport 1(3): 233–251.

Wang, D. Z. W. and H. K. Lo. 2010. Global optimum of the linearized network design problem with equilibrium flows. Transportation Research Part B: Methodological 44(4): 482–492.

Wang, S., Q. Meng and H. Yang. 2013. Global optimization methods for the discrete network design problem. Transportation Research Part B: Methodological 50: 42–60.

Watling, D. P. and G. E. Cantarella. 2013. Modelling sources of variation in transportation systems: theoretical foundations of day-to-day dynamic models. Transportmetrica B: Transport Dynamics 1(1): 3–32.

Yang, H. and M. G. H. Bell. 1998. Models and algorithms for road network design: A review and some new developments. Transport Reviews 18(3): 257–278.

Index